No. 8434
$31.90

DESIGNING & BUILDING YOUR OWN STEREO FURNITURE

BY CARL W. SPENCER

30 PROJECTS TO IMPROVE YOUR STEREO SYSTEM

BY DAVID B. WEEMS

FIRST EDITION

FIRST PRINTING

MARCH 1981

Printed in the United States of America

Library of Congress Cataloging in Publication Data

Weems, David B
30 projects to improve your stereo system.

Includes index.
1. Cabinet-work. 2. Loud-speaker cabinets. 3. Stereophonic sound systems—Amateurs' manuals. I. Spencer, Carl W. Designing and building your own stereo furniture. 1981. II. Title.
TT197.W43 621.389'334 80-28747
ISBN 0-8306-8434-4

No.1334
$14.95

30 Projects To Improve Your Stereo System

by David B. Weems

FIRST EDITION

FIRST PRINTING

MARCH 1981

Printed in the United States of America

Library of Congress Cataloging in Publication Data

Weems, David B
30 projects to improve your stereo system.

Includes index.
1. Stereophonic sound systems—Amateurs' manuals.
I. Title.
TK9968.W436 621.389'334 80-28731
ISBN 0-8306-9631-8
ISBN 0-8306-1334-X (pbk.)

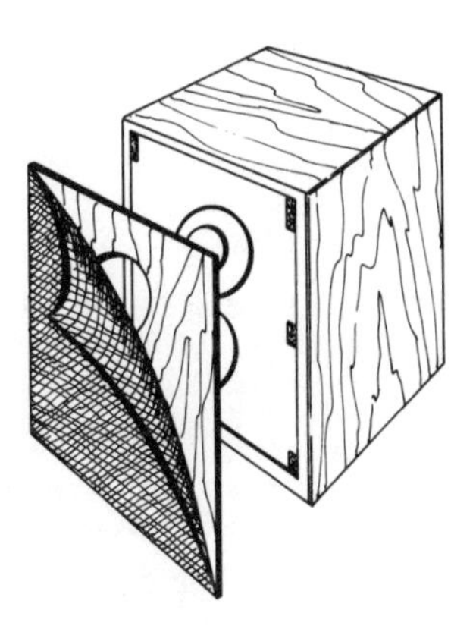

Contents

Introduction

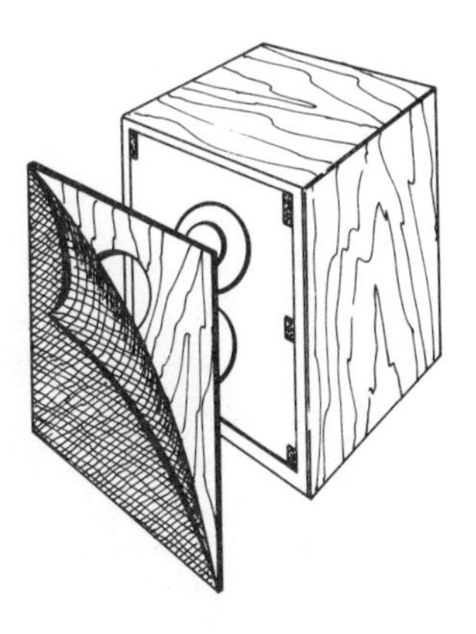

Almost any audio system can be improved, but some improvements are expensive. The projects in this book were chosen with one eye on performance and the other eye on cost. After examining several typical audio systems on that basis, we found that there are two places where small changes can pay big dividends in better sound: at the antenna and at the speaker. That's why you will find more antenna and speaker-related projects in this book than any other kind of projects.

So take a look. You will surely find some projects here that can significantly improve your stereo system. And they won't strain your budget.

David B. Weems

1
FM Antennas

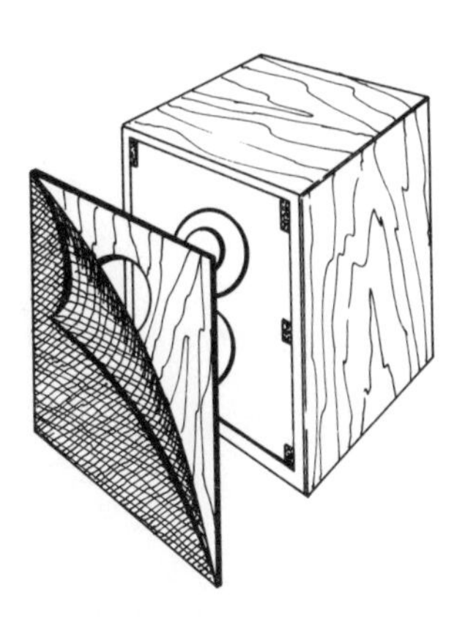

The FM antenna is probably the most overlooked component in most stereo systems. Many stereo fans invest hundreds of dollars in components, yet resist the cost of a decent antenna or the time required to make one.

This pattern was highlighted recently when I was asked to solve an audio mystery. The problem, the owner said, was in the speakers. He had just bought a new receiver which worked fine with his old speakers, but the system developed some odd habits after he installed new speakers. When listeners moved around the room, the sound sometimes went dead. One guest discovered that he could mute the speakers by simply lifting his leg—very strange!

After considerable questioning, one important fact surfaced: When the owner got the new speakers, he moved the receiver to a new location. And what kind of antenna and lead-in did he use? None. This particular receiver had a built-in antenna in the AC cord, which the owner had assumed would be adequate. A simple *folded dipole* made up from a few cents worth of *300-ohm lead-in* eliminated the problem.

That installation may not have exhibited the typical symptoms of a poor antenna system, but it shows that antenna problems are sometimes blamed on other components. Almost any receiver in almost any location can be improved by a better antenna. Regardless of receiver quality, there is no substitute for a good antenna.

ANTENNA FUNCTION

The antenna picks up the electromagnetic waves sent out by the station transmitter, converting them into electrical current that flows through the antenna. This electrical energy is extremely weak at the antenna—so weak, in fact, that the voltage that sends the current through the antenna and lead-in must be measured in *microvolts.*

Every FM stereo receiver has a characteristic—also measured in microvolts—that indicates how much signal the tuner must have to perform adequately. The lower the microvolt figure for full quieting is, the more sensitive the receiver is. Any increase in the voltage delivered to the receiver input will help it to give better stereo reception.

Fringe area stations can often be received well enough for good performance when the receiver is switched to the MONO mode, but in STEREO the reception deteriorates. This symptom shows that the voltage delivered to the receiver by the antenna system is marginal. Other symptoms of a weak signal are rhythmic fading when aircraft are overhead or a clicking sound from nearby car engines.

When a particular fringe- area station is desired, it is possible to tune the antenna to the frequency of that station as well as orient it toward the direction of the station. To do this, the antenna element dimensions can be altered to match the frequency of the carrier wave of the station. Formulas for doing this will be given later.

DIPOLE ANTENNAS

Most FM antennas are based on the simple *half-wave center-fed, dipole* (Fig. 1-1A). The length of the dipole, from end to end, is equal to one-half that of the radio frequency carrier wave sent out by the station. The dipole accepts the maximum energy at this frequency.

For several reasons most practical antennas contain a folded dipole (Fig. 1-1B). The folded dipole has a broader resonance, so it can pick up a wider range of broadcast frequencies. This makes it practical to use a single dipole for the entire FM band. Also, the folded dipole has a higher characteristic impedance which makes it a better match for the usual 300-ohm lead-in. The simple dipole has a typical impedance of about 70 ohms, but when additional elements are added to the antenna, the impedance drops. Any mismatch between antenna and lead-in, or lead-in and receiver will cause a significant loss of energy to the receiver input. If a single

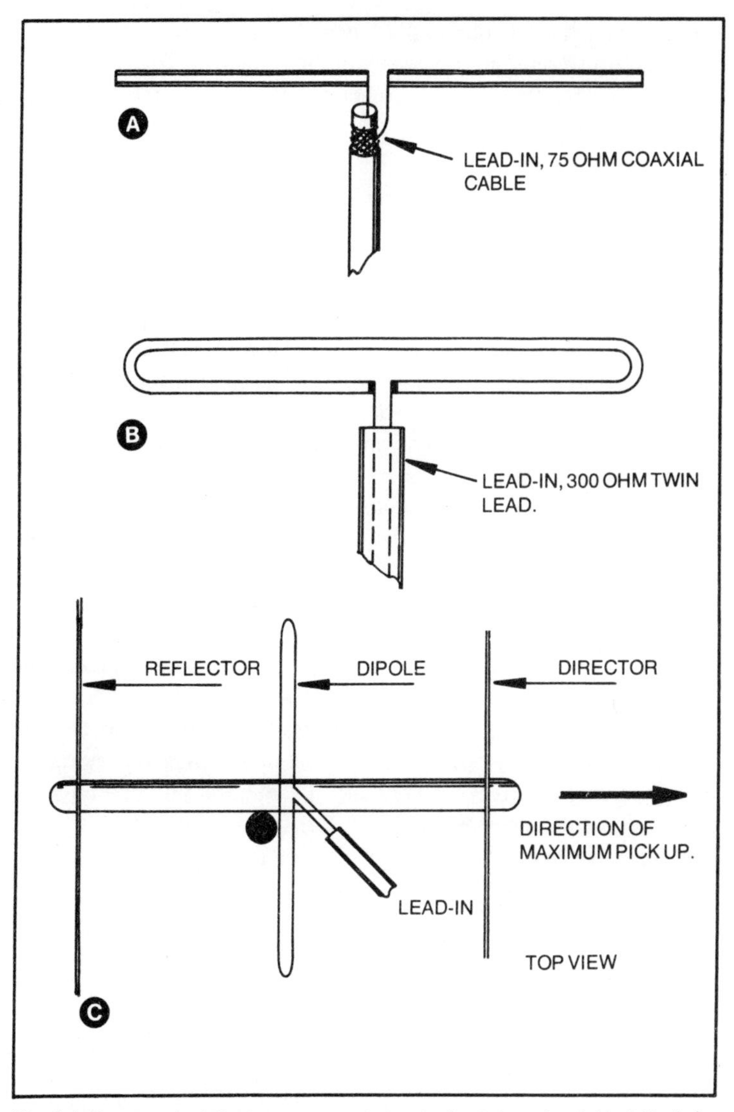

Fig. 1-1. Some typical dipole antennas. A simple dipole is at A, a folded dipole is at B, and a multielement dipole antenna is at C.

folded dipole is made from small tubing of uniform diameter, its impedance will be about 300 ohms.

The folded dipole is often used in a multielement array consisting of the dipole and one or more parasitic elements, such as a reflector and a director (Fig. 1-1C). The reflector is slightly longer than the dipole and is usually spaced about one-fifth wavelength to one-quarter wavelength behind it. The director is slightly shorter than the dipole and is placed an equal or shorter distance in front of it.

The pick-up directivity of the antenna will vary greatly according to its construction and how many elements are used. A simple dipole has a relatively broad directivity, picking up most effectively from stations whose direction is at right angles to the dipole (Fig. 1-2A). As the pick-up pattern shows, a dipole will pick up little energy from its ends.

When two dipoles are crossed, the pick-up pattern is almost omnidirectional (Fig. 1-2B). This method is used in some commercial FM antennas, usually called *turnstyle* antennas.

In some cases you may want to pick up a station that is, say, east of your location, but a more powerful station to the west interferes. You can add a reflector to your dipole so that its pick-up pattern becomes more directional (Fig. 1-2C), by increasing the *front-to-back* ratio of signal strengths. Adding a director, instead of a reflector, can also increase directivity.

For a still more directional antenna, you can add both a reflector and a director to the dipole (Fig. 1-2D). As more elements are added, the antenna becomes both more directional and narrower in its frequency range. Extra directors can be used to increase directivity, but since the outside reflector shields the other elements, it is customary to use only one reflector. Multielement antennas based on the dipole are known as Yagi antennas. Because they are strongly directional, they often require a rotator for a reception of stations at various points of the compass.

If you decide to use some kind of dipole antenna, the best way to find out whether extra elements are needed is by trial and error. If the simple dipole doesn't give satisfactory performance, you need another element—or two. Here is a rough guide to what you may need. If the stations you want to receive are less that 15 or 20 miles from your receiver, you can probably get by with a simple dipole. For 20 to 30 miles you may need an extra element, and for 30 miles or more you will probably decide to build a three-element array. Local topography or buildings can reduce these figures by blocking the signal.

Various kinds of antennas are often described by assigning gain figures, rated in decibels (dB), to them. These figures show how much gain a particular antenna will have when compared to a

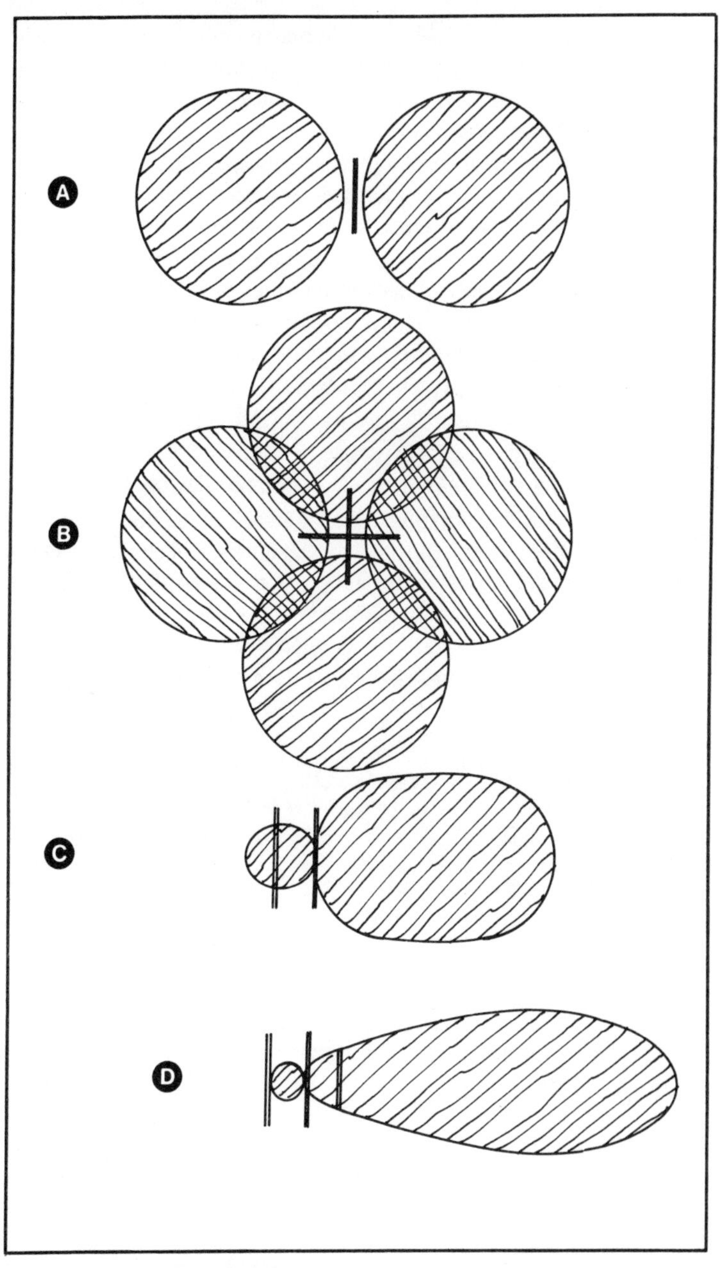

Fig. 1-2. Pick-up patterns of various dipole antennas. Pattern of a simple dipole is at A, of crossed dipoles at B, of a dipole plus one element (reflector) at C, and of dipole plus two elements at D.

simple dipole. For example, a dipole plus one element is typically rated at 3 dB and a three-element array is typically rated at 6 dB. More gain can be obtained from a dipole antenna by *stacking*, which is placing one antenna above another. A pair of stacked three-element antennas would probably have a gain figure of about 10 dB, but the antennas must be identical and should be spaced at a distance of one-half wavelength from each other. Estimates of gain are based on theory and are not always achieved in practice.

In addition to dipoles, many other kinds of antennas have been used for FM or TV reception. Some examples of antennas that offer features attractive to builders are given in some of the antenna projects in the next chapter.

Dipole Dimensions

The dipole is called a half-wave dipole because its length is equal to one-half of a wavelength at the station carrier frequency. When a single dipole is to be used for the entire FM band, it is cut to give optimum reception at the midpoint of the band. Because the FM band extends from about 88 Megahertz (MHz) to 108 MHz, the midpoint is about 98 MHz.

The formula that gives the wavelength for any given frequency is:

$$\text{Wavelength (inches)} = \frac{\text{Energy velocity (inches)}}{\text{Frequency (Hz)}}$$

While this formula is correct, using it directly would involve some unnecessary work with large numbers. By changing the frequency to values in MHz and making the appropriate adjustment in the velocity figure, the formula is much simpler:

$$\text{Wavelength (inches)} = \frac{11{,}800}{\text{Frequency (MHz)}}$$

But to use the formula for antenna design we really want to find the distance that is equal to a one-half wavelength, so the formula becomes:

$$\text{Dipole length (inches)} = \frac{5900}{\text{Frequency (MHz)}}$$

This suggests that the correct length for a dipole that is calculated to receive the FM band should be about 60-¼" long

$$\left(\frac{5900}{98}\right),$$

but there is one correction to be made. The velocity of electromagnetic energy in the metal used for antenna construction is slightly lower than in space. It is about 94 or 95 percent of the velocity in space. To compensate for this difference, we must multiply the adjusted velocity figure, 5900, by 0.95. Thus the corrected formula is:

$$\text{Dipole length (inches)} = \frac{5600}{\text{Frequency (MHz)}}$$

or for 98 MHz:

$$\text{Dipole length (inches)} = \frac{5600}{98}$$
$$= 57\text{-}\tfrac{1}{8}'' \text{ or } 57''$$

This formula can be used to design a dipole to give peak reception for any station in the FM band. Simply divide 5600 by the frequency in MHz of the desired station. Even if you are making a dipole to cover the FM band, you might want to tune the antenna to the frequency of a distant station, particularly if the other stations you want to receive produce a strong signal at your location.

Dipole Impedance

For minimum energy loss, the impedance of the dipole should match that of the line and the line impedance should match the input impedance of the receiver. Most receivers have an input impedance of 300 ohms, the typical impedance of antenna lead-in. If there is to be a mismatch anywhere in the system, it is less harmful at the antenna end than at the receiver end where it can cause energy reflection throughout both line and antenna. Minor impedance variations in home-built antennas will not affect the performance of a receiver as seriously as using a lead-in of the wrong impedance with no provision for proper matching.

As mentioned earlier, a straight dipole has an impedance of about 70 to 75 ohms. If the dipole is folded, using wire or tubing of the same diameter throughout, the impedance is increased to about 300-ohms. This makes the simple dipole an almost perfect match for 300-ohm lead-in and your FM receiver.

If you add any parasitic elements to your dipole, you will lower the impedance of the dipole. The amount of impedance reduction depends on the spacing of the elements; the closer they are installed to the dipole, the more they affect its impedance. This means that the impedance of the original free dipole should be increased if additional elements are used with it.

There are two easy ways to increase dipole impedance: by making a three-conductor dipole or by using conductors of different diameters in the upper and lower arms of the dipole. The second method is easier.

To understand how dipole tubing diameter affects impedance, you must be aware that impedance varies inversely with the square of the current. To increase the impedance in the section of the dipole that is connected to the lead-in, therefore, you only have to decrease the current in that arm. To decrease the current in the lower arm of a folded dipole, you can simply make that arm from smaller tubing than that used for the upper arm. Since the impedance varies inversely with the square of the current, you can quadruple the impedance by reducing the current in the lower arm to one-half that of the upper one. If you want to make the impedance nine times that of the upper arm, decrease the current to one-third of the upper arm, and so on.

Note that antenna elements behave somewhat like pieces of wire. The smaller the cross-sectional area of a wire is, the higher its resistance is to the flow of an electric current. But with the antenna, it is the ratio of low arm tubing diameter to that of the upper arm that counts. The spacing between the arms also has an effect on impedance. Keep this in mind if you decide to build the three-element antenna project. Don't be tempted to substitute tubing of equal size throughout the dipole or to change dimensions significantly.

ANTENNA LEAD-IN

In some ways the transmission line, or lead-in, that conducts the signal from the antenna to the receiver is as important as the antenna itself. The lead-in must match the impedance of the antenna at one end and the impedance of the receiver at the other. Most receivers have a 300-ohm input, so 300-ohm lead-in is the standard.

The ideal lead-in would produce no loss of signal between the antenna and the receiver, and it would pick up no external spurious radio frequency energy, such as ignition signals from passing cars.

All lead-in causes some loss, particularly if the line is long, but there is a significant difference in the amount of spurious energy picked up by various kinds of lead-in.

In addition to the above desirable electrical characteristics, a lead-in should be durable. This is particularly important if your antenna is to be mounted outdoors. An additional characteristic that most people consider is price.

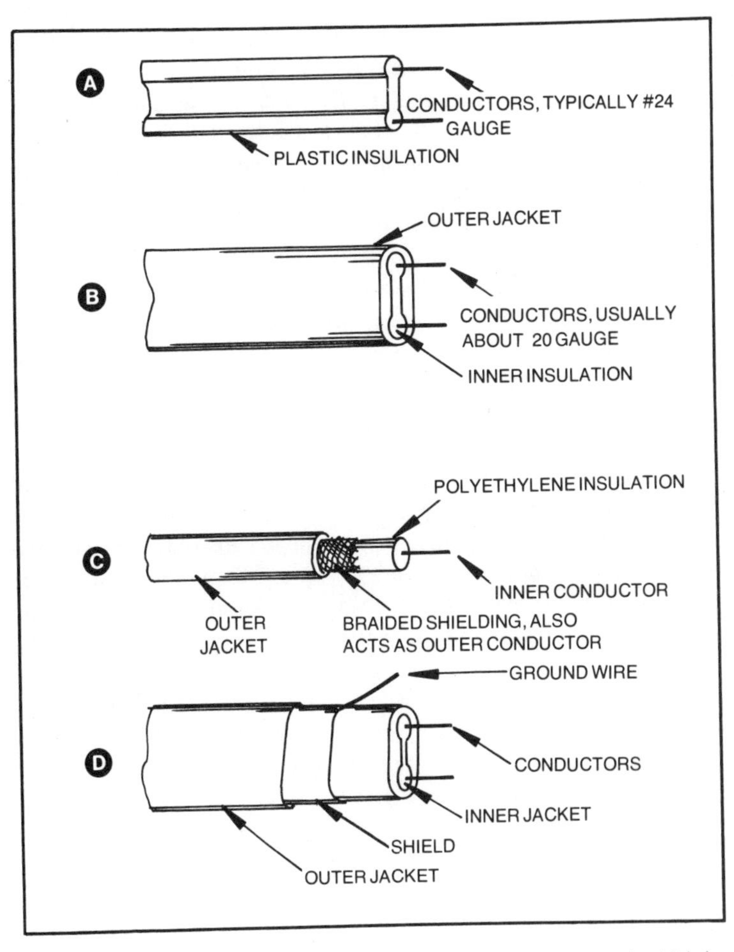

Fig. 1-3. Types of lead-in generally available. At A is flat twin-lead, which is recommended for strong-signal areas; it is relatively inexpensive. At B is heavy-duty twin-lead, which is excellent except where interference is a great problem. At C is coaxial cable, which requires a transformer for impedance matching. At D is shielded twin-lead, which is the best choice for areas where signal interference is a problem.

Although about 10 kinds of lead-in are manufactured, most electronic stores carry only a few of the most practical and widely used kinds. The most available and commonly used lead-in is flat twin-lead (Fig. 1-3A). Its 300-ohm impedance gives good matching to antennas and receivers, and it is inexpensive. But it produces high signal losses in long leads and is subject to extraneous energy pickup. If you live in a high signal area, it will work fine. For indoor use you can get the kind with white or clear polyethylene insulation. Sunlight will cause this kind to deteriorate, so the brown covering is more durable for outdoor lead-in.

Where the antenna must be placed more than a few feet from the receiver, heavy-duty twin-lead is a better choice (Fig. 1-3B). It has heavier gauge conductors and so produces lower loss.

If you have problems with interference noise, choose a shielded lead-in. In the early days of FM and TV, the only kind of shielded lead-in generally available was coaxial cable (Fig. 1-3C). The outside conductor of coax is grounded, so this lead-in is unbalanced. Also it has a typical impedance of 75 ohms, requiring transformers to match it to the antenna and the receiver. These transformers produce loss of signal energy, making coax cable of questionable value, since the transformers add to the cost of installation.

The best lead-in for locations with weak signal energy and interference is probably a shielded twin-lead (Fig. 1-3D). This lead-in combines the good matching and low loss characteristics of heavy-duty twin-lead with the shielding feature of coax. It is somewhat more expensive per running foot of lead-in than coax, but because it requires no transformers, the total cost may be no greater.

To summarize your choice of lead-ins, use cheap flat twin-lead for simple indoor antennas; choose heavy duty twin-lead for long runs and outdoor use where noise is no problem; and if you live in a weak-signal area and are subject to heavy traffic or other noise-producing signals, go for the shielded twin-lead. Regardless of the kind of lead-in, it should be installed with little slack and with a minimum of horizontal runs.

ANTENNA HEIGHT

A rule of thumb used by antenna technicians is "the higher the better." But the rule holds only in regard to significant changes in elevation. One man who moonlights by installing antennas for friends and neighbors told me that he never tries for additional height because an antenna can be too high. He learned that fact when he accidentally lowered an antenna slightly during testing and the TV picture improved tremendously. What his experience proves is that minor changes in elevation sometimes affect reception unpredictably because the height above ground, in wavelengths, affects antenna impedance. The best way to establish the right height for your antenna is by experimenting. As a rough guide, you can expect twice the signal voltage at your receiver when you double the height of the antenna above ground.

Outdoor antennas almost always work better than indoor models. Good performance can often be obtained with attic antennas unless you have a roof made of metal or slate. House walls often shield your antenna, especially houses with metal plaster lath or wire-reinforced stucco.

SAFETY PRECAUTIONS

Thousands of people have been killed while erecting outdoor TV or FM antennas by permitting the antenna or the mast to fall against or near a high-voltage power line. Make sure your outdoor antenna is far enough from any power lines that it can't possibly come within 10′ of the line. In some cases this may limit the possibilities for antenna installation, but under no circumstances would it be prudent to ignore the danger. Even if the antenna is securely installed, a storm with high winds can bend the antenna or mast into the power line. This could produce a serious hazard to other people as well as yourself.

Another possibility of injury by an antenna is inadequate mounting procedure that permits the antenna to fall and hit someone. Even if no one is injured by a falling antenna, the tubing can cause serious property damage. Antennas are frequently mounted on short masts and attached only to chimneys. This kind of installation is adequate if the antenna is not much higher than the chimney and if chimney straps are used. Roof-mounted antennas will require guy wires or cables that are securely anchored. Another possible mounting procedure is to anchor the base of a free-standing mast at the foundation of the building and again at the roof level with a U-bolt or to a chimney by straps. You can test the stability of the mast by shaking it with considerable force and watching for excessive movement at any point.

Even if you make an installation that is perfect mechanically, you can endanger your receiver or even your home by failing to give proper attention to lightning hazards. The possibility of receiver damage is the more likely one, and it can happen even if you never experience the ear-splitting crack of a direct hit. Charges can build up in an exposed antenna and leak off. The leaking away may prevent the more dramatic event, but it can burn out the fine wires in an antenna coil or other parts.

For lightning protection, you should combine a lightning arrestor with a good ground rod (Fig. 1-4). An arrestor will probably prevent damage to your set from the kind of leading charges just mentioned. It will do little to prevent a direct hit except that any provision that encourages the leaking of charges will help to prevent the massive build-up that precedes a strike. One major precaution is that the mast should be grounded to a copper rod driven into the ground. The required depth of a ground rod will depend on soil conditions, so it varies from site to site. Soil moisture is more important than depth, so you should soak the ground around the rod, particularly during storm seasons. Place the ground rod at least a foot from the foundation, where the soil is more likely to be moist. If you live in a rural area with few tall structures or other antennas near, you must give far more attention to lightning protection than someone who lives in a city apartment house. Notice that if you choose shielded lead-in, you can ground the shielding and will need no lightning arrestor. For all ground connections use a heavy aluminum wire; No. 8 wire is a frequent choice.

CHOOSING AN ANTENNA

The question is often asked, "Which kind of antenna is best?" If there were a single type that could excel in every respect, it could be recommended to everyone. Some antennas are obviously superior to others, but each one has certain disadvantages. One disadvantage that can be easily appreciated is a high price. One shouldn't skimp on getting a good antenna, but no one should get an expensive antenna without some indication that it will make a significant difference in FM reception. To make an intelligent choice, consider your location and the location and signal power of the FM stations you will want to receive.

One obvious method of gathering information on these factors is to check with your neighbors to see what kind of antennas they use and how well they work. Even if there are no stereo fans in your area, you can gain useful information from observing the kind of TV reception available. If your neighborhood is in a fringe area, you will need an antenna that offers high gain, and this means one that is

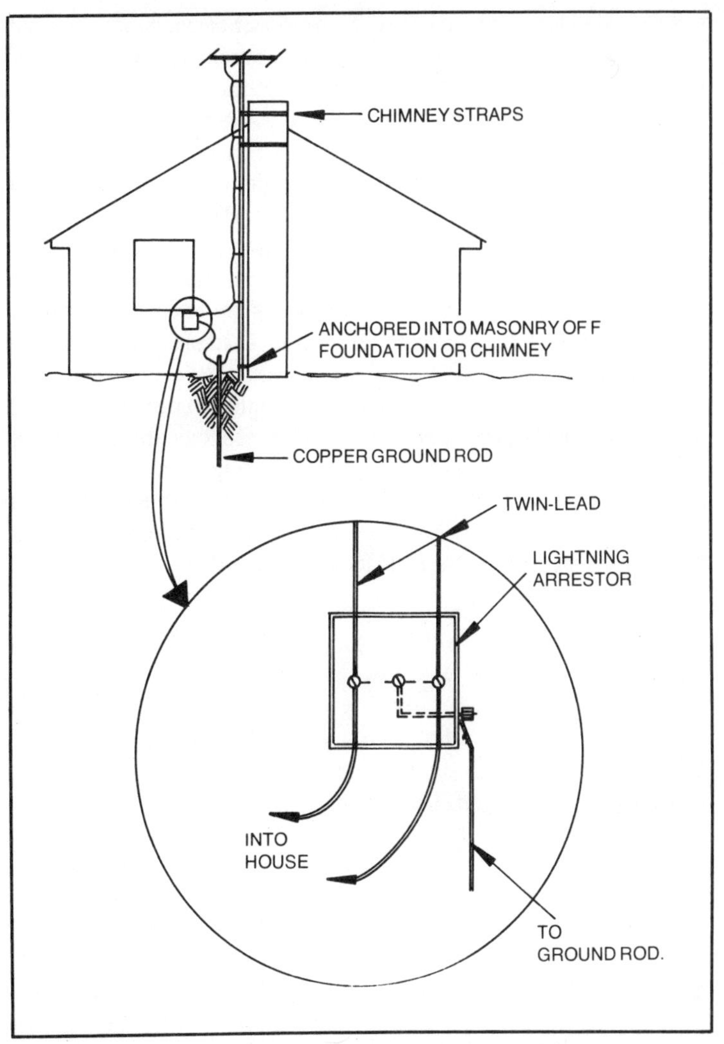

Fig. 1-4. Proper outdoor antenna installation. Note that the lightning arrestor is placed on the lead-in near where the lead in enters the house and is connected to a good ground.

highly directional. The more directional antennas would be appropriate for most fringe or near-fringe areas, but they are most practical where the desired FM stations are located in the same city or in cities on the same compass point. Of course, this situation rarely occurs. If you choose a directional antenna and want to receive stations from more than one direction, therefore, you must either install the antenna so you can rotate it or use more than one antenna with a switch to permit you to select one that is aimed toward the station you want to receive. If the number of stations you want is limited, you can make up antennas to match the frequency of each one.

For more typical situations, such as an urban residential area where signal strength is high, a highly directional antenna would probably be an impractical choice. Here, good reception is usually possible with a single omnidirectional antenna, such as Project No. 7 in Chapter 2. In many such locations an indoor antenna is adequate, although outdoor models almost always give improved reception.

A logical procedure to follow in choosing an antenna is to try a simple model first, one that requires little expense or construction time. Then if its performance leaves something to be desired, go to one with higher gain, such as Projects No. 6, 8 or 9. Regardless of your final choice, you should expect to do considerable experimenting to get the best performance it can give you.

ANTENNA EXPERIMENTS

Even if you live in a high-rise apartment, a few simple experiments will show where and how to install your antenna. For indoor models, you will probably want to give some weight to appearance, so it will be desirable to hide or camouflage the hardware. Because this greatly limits the possible sites, it is all the more important that you choose the best one.

To help you pinpoint the best location, try to find an old set of TV rabbit ears. You can often pick one up for pennies at a flea market or garage sale. Connect the lead-in to your receiver, and start with the rabbit ears in a horizontal position, set at about 57" from tip to tip (Fig. 1-5). Move the antenna through various positions behind large pieces of furniture and note the positions that give superior reception. Sometimes you will find that even a minor change of angle or location will make a significant difference. In some cases the house wiring or plumbing can produce reflections that can cause distortion and even reduce stereo separation. While moving the antenna, vary the angle of the elements from 0 degrees to 180 degrees. At the same time, try various unbalanced arrangements, with one element extended farther than the other one. While most people use rabbit ears by placing the rods in a vertical V, you may be able to get adequate reception by rotating the V to a horizontal position in a corner of the room. This would suggest the possibility of making a folded dipole from 300-ohm

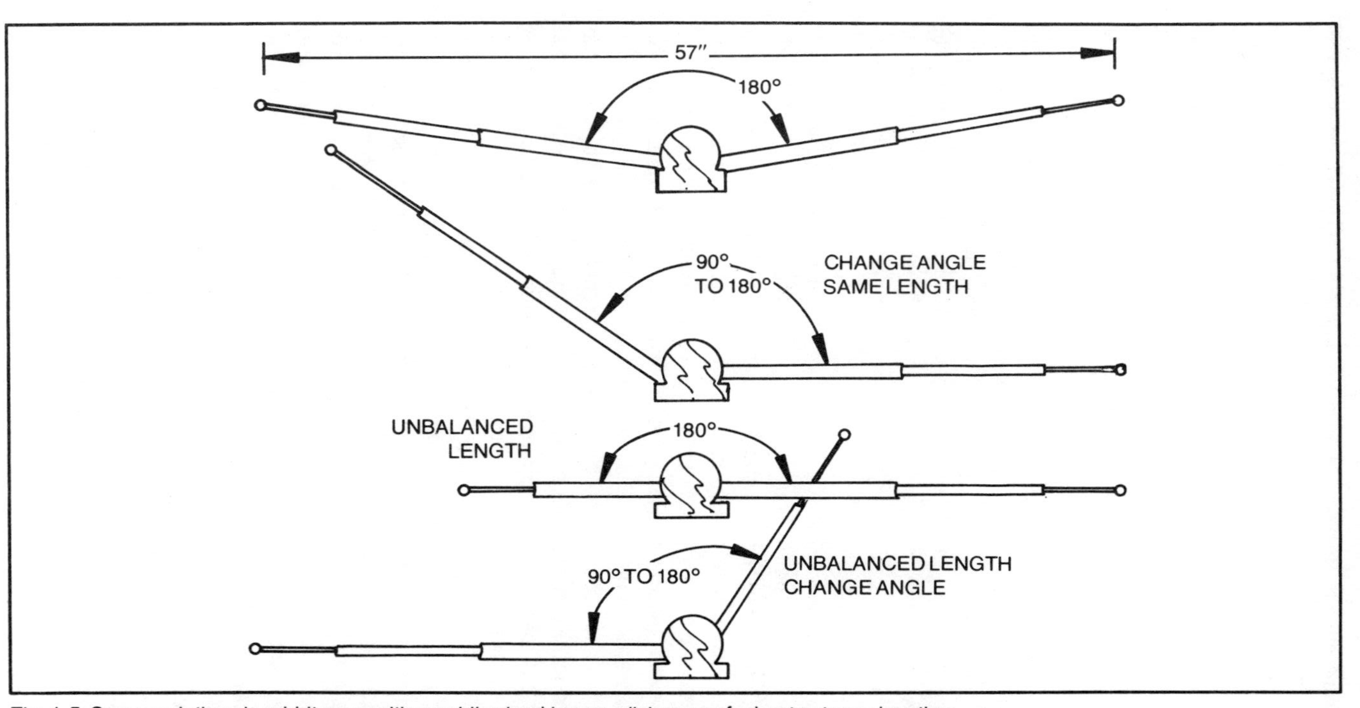

Fig. 1-5. Some variations in rabbit ear positions while checking your living area for best antenna location.

transmission line and taping the dipole to adjacent walls in that room corner, an inconspicuous arrangement.

HOW TO REDUCE STUBBORN REFLECTIONS

The kind of experiments mentioned earlier will usually solve minor reflection problems, but further steps are sometimes necessary. One of the best ways to analyze what you are doing is to have a visual display of the symptom to verify what you hear. If there is a VHF television station transmitter located in the same direction from your receiver as the FM station you want, you may be able to use a TV set to monitor what you are doing. Connect your FM antenna to the TV set. TV reception with FM antenna may be marginal, but any reflections will be obvious on the picture tube as *ghosts*. While making subtle changes in your antenna, watch the ghosts and select the antenna position and orientation that best eliminated them.

One way to block the reflected part of a dual-path signal is with an extra dipole. You can use the extra antenna as a *parasitic* element with no electrical connections to the driven antenna; in fact, after installing a 300-ohm lead-in dipole, you can use your old rabbit ears as the reflector. The chief problem here is to get the proper spacing and yet keep the hardware out of sight. About 20″ between dipole and reflector is usually about right.

The lead-in can be a source of trouble. Keep it as short as possible and check its position in relation to metal objects, such as metal windows or appliances. It should not touch any of these; sometimes even a minor separation will work wonders.

You can try a shorted stub, tuned to remove the reflection by trial and error. This stub can be connected to the antenna terminals of the receiver in the same way that the stub for Project No. 2 is connected. For this purpose the stub is shorter, about 6″ long. Shorten it until you find the length that gives best reception. You can install a SPST (single-pole-single-throw) switch in one side of the 300-ohm line of the stub to switch it out of the line when not needed.

Reflections are sometimes especially troublesome in areas of high signal strength. If this is your situation, you can employ a signal attenuator that will reduce the strength of the primary signal and the reflected signal. An attenuator sometimes effectively removes the reflection by dropping its strength below the threshold of your receiver. This method is demonstrated by Project No. 4.

If none of these measures work, you might have to go to a more directional antenna, one that rejects reflections by its high front-to-side and front-to-back pick-up ratio.

ANTENNA MAINTENANCE

Indoor antennas need little maintenance, but if you install an outdoor antenna, you should inspect it periodically for damage or gradual deterioration. This inspection should include both mechanical and electrical tests.

Your first mechnical test should be to the antenna mounting system. Shake the mast and see if the mounting hardware has loosened. Sometimes the mast is held fairly well but can twist in its mounting, changing antenna direction and affecting reception. If the antenna is a multielement design, such as a Yagi, check to see that each element is securely in position and that none are bent or cracked. If you live in a region of frequent high winds or of ice buildup, you might have to rework the antenna at regular intervals.

In some industrial areas, corrosive fumes may cause problems. Here, the antenna and even the lead-in should be wiped clean occasionally. Such problems are not confined to urban population centers; the recent shift to wood stove heating in many rural areas has increased the soot particles in the air. This will be particularly true if you have a fireplace or wood stove in your house.

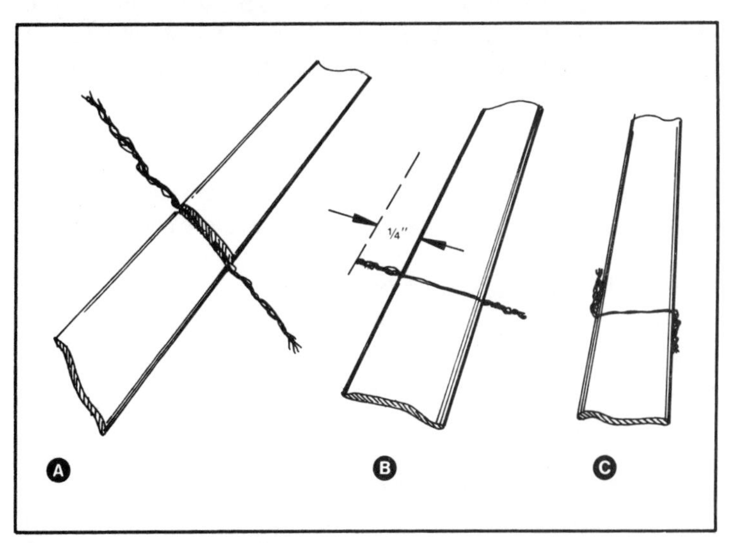

Fig. 1-6. How to splice 300-ohm twin-lead: twist leads together at A, solder and cut off to ¼" at B, and bend leads against the edge of the line and tape at C.

For a quick electrical check, disconnect the lead-in at the receiver and measure the resistance through the lead-in and antenna. Note that you will obtain the actual resistance of the lead-in, connections, and dipole with a folded dipole. If you have another type of antenna with a noncontinuous element connected to the lead-in, you must short across the elements to get an accurate reading. The resistance should be no greater than a few ohms; the exact value will depend on the length of lead-in. If the reading you get is higher than expected, there are probably resistance points where the lead-in connects to the antenna. Remove each connection and polish the contact points with fine sandpaper or emery cloth. Replace rusted connector bolts with new ones. After making the connection, it's a good idea to spray the contacts with a protective plastice spray, such as auto ignition spray.

In some cases the electrical test will show an open circuit. One likely cause is a break in the lead-in, probably caused by high winds whipping the antenna. Lead-in can be patched by twisting the conductors tightly together and soldering them. The soldered leads should be bent flat against the edge of the ribbon and taped (Fig. 1-6). To prevent future breaks, leave some sag in the line near the antenna so that the antenna can move without putting undue stress on the line.

If FM reception problems arise at intervals between antenna inspections, first eliminate the antenna as a possible cause before suspecting your receiver. You can easily check the performance of an antenna system by substituting another receiver known to be in good working order. If another receiver is not available, use a substitute antenna and compare the performance of your receiver with each antenna. A simple indoor antenna should not be expected to even approach the performance of an outdoor directional model, so if it does, you know something is wrong out there on the roof.

TV ANTENNAS OR CABLE HOOKUPS FOR FM

In areas of high signal strength, you can often use a single antenna for your TV and your stereo system, but only if your TV antenna has no FM trap. You can make a simple resistive splitter (Project No. 3) that will permit you to connect your TV set and FM receiver to one antenna. The resistive splitter causes some loss of signal, so to be successful, it requires a strong TV signal as well as a strong FM signal.

If you live in an apartment building that has a master antenna on the roof, you can probably forget about using the signal from it.

Because more people are interested in a good TV signal than in high-quality FM sound, these antennas often are designed *strictly for TV*. In such situations, you can rarely mount any kind of external antenna, so you must be satisfied with one of the indoor models. Some of the projects in the book will offer good performance, particularly if you experiment thoroughly to find the best location and orientation after you finish the antenna.

Salesmen for cable TV almost always contend that their systems will give good FM reception. Unfortunately, they usually exaggerate the value of their cable system for FM and are often even unaware of the kind of reception possible with a good antenna. Like master TV antennas, cable systems are designed almost exclusively with providing TV reception. In our tests, we found that even the simplest antenna projects in this book gave better FM performance than a cable connection. It is said that some cable systems offer good performance on FM, but not the ones we checked.

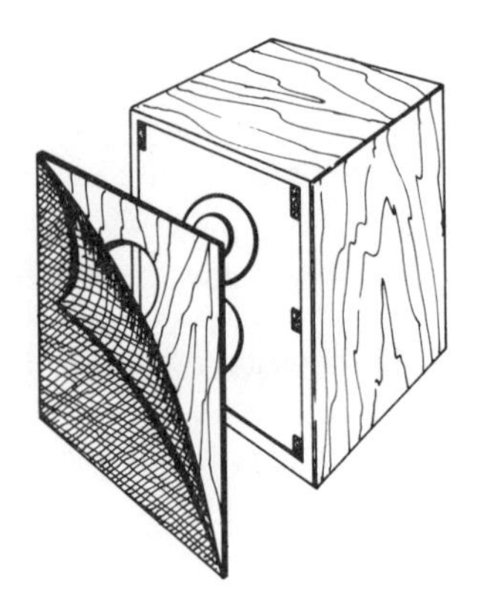

2
Antenna Projects

As you glance through the antenna projects in this chapter, you will see that some of them are a bit unusual, both in design and material. Don't underrate these projects because of their somewhat bizarre appearance. They all work well in most situations, and each project is based on a solid theoretical foundation, not on magic.

Before you begin, here's a word of advice: Expect to spend a considerable amount of time experimenting with your antenna if you want optimum performance. Antenna behavior in any specific location defies analysis by theory alone because there are so many variables, such as topography, neighborhood structures, kind of dwelling, and many others. Many installations represent a rough guess on the part of the technician as to the best location for the antenna. No one wants to pay for exhaustive experiments in addition to the usual cost of labor and material. By experimenting, you can obtain superior performance from your antenna system as well as having some fun.

PROJECT 1: CAPACITIVE LINE LOADER

This project looks like a Mickey Mouse job, but don't sell it short. While it takes only a couple of minutes to complete, it can sometimes produce results that approach wizardry. In other situations, it will make no discernible difference. But with practically no cost or time involved, it's always worth a try.

The only material required is a 2″ x 2″ or slightly larger piece of aluminum foil. To thoroughly test it out, try at least three pieces, one 2″ x 2″, one 2-½″ x 2-½″, and one 3″ x 3″. Wrap the foil around the lead-in near the receiver (Fig. 2-1). Then slide the foil wrapper up and down the line. The best position is usually found within an inch or so from the set to about a yard away. Obviously you will want to test this while tuned to a problem station, one in which you can hear slight distortion or the fizz of inadequate quieting. When you reach the position of minimum noise, leave the foil in that location.

If you have any TV reception problem, try this project on your TV antenna lead-in. Again, it is useful only if you have marginal reception on at least one station. With a TV set, you can often see the picture change contrast from a washed-out image to deep tones when the foil is moved to the best position on the lead-in. Even though it doesn't work in all situations, the line loader is worth a try for reception problems.

PROJECT 2: TUNABLE STUB

Many of the statements used to describe Project No. 1 could be used again for Project No. 2. Here again, you use an aluminum foil slider to tune the system, but the slider is on a short length of 300-ohm line that is connected in parallel to the antenna lead (Fig. 2-2). Such a piece of line is called a stub, or more specifically, a quarter-wavelength stub. Theoretically the stub should be equal to one-quarter wavelength at the frequency to which it is tuned. For the FM band, you can start with a 30″ stub and reduce the length, in

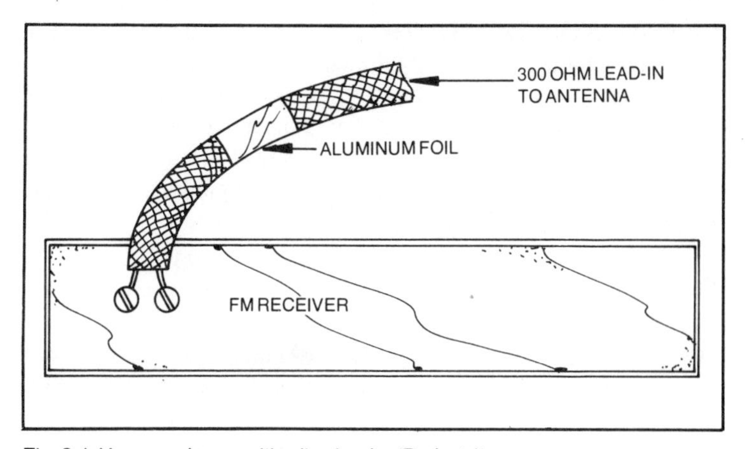

Fig. 2-1. Homemade capacitive line loader (Project 1).

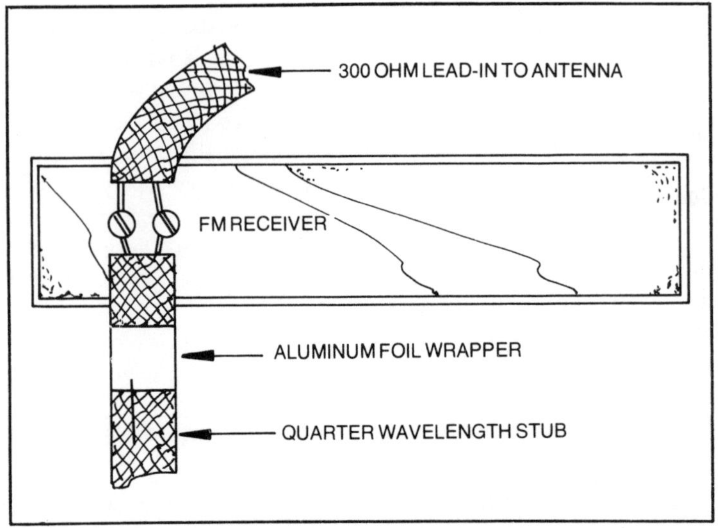

Fig. 2-2. Tunable stub used to reduce interference and reflections (Project 2).

small increments, until maximum effect is obtained. At each length try moving the slider over the entire length of the stub.

Again, this is a project that can also be adapted to your TV set. It is an excellent method of reducing reflections, or ghosts. As your ear becomes tuned to the difference that even minor reflections can make in your sound system, you will begin to appreciate the fact that on FM, as well as TV, it pays to prevent reflections. As in Project No. 1, one of the main ingredients to success for this project is patience.

PROJECT 3: RESISTIVE SPLITTER

If you have a suitable TV antenna in a good signal area, you can use this splitter to connect your FM to the same antenna as your TV set. Note the two qualifications. Some TV antennas have FM traps or are just not very sensitive in the FM band. But if you live in a high strength signal area, you can use many kinds of TV antennas this way to good advantage.

While some splitters, or couplers, are designed to split the TV signals between VHF and UHF, this simple resistive splitter is not frequency sensitive (Fig. 2-3). It can be used for two FM sets or two TV sets, as well as for a TV and an FM set.

If you live in a fringe area, this splitter is not recommended. Since it is resistive, it produces some loss of signal strength. In fact, the expected loss will be equal to about one-half the strength of the signal without the splitter. This doesn't make as much difference in performance as you might think, but in marginal areas that much loss would be noticeable. Another reason that is doesn't make sense for fringe locations is that in such areas you would want an antenna designed specifically for FM anyway.

Construction

There are many ways to accommodate such a small gadget as this splitter. It can be built into a small plastic hardware container, like the one shown in Fig. 2-4, or on a piece of hardboard. If you prefer to use binding posts, instead of the screw-type terminals, you can mount the binding posts on a piece of perfboard or other nonconductor. Regardless of the method of construction you follow, try to keep the resistor leads as short as possible (Fig. 2-5).

The splitter shown here was built on a 2″ x 2-½″ hard plastic container in which small brads were originally packaged for sale. Many hardware items are offered in such containers. You can buy small project boxes, but their cost is usually more than that of all the other parts used in this one project.

To help locate the holes you will drill, it's a good idea to temporarily cover the top of the box with paper. If you have any adhesive-backed labels, they are ideal for this purpose. Simply cut out a piece of label large enough to cover the top of the box and stick it on. Then set the terminal strips in place and mark the proper location of the holes for mounting bolts and terminal lugs. Drill the mounting holes with a ⅛″ drill bit and the holes for the lugs with a ¼″ bit. You will have to elongate the ¼″ holes slightly to allow the terminal strips to fit properly.

Install the terminal strips with #6-32 x ⅜″ bolts and nuts. Mount the resistors between the terminal lugs, as shown,

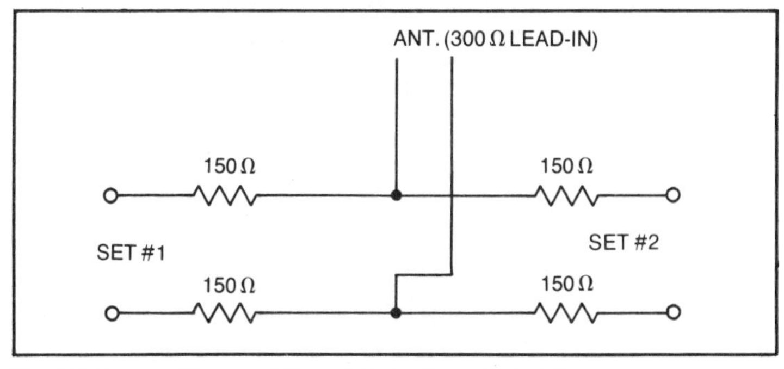

Fig. 2-3. Schematic diagram for resistive splitter (Project 3).

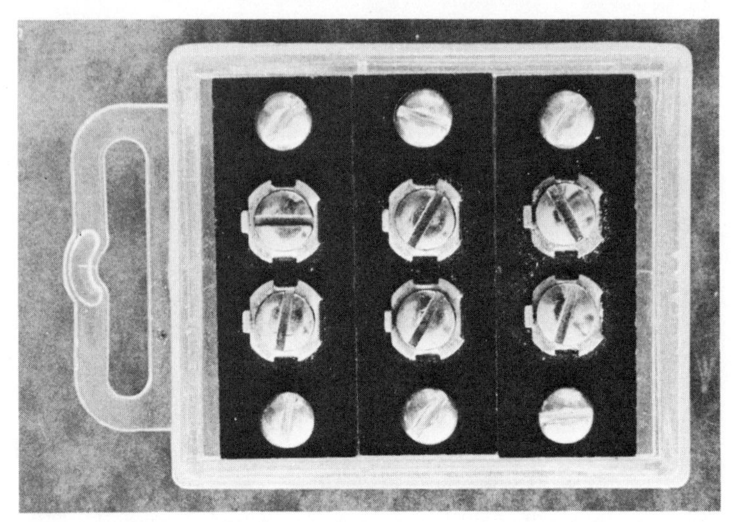

Fig. 2-4. Resistive splitter (Project 3) can be built on a small plastic hardware box.

and solder the connections. Cut off the excess leads. Connect your antenna lead-in to the middle terminal strip. You can then connect your TV set to the terminal at one side and your FM receiver to the opposite terminal.

PROJECT 4: SIGNAL ATTENUATOR

Most of the antenna projects in this book are designed to provide your receiver with a stronger signal. This project does the opposite: It reduces the signal strength. Why? There are at least two situations in which an attenuator can provide you with better FM reception. The obvious case is where you live near a powerful transmitter that overloads the front end of your receiver and produces distortion. But there is another kind of problem that can sometimes be cured by reducing signal voltage: reflected signals. Because bad reflections often occur in cities with tall buildings, they frequently go hand in hand with a healthy primary signal. In such cases, you can sometimes reduce the input from the antenna to your receiver just enough to eliminate the reflections, but not the primary signal.

In many situations, the attenuator will be needed for only the most powerful stations. If this is true, you can prepare a short piece of 300-ohm lead-in with alligator clips at one end. Connect the other end to your set. With this arrangement, you can short out the attenuator when it isn't needed.

The Table in Fig. 2-6 shows the correct value resistors to use for various amounts of attenuation. If you have no idea how much attenuation to apply, start with the 6 dB pad. This will cut the signal voltage in half. If you need more attenuation, go to the 10 dB or 20 dB values. If you are in doubt about what values to use, don't solder the connections until you have tried the attenuator in your system.

Construction

This attenuator, like Project No. 3, can be built on a small plastic box or on a piece of perfboard (Fig. 2-7). Because there is no need for screw terminals or other connectors, the perfboard was chosen here. The 1″ x 1-⅜″ perfboard makes for a more compact unit in this project.

The only precaution to observe is to keep the leads short. Install the resistors on the perfboard, choosing holes that make for short leads. Twist the leads together under the perfboard and solder the connections. Cut off the excess leads. Connect your lead-in at one end and two spade lugs at the other end. Then solder the connections. That completes the project.

PROJECT 5: 300-OHM DIPOLE

This project is identical to those antennas made of lead-in that are often included free with receivers. A slight difference exists, however. The antenna here is mounted on a board to hold it in

Fig. 2-5. Bottom view of Project 3. Keep leads short.

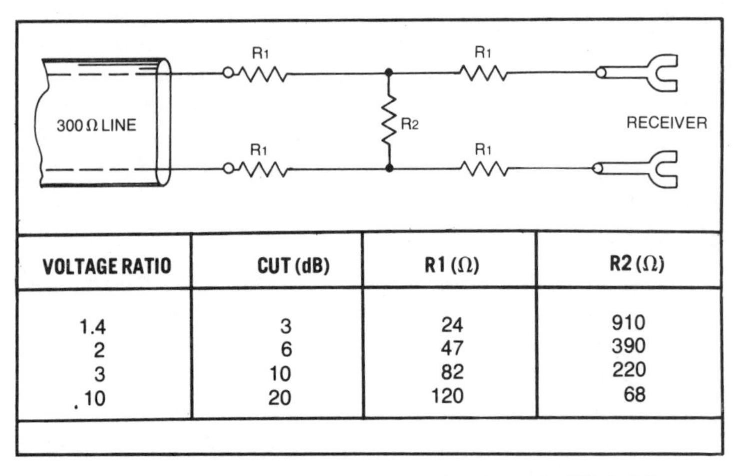

VOLTAGE RATIO	CUT (dB)	R1 (Ω)	R2 (Ω)
1.4	3	24	910
2	6	47	390
3	10	82	220
.10	20	120	68

Fig. 2-6. Schematic diagram and resistor values for attenuator (Project 4).

position and with a clamp that permits you to experiment by moving it around the room and clamping it on various pieces of furniture. After you have fully explored your living quarters for the best location, you can make the installation in a more permanent and inconspicuous way.

Although the project pictured uses a lamp clamp, you can employ various kinds of clamps. In some cases, however, you will have to cover the jaws of the clamp with tape to protect furniture. A car-battery booster clip, for example, will work if you are clamping the antenna to thin objects such as bookshelves. The lamp clamp is more versatile.

You can probably find an old lamp clamp at a flea market, but I bought a new clamp from which the reflctor and socket had been lost or damaged for $1.50. The only other hardware needed is a couple of No. 8-32 x ¾″ bolts, a corner brace, two yardsticks, and 300-ohm lead-in.

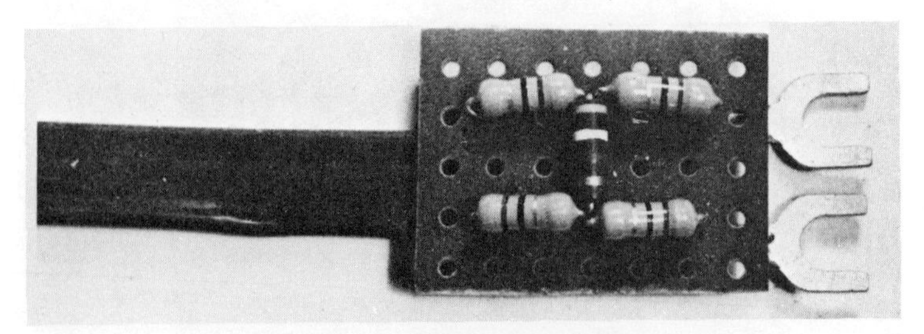

Fig. 2-7. Attenuator constructed on perfboard. Keep leads short.

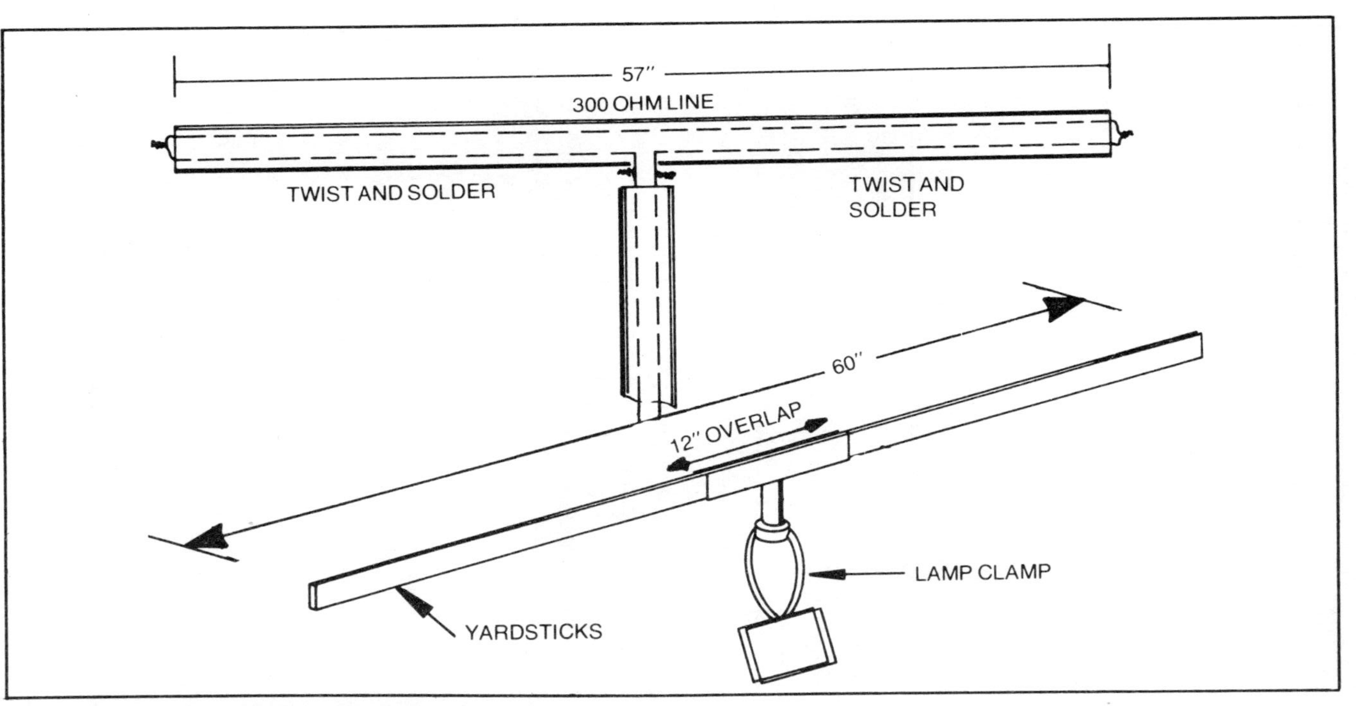

Fig. 2-8. Dimensions for 300-ohm dipole (Project 5).

Construction

Apply wood glue to a 12″ length of each yardstick and clamp them together until the glue sets. If you have no clamps, you can use weights to hold the yardsticks together. Prepare a 58″ length of 300-ohm lead-in by stripping ½″ of insulation from each end and twisting the leads together (Fig. 2-8). Cut one conductor at the halfway point and strip each cut end back ¼″. Prepare a piece of lead-in long enough to permit you to move the antenna around the room, propbably at least 10′ or 12′ long, and bare the conductors at one end. Twist these to those at the midpoint of the dipole and solder them. Solder the twisted conductors at the end of the dipole. Prepare the other end of the lead-in by splitting it for an inch or two and stripping ½″ of insulation from each conductor. Install spare lugs on each conductor, or solder each one and bend it into a "J."

When the glue has set on the yardsticks, drill a 3/16″ hole at the center of the combined sticks. You can locate this point easily by looking for the 6″ mark on the overlapping part of the sticks (Fig. 2-9). Bolt a corner brace to the lamp clamp. Then bolt the corner brace to the yardstick beam. In some cases you may not need the corner brace. Some lamp clamps have a mounting arrangement that is different from the one shown in the photos, and many have a ball joint for more flexible orientation. The brace was used here to make the yardstick beam parallel to the piece of furniture to which it was clipped, rather than at right angles to it.

Tape the 300-ohm dipole to the yardstick beam so that it rests flat against the yardsticks without drooping. You can now clamp the antenna to any chair back, cabinet door (Fig. 2-10), or other projection that appears to be in a suitable location for your antenna.

You will find that this versatile antenna will often give far superior performance to the throw-on-the-floor-behind-the-sofa or tape-on-the-wall twin-lead dipole that many people use. It is excellent in the attic where you can clamp it to the upper edges of ceiling joints or to cross braces to find the best position.

PROJECT 6: BIKE WHEEL ANTENNA

Here is an antenna that offers an amazing combination: good performance at lowest possible cost. It has an amazing bandwidth, working well on FM and TV channels 7 through 13. Although you might expect it to be omnidirectional by its appearance, it is really quite directional. This makes it a good choice where reflections are a problem. And it is so simple and light that you can easily set it on a wood dowel and rotate it for maximum signal strength from various directions.

Fig. 2-9. Use small corner brace to bolt lamp clamp to yardsticks.

Fig. 2-10. Project 5 in use.

The low cost is possible because the antenna can be made from little more than junk. All you need is an old bicycle rim, one resistor, a few scraps of wood, and enough 300-ohm lead-in to span the distance to your receiver. The bicycle rim can be a damaged rim which you can probably get free from your local bicycle shop or a neighbor's trash can. If you can't find a bike wheel, use a length of aluminum or copper tubing, or even just a piece of heavy copper wire.

You will note that each half of the rim should measure about 28″ or 29″ long for a total wheel circumference of about 57″, which is one-half wavelength (Fig. 2-11). The wheel that comes nearest to this size is the rather rare 20″ x 1-3/8″ wheel. If you can find an old one of these, you can use the rim as is. But any larger wheel will do; simply cut out enough rim to make the length right, and bend the remaining portion into two semicircles. The most common sizes of wheels these days are those used on 10-speed bikes, 26″x1-3/8″ and 27″ x 1-1/4″. These measure about 73″ and 79″ in circumference. If you choose one of these, you should cut out a section of rim, 16″ from the 26″ wheel and about 21-1/2″ from the 27″ rim.

Don't substitute other sizes of 26″ rims unless you measure the circumference and adjust the cutout to make a final circumference of 57″. In the strange world of bicycling a 26″ x 1.375″ tire won't fit a 26″ x 1-3/8″ rim. And a 20″ x 1.75″ or a 20″ x 2.125″ rim will have smaller diameters than the 20″ x 1-3/8″ size that is preferred. The tire size is measured from the outside of the tire to the outside of the tire on the opposite side of the wheel, so the rim sizes can vary considerably according to the width of the tire. The measurements given here for each size were made around the drop center of the rim; the external rims that hold the tire bead would have a slightly greater diameter.

One advantage of using the 26″ or 27″ size rims, aside from their easy availability, is that you can often cut away the damaged part of a ruined rim. For an attic or for any location where the antenna isn't exposed to view, the damaged rim will work just as well as a new one. If you plan to install the antenna outdoors, try to

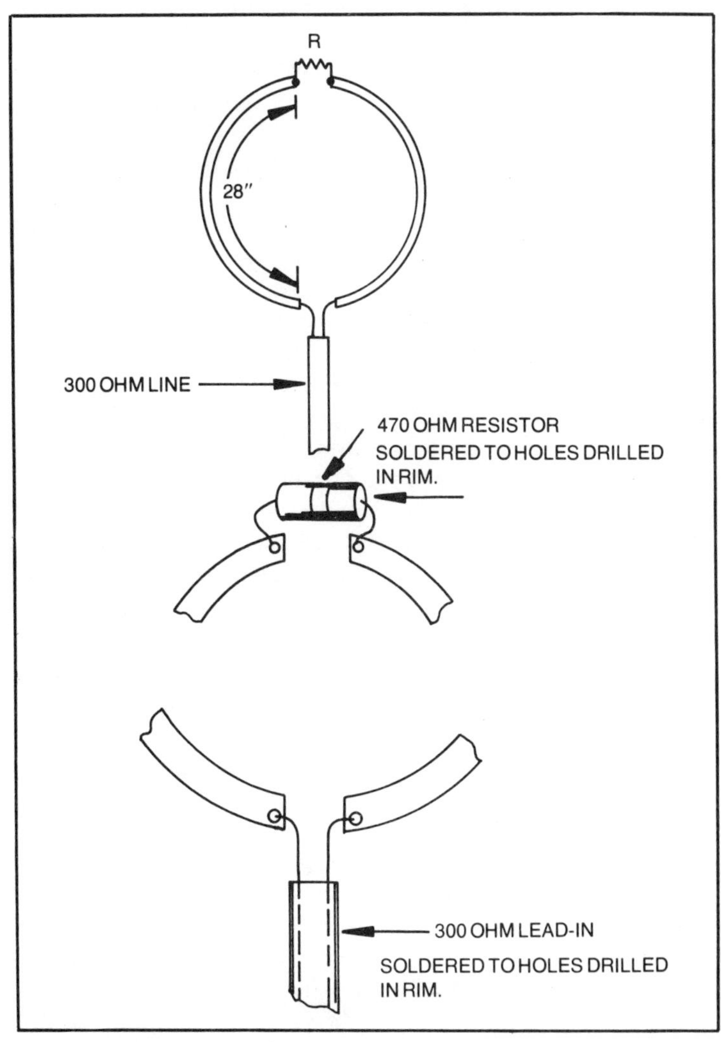

Fig. 2-11. Details of bike wheel antenna (Project 6). Antenna is directional. Use with capacitive line loader (Project 1) for best performance.

get an aluminum alloy that will not rust. The only size you are likely to find in aluminum alloy is the 27″ size.

A special kind of 27″ rim is the kind made for sew-up tires. Sew-up rims invariably are made of aluminum alloy. These are often available because some cyclists find that the risk of flats with the thin-walled sew-up tires outweighs the potential reward they offer in easy pedaling, so they convert to conventional tires. Because of this, used sew-up rims are frequently available in good condition at low cost unless your bike shop caters to racers.

To adapt the 26″ or 27″ rims, you need not even measure the rim if it has 36 spoke holes, the most common type. For the 26″ x 1-3/8″ size, simply remove eight spoke intervals. You will then have about 56″ of rim left. For the 27″ size, remove 10 spoke intervals.

Construction

When you get the rim, it will likely be in a damaged wheel, complete with spokes and hub. Cut out the spokes with wire-cutting pliers. If the rim is a steel one, clean it with steel wool to remove the rust. Count the number of spoke holes in the rim. If you have a 20″ rim, it may have as few as 20 or as many as 28 holes. Choose a spoke hole as a starting point. With a hack saw, cut the rim at that point. Count around the rim to the spoke hole that equals half the total number. When you do this, don't count the hole you have sawn through. When you locate the opposite hole, cut the rim at that point.

The procedure above assumes that you have a 20″ x 1-3/8″ rim. For the 26″ or 27″ rim, simply remove the number of spoke intervals mentioned earlier. Then cut the remaining rim in half. But beware of the occasional rim that has a number of spoke holes different from the usual 36 holes. After cutting out the sections you will use, bend each into a semicircle.

Set the rim halves on a table or work bench with the ends about ½″ to ¾″ from each other. Count to the third spoke hole from each cut and mark these. Then measure the distance between opposing marked holes; this is the length for the cross braces that hold the rim halves in proper relation to each other. If you are using a 20″ rim, you might be able to use the second hole from each cut. The exact spacing is unimportant, but attach the braces to the rims at a distance of about 5″ or 6″ from each cut end, so the braces form an "X" rather than a cross.

Cut braces from 1″ x 2″ material or from wood dowels to match the lengths measured. Set one brace in place and put the other one over it to mark the area on each brace where they cross. Remove half the thickness of each brace so they will fit together with flush edges and permit the ends of the braces to align with the holes in the rim. Install the braces by screwing a No. 8 x ¾″ sheet metal screw through each of the four chosen holes in the rim into the end of the braces (Fig. 2-12). Then nail the two braces together where they cross with a nail 1-¼″ long.

Drill small holes at the upper edge of each rim, about ⅛″ from the cut end. Wire a 470-ohm carbon resistor between the two rim halves at one side of the antenna, using the drilled holes to receive the resistor leads. Solder the resistor leads to the rim. Wire a piece of 300-ohm lead-in, long enough to reach from the permanent location of the antenna to your receiver, to the holes at the other side of the rim. If you are unsure about the final position of the

Fig. 2-12. Screw bike rim to cross braces.

antenna, experiment with position and orientation before you solder the lead-in to the rim. This will insure that you use enough lead-in to reach your receiver without splicing it.

If you are constructing an outdoor antenna, you might want to find an aluminum alloy rim, as mentioned earlier. In that case you won't be able to solder the resistor or the lead-in to the rim. Instead you can install No. 6 x ⅜" sheet metal screws to bind the leads to the rim. Spray the connections with a plastic spray before putting the antenna outdoors. For greater gain, you can use two rims—one placed 57" above the other. In this arrangement, change the resistor in each rim to 1000 ohms.

This antenna is more directional than a simple dipole, but with improved performance. Theoretically, the resistor should face the direction of the incoming signal, but that wasn't true in our experiments. Instead, the best reception occurred when the side with the lead-in attached to the rim was aimed at the transmitter (Fig. 2-13).

This antenna may vary somewhat in impedance. It can be fine-tuned by using it in conjunction with Project No. 1. If you are using a simple dipole and are not satisfied with the performance, this may be just the ticket.

PROJECT 7: AN OMNIDIRECTIONAL ANTENNA

Here is another antenna project that costs very little and is exceptionally easy to make. It is ideal for an attic position where there will be no provision for rotating it, and it can easily be adapted to outdoor use.

Although this antenna is called omnidirectional, it really gives its best performance to any four points of the compass. Its maximum sensitivity is to the four directional lines that make a 90-degree angle with each of the four sides. Compared to the other antenna projects, however, it is essentially omnidirectional.

This antenna can be made from a 10' length of wire, a little over 6' of 1" x 2" pine, and a couple of small bolts (Fig. 2-14). Almost any kind of wire will do; the antenna pictured here was made from some junk magnet wire removed from a burned-out car alternator. The enamel insulation on the wire was scorched, but for our purpose that was of no significance. If you want to construct an omnidirectional antenna for outdoor use, you can substitute aluminum tubing for the wire. In that case, you can leave out one of the cross braces shown here because the aluminum tubing would be self-supporting.

Fig. 2-13. Bike wheel antenna in use near an attic ceiling.

The design can be considered an *expanded dipole*. Each side of the diamond should be about one-quarter wavelength for good performance. With the dimensions shown in the plans, the antenna is tuned to about 98 MHz, or for the middle of the FM band. If you want to tune the antenna to receive a particular station, one that produces a weak signal in your area, you can use Table 2-1 as a guide.

Construction

Cut the cross braces to the lengths indicated for the frequency being received. For a general FM band antenna, use the dimensions for 98 MHz (Table 2-1). Notch the braces where they cross so that the edges will be flush (Fig. 2-15). At one end of the shorter brace, drill two holes just large enough to receive No. 6-32 x 1″ bolts. Cut notches at the edge of the other three brace ends to hold the wire in place. Make sure these notches are not deep enough to effectively shorten the length of each brace.

Glue and nail the braces together. Install the bolts, facing them in opposite directions. Use sandpaper to clean one end of the wire. Then loop it around one of the bolts and install and tighten the wing nut. Run the wire around the frame, pulling it tight at each brace end. When you get back to the starting point, cut off the excess wire. Sand off the section of wire that will contact the bolt. Then loop it around the second bolt and install the wing nut.

Install the antenna in your attic or upper part of a closet near your receiver. Prepare some 300-ohm lead-in and—one at a time—loosen the wing nuts enough to wrap the conductor of the lead-in around the bolt. Retighten the wing nuts. Connect the lead-in to your receiver and rotate the antenna slightly until you find the direction that gives the best performance on the stations you want to receive. Drive a couple of nails into a ceiling joist, or other house beam, and bend them over the braces at two points near the center of the antenna.

The omnidirectional antenna is an ideal choice if you live in a building where you have access to a closet or attic but no way to install an antenna that can be rotated.

PROJECT 8: A MULTIELEMENT ANTENNA

This antenna is of more conventional design than the two preceding models (Fig. 2-16). It can be built from pieces of old TV antennas and is suitable for either attic or outdoor installation.

The design consists of a folded dipole with a reflector and a director. When extra elements are added to a simple dipole, the

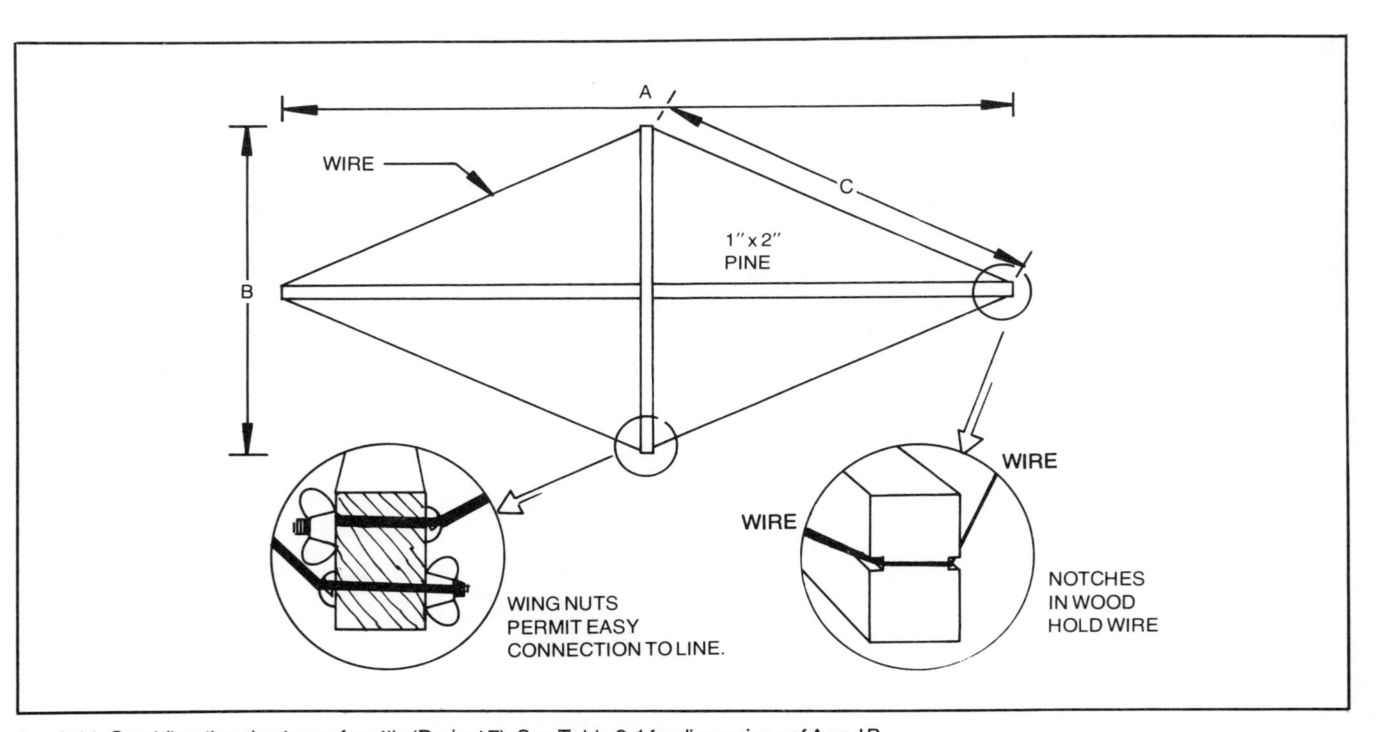

Fig. 2-14. Omnidirectional antenna for attic (Project 7). See Table 2-1 for dimensions of A and B.

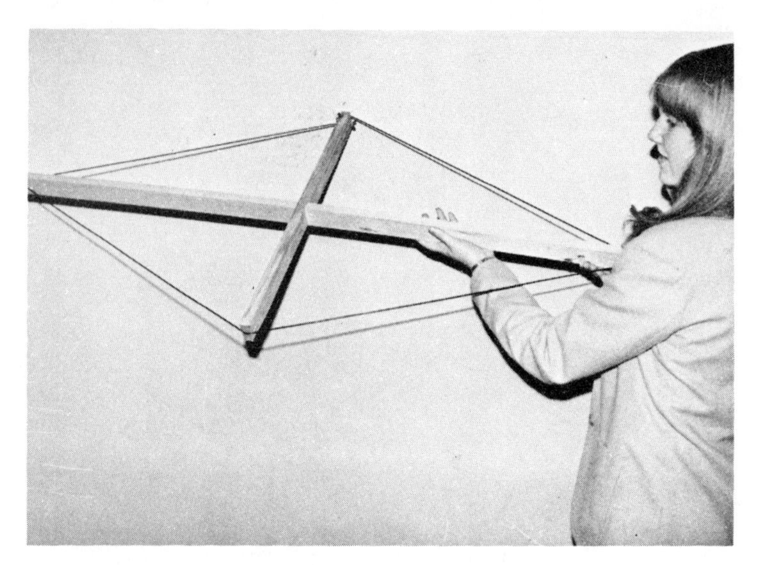

Fig. 2-15. Omnidirectional antenna is easy to construct.

impedance of the dipole is reduced, causing an impedance mismatch between the antenna and line unless some kind of matching device is employed. Hre the matching is accomplished by using dipole tubing of two diameters. The small tubing, having a higher impedance than the large tubing, raises the net impedance. Most old TV antennas will have enough ⅜″ tubing in the elements and enough 1″ tubing in the boom to provide the material for this project.

Table 2-1. Dimensions of Cross Braces and Length of Wire per Side of Omnidirectional Antenna for Various FM Frequencies.

f (MHz)	A (in.)	B (in.)	C (in.)	
88	56-½	26-¼	31-¾	
90	55-¼	25-⅝	31-⅛	
92	54-⅛	25-⅛	30-½	
94	52-¾	24-½	29-¾	
96	51-¾	24	29-⅛	
98	50-⅝	23-½	28-½	FM BAND MIDPOINT
100	49-½	23	28	
102	49	22-¾	27-½	
104	48	22-¼	27	
106	46-¾	21-¾	26-½	
108	46-¼	21-½	26	

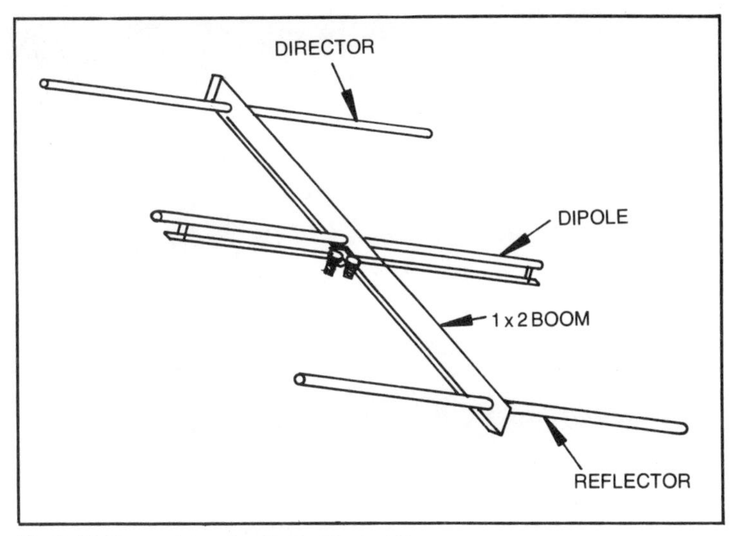

Fig. 2-16. Three-element antenna (Project 8).

The dimensions shown in the drawing are right for the middle of the FM band (Fig. 2-19). If you want to favor a particular frequency, you can use the formula in Chapter 1 to cut the dipole length to match your station. You should also alter the director and reflector by a similar proportion. In fact you can just scale the entire antenna, including the spacing of the elements, up or down to give optimum performance in any part of the FM band.

Here is another way to favor one station without changing the antenna measurements. If the station you want to boost is a distant

Fig. 2-17. Use binding posts for dipole connectors.

Fig. 2-18. A 7/8″ piece of 3/8″ tube acts as a spacer for dipole tubes.

one but is located at a different point of the compass from other, stronger stations, you can aim the antenna at the more distant station. In some cases this will work well; the final performance depends on just how strong the signal is for the other stations.

While pine was used for the boom in this antenna, hardwood is a better choice for outdoor installation. Also, the thin perfboard might be subject to damage if the antenna were whipped by the wind. A piece of 1/4″ plastic or tempered hardboard would make a stronger bracket. To maintain the same center-to-center distance between the dipole tubes, the bracket should be set flush with the bottom edge of the boom. You can remove 1/4″ of the boom with a couple of saw cuts and wood chisel, but you should reinforce the section of the boom near the dipole with 3/4″ wood braces on each side. The perfboard, as shown, is thoroughly adequate for an attic antenna.

This antenna is more specific in its performance than others shown in this chapter; that is, it is designed for the FM band and it works well for FM reception, but it is not a broad-band antenna. Like the other antenna projects, it was tested on TV as well as FM. It gave poor performance on most TV channels. It is also quite directional. It gave good quieting for distant FM stations when it was aimed in the proper direction. But its side pickup was poor. This, of course, is a great advantage if you are bothered by severe reflections. The only disadvantage is that you must make some provision for rotating the antenna if you want to receive stations that are different points of the compass from you home.

Construction

When you have found an old TV antenna to supply the parts, break it down and remove any pieces of extraneous hardware from the tubing. Cut the tubing to the proper length. For the longer pieces, such as the reflector, you may have to bolt together two pieces of tubing. If you do, flatten a section of each tube, drill matching holes, and use No. 8-32 x ¾" bolts and nuts to hold them together.

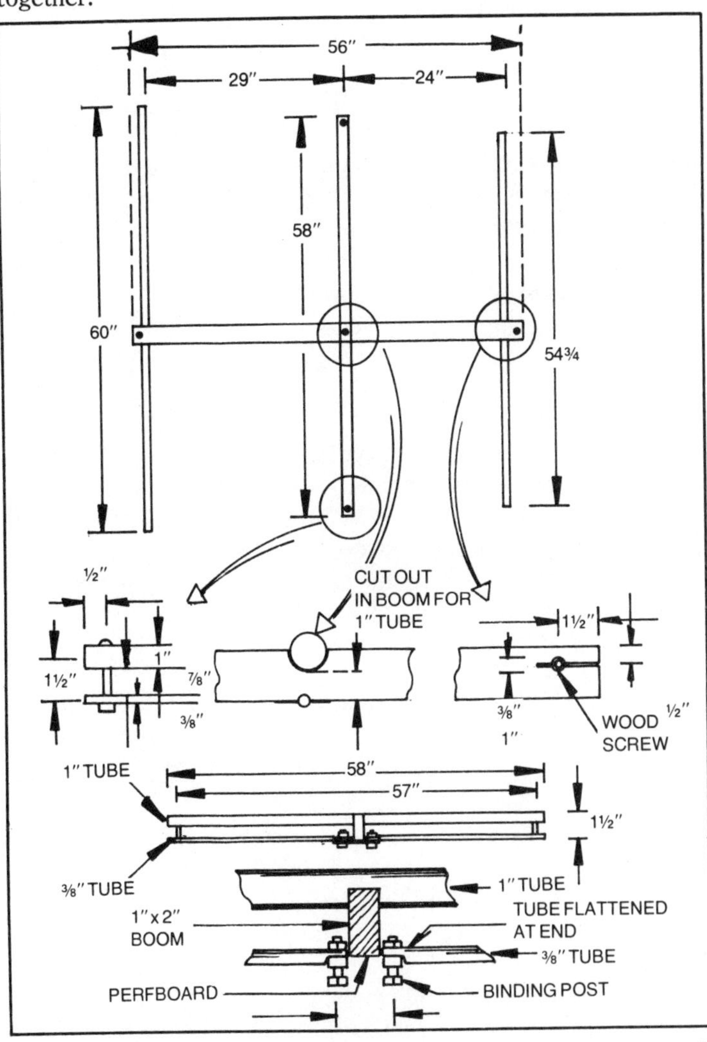

Fig. 2-19. Dimensions for three-element antenna (Project 8). Length of each ⅜" tube used is 25-⅝".

Cut the pine boom to length and mark the position of each cutout or hole. Drill the ⅜" holes for the director and reflector. Then saw a straight cut through the center of each hole (Fig. 2-19). Drill a guide hole and a pilot hole for the screws as shown. Use an 11/64" drill bit for the shank hole in the upper section, above the saw cut, and a 5/64" bit for the pilot hole below the cut.

Clamp another piece of wood flush against the boom at the point where you need a cutout for the large dipole tube. This cutout will be semicircular hole, so the extra piece of wood must contact the upper edge of the boom but set flush with the side of the boom. Drill a 1" hole at the line where the two pieces of wood touch, so that half of the hole will be located in the boom wood. After drilling this hole, ream out the cutout, if necessary, to permit the 1' tube to rest in the boom at the proper position so that its lower surface is 1" form the bottom surface of the boom.

Locate the tubing for the dipole, the 58" large tube, the two 28-⅝" lengths of ⅜" tubing, and the two ⅞" pieces of ⅜" tubing. Clean the tubing with steel wool or sandpaper, especially at contact points. Use a center punch or a large nail to make depressions in the tubing where it will be drilled. To drill the holes, clamp or hold the tubing firmly against a wood block. The size of the holes at the outer ends of the dipole will depend on the size of 2-½" bolt you use. The preferred size is No. 8-32 x 2-½", but that size is harder to find than the 3/16" diameter. Flatten the inner ends of the ⅜" dipole tubes and drill holes at the very end just large enough to accommodate the posts on the binding posts you select. Drill matching holes 1" from each other in the perfboard.

Begin assembly by bolting the binding posts to the perfboard. Then add the flattened tubing and use the extra nut on the binding post to hold it (Fig. 2-17). Because the tubing is likely to have rough burrs on it, no lock washer is necessary.

Screw the perfboard assembly to the underside of the boom, opposite the cutout for the large tube. Install the large tube in the cutout, using a No. 8 x 1-½" wood screw to hold it there. Install the 2-½" bolts through the outer ends of the large tube, then the ⅞" spacers, and finally the ⅜" tubes (Fig. 2-18). Tighten the nuts to make good contact between the members.

Next, insert the ⅜" tubing for the director and reflector through the holes made for them. Install a No. 8 x 1" wood screw in the prepared holes to tighten the split boom over those tubes.

If you want to use this antenna outdoors, clamp it to a mast with a U-bolt. Don't forget to adequately support the mast.

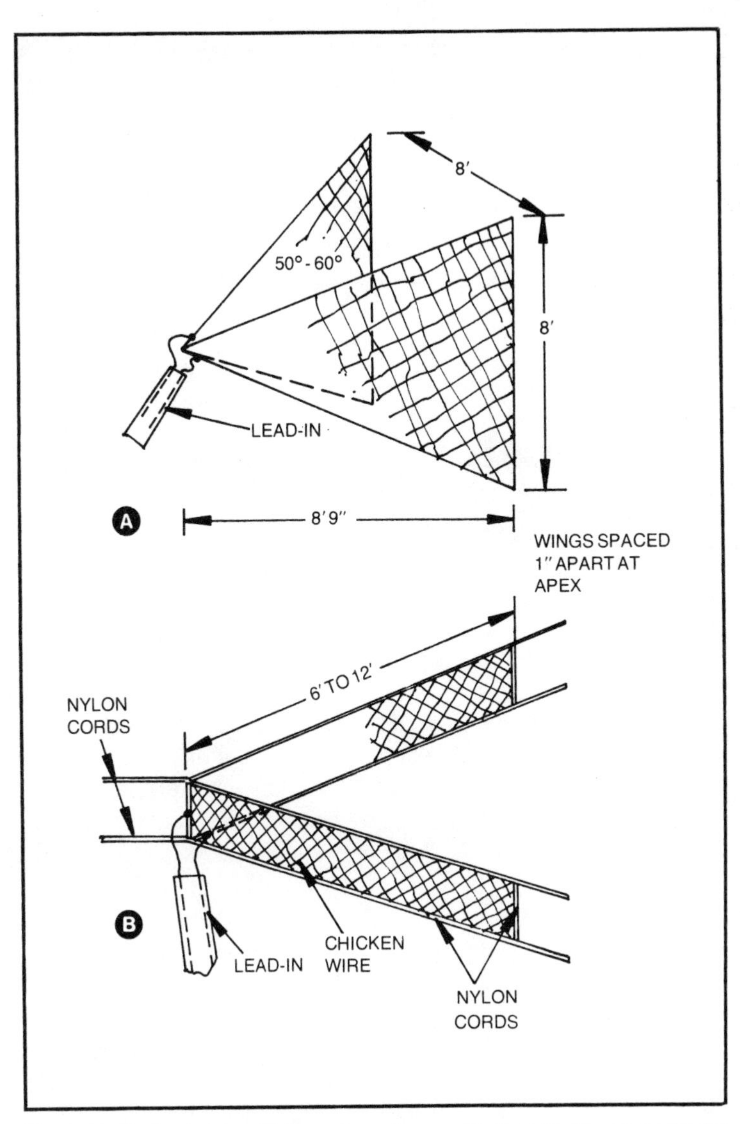

Fig. 2-20. Basic plan for chicken wire antennas (Project 9). Flared model for outdoor use is at A, and V model for attics is at B.

PROJECT 9: THE CHICKEN WIRE SPECIAL

Here is an antenna that is extremely wideband, but it works best from the FM band up to UHF. When it was installed on the roof of an old shed (Fig. 2-22), it gave good FM stereo performance, as well as good TV reception, on stations with transmission towers 75 miles from the site. How much farther it might have reached is anybody's guess because the next distant stations were located more than 100 miles farther away.

While the chicken wire special is wideband, it is directional. This makes the antenna most practical for situations where the most distant stations are located at the same compass point, unless you use it with a rotor. It is obviously not for everyone, but if you

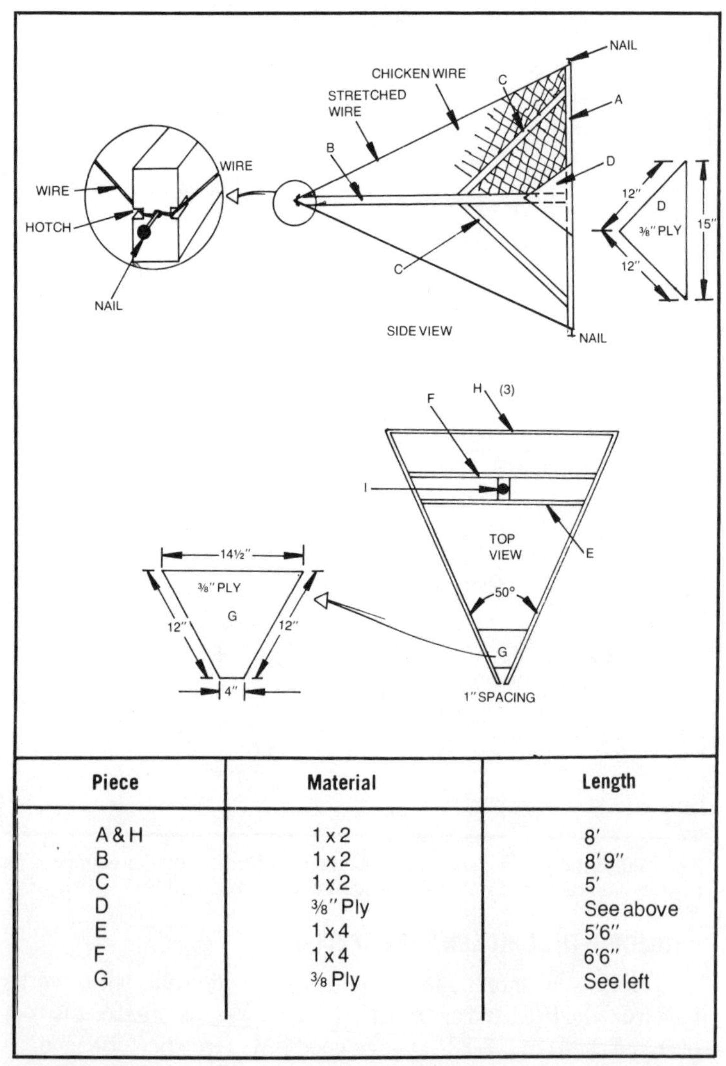

Piece	Material	Length
A & H	1 x 2	8′
B	1 x 2	8′ 9″
C	1 x 2	5′
D	⅜″ Ply	See above
E	1 x 4	5′6″
F	1 x 4	6′6″
G	⅜ Ply	See left

Fig. 2-21. Dimensions of wood parts for outdoor flared antenna (Project 9).

Fig. 2-22. Chicken wire antenna mounted on the roof of an old shed.

live in a rural area or if you want to install an antenna at a summer cabin to pull in your favorite station, give this one a try.

Although this huge net might seem out of place in some neighborhoods, if you are lucky enough to have a full-sized attic, there is hope. For an attic it is possible to eliminate the wood frame and use tautly stretched nylon cord or rope to hold the elements in place. If your attic is too shallow for the 8′ model, as shown, you can get 3′ chicken wire and make the height of the mouth only 6′. For extremely tight attic crawl spaces, you can convert the design to a simple V pattern instead of the flared mouth design (Fig. 2-20). For the V, simply stretch two pieces of chicken wire so they make a large V. The two pieces should be spaced about 1″ apart at the apex of the V. The longer the sides of the V are, the sharper should be the angle, but this is best determined by experiment.

Construction

As mentioned earlier, this design can be adapted to a much smaller attic antenna with similar characteristics. But because the attic model must be varied to fit the conditions in your attic, the construction plans here will consider only the more demanding outdoor antenna.

Start construction by cutting out the wood frame parts. It's a good idea to label the various parts with letters corresponding to those in the plans. Measure the center points on the two A members and mark them. Assemble A and B parts, first by driving a nail through the A members into the B midribs, then by nailing the triangular piece of ½″ plywood over the two pieces. Use marine plywood if possible. As you nail down the plywood, make sure that A is at a right angle to B. An easy way to do this is to drive a nail into the center of the far end of B. Make a loop in a piece of twine and place the loop over the nail. Measure the distance from the nail to the center of each of the ends of A. Adjust the angle until this distance is equal. Then nail down the plywood bracket.

Nail the 1″ x 2″ pine braces on next, placing them so that they reach from a point about 10″ from the end of A to B. These braces, if they are 5′ long, as shown, will reach to a point about 44″ from the mouth of the antenna.

Make shallow saw cuts into both corners of B at the end opposite from A. Make a single cut at the inner corner of each end of A. These notches are to hold the stretched wire to which the chicken wire will be fastened. Drive a nail into the center of each end of A. String a wire from the nail at one end of A to the nail

Fig. 2-23. Stapling the chicken wire to the "yard arm" frame.

driven in B earlier. Pull this wire taut and wrap it around the nail in B for at least one loop. Then proceed to the other end of A. Again, pull the wire taut and make it secure by tying, twisting, or twisting and soldering. Note that you will need 40′ of wire altogether, 20′ for each wing of the antenna. You can use copper or steel wire. I used electric fence wire because it was easily available at a building supply dealer.

Roll out your chicken wire next. Place one end at the yard arm end (A) and cut it at the end of B, but leave a couple of inches to spare. Staple one side of the wire along the midrib (B). See Fig. 2-23. Now you can cut the chicken wire along the path made by the stretched wire. The space piece of chicken wire left over from this cut can then be flopped over to make the other half of the wing. Staple it along B. Then twist the loose ends of the chicken wire around the taut wire. Lace a loose piece of wire between the two sections of chicken wire so they will be electrically connected as one.

Cut three boards 8′ long for crosspieces (H) at the mouth of the antenna. Saw or rasp the ends of these boards to a 25-degree angle. These boards can be either 1″ x 2″ or 2″ x 2″ pieces. Mount one of these across the center of the antenna mouth, using wood screws. It's a good idea to reinforce these joints with a triangular bracket like those used to brace the A-B joints.

Cut two pieces of ⅜″ or ½″ marine plywood to the dimensions shown (G). These pieces can be screwed to the upper and lower edges of the two B members. When you do this, make sure there is a gap of about 1″ between the chicken wire or stretched wire of each wing at the point where they are nearest to each other. At any time you can install the two remaining cross braces at the mouth.

Cut the interior braces, E and F, from 1″ x 4″ plywood. If you intend to mount the antenna on a mast, these braces should be stiffened by cutting two extra pieces and installing them vertically under E and F to make a beam with a T-shaped cross section. Install E and F to the two B members with wood screws or with 3/16″ bolts.

The last piece to install is a short length of board (I) that will reach from E to F. Since there will be some variation in the distance of E to F, this board should be cut to fit the space you find available. If you plan to mount the antenna on a mast, this piece should be made from 2″ material, such as a piece of 2″ x 2″ lumber. For a mast mount, you will want to find the center of gravity of the antenna along this final piece (I). Then clamp the antenna to the mast at this point, using a U-bolt or other suitable hardware. If the antenna lacks rigidity, use either wood or nylon rope "X" bracing across the mouth.

Connect the 300-ohm lead-in to the chicken wire at the apex of the antenna. Orient the antenna for best performance, and the job is completed. Happy listening to you.

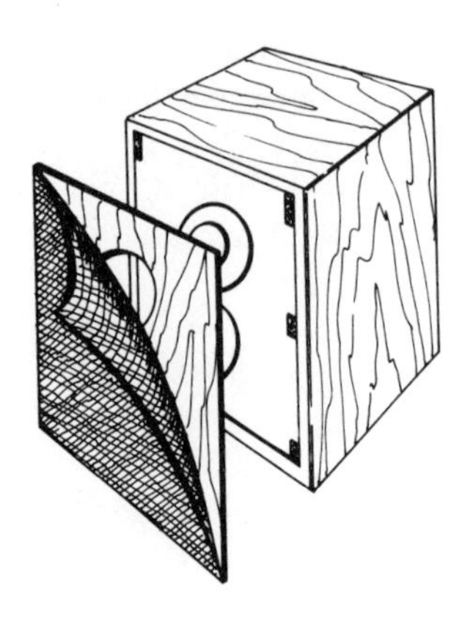

3
An Amplifier Project

Modern amplifier design and manufacturing has progressed to the point that it isn't easy to improve the typical commercial product. In the early days of hi-fi, an amplifier that could supply 10 watts of power at less than 1 or 2 percent distortion was considered to be a high-quality component. Now we measure intermodulation and third harmonic distortion in *hundredths* or even *thousandths* of a percent. And these figures are often made at output power levels of 50 to 100 watts per channel.

Not that every amplifier can meet these standards. Many relatively simple amplifiers and receivers in use today give good performance but don't have either the blueblood specifications or the versatility of more expensive models. While most listeners could do without many of the controls on current high-priced stereo components, there is one control that should be used more often: an infrasonic—sometimes called a subsonic—filter.

When that control is mentioned, one often hears a protest. The usual complaint against infrasonic filters is that they cut the bass response of the system, and the protester goes on to say, "I like bass." This line of reasoning, if carried to its logical conclusion, insists that the quality of music reproduction will always be improved by widening the frequency range of the equipment. After all, the argument goes, that's what high fidelity is all about.

A moment's reflection will show that this path can sometimes lead to problems. Suppose you have a speaker system consisting of a single driver in each stereo channel and for some reason—say, cost—you could not make any changes in those speakers. If the speakers had a fairly smooth frequency response up to about 10,000 Hz, but a shrieking resonance in the small whizzer cone at some point above that, would you want to continue to feed wide-range signals to those speakers? The truth is that you would have far more *musical* sound if you could impose a sharp cutoff at 10,000 Hz. The quality of sound within a frequency range is a more dependable trait of high-fidelity performance than is the extent of the frequency range.

The same kind of violence to good reproduction can be done by trying to extend the frequency response too far into the "basement" of the tonal range. By trying to coax infrasonic lows out of your speaker system, you can increase distortion by many times while obtaining no more usable music range. Recording engineers say that very little musical material under about 40 Hz ever gets onto tapes and records. If true (and why should we doubt them?), any effort to extend the bass range of a speaker below 20 Hz, at most, becomes questionable. In fact, there is good reason to believe that the cutoff point should be significantly higher than that.

Take a look at your amplifier or receiver. Does it have an infrasonic filter? If it has none, the equalizer project described in this chapter, which boosts the usable bass but sharply cuts off the lowest frequencies, can make a significant improvement in your stereo system. And it's not the kind of improvement that you need

Fig. 3-1. The completed equalizer built on a circuit card available from Speakerlab.

sophisticated measuring equipment to detect; you will hear the difference immediately. Even if your amplifier or receiver has such a filter, this bass equalizer (Fig. 3-1) can boost a sagging bass response or extend the low-frequency range of almost any stereo rig.

WHY AN EQUALIZER?

You may wonder, if you have a top-notch amplifier or receiver that can produce a frequency response curve as flat as any ruler, why you would need any kind of equalizer. The equalizer works in the amplifier circuit, but its usual function is to equalize acoustical deficiencies: those that arise in the speaker system or the room. As mentioned earlier, amplifiers are good, but rooms and speakers are always less than perfect. Electronic gadgets are much more predictable than those at the acoustical end of the stereo reproduction chain. To improve the bass response of an amplifier, the designer must invest some design time and money; after all, it must have higher quality throughout to avoid the problems of noise and hum, but the problem is a straightforward one. For example, by increasing the value of the capacitors, relative to resistance, the designer gets more extended bass. But with speakers, the problem is often one of space. Big bass often requires bigger drivers as well as larger cabinets in a room that may already be too crowded.

Because of these acoustical problems, most stereo systems are compromised more in bass reproduction than in any other way. The typical speaker system has a bass response that falls off rapidly below about 100 Hz. One way to attack this problem is to add an equalizer to your amplifier that will lift the lower bass. A control on the equalizer permits you to adjust the bass until the sound is balanced.

At this point, some will wonder why that same effect cannot be achieved with the bass control on the amplifier or receiver. An ordinary bass tone control will boost the bass all right, but it boosts the middle bass right along with the low bass. Many speakers already have too much middle bass, so when the bass is rolled up, they sound *tubby*, or dull and without proper resonance. An equalizer, such as the one described in this chapter, can boost the bass below 100 Hz. Yet it will have little or no effect on the middle or upper bass. You can even tailor this project to your system by choosing the frequency at which the maximum bass boost will occur.

HELP FOR REFLEX WOOFERS

Equalizers can be used with any kind of speaker system for improved bass performance, but they are especially desirable for reflex speakers. Bass reflex cabinets have ports in the boxes that couple the air in the cabinet to the room. The air flow in the port of a well-designed reflex enclosure damps the woofer cone at its resonance, adding bass but reducing distortion. A reflex also permits a more extended low-frequency response than is possible from the same driver in a closed box. Unfortunately, there is a price to pay for these advantages, and the price can be the woofer cone itself. At frequencies below the tuned frequency of the box, the air in the port fails to damp the woofer cone because instead of moving in phase with the cone and compressing the air in the box with the woofer, it moves out of phase with it, leaving the cone free of box pressure. If you have a reflex speaker, you can easily see this movement. Have someone tune your FM receiver from station to station with the volume control well up. At each station, you'll think the cone is going to leap right out of the box.

This peculiar behavior can easily lead to excessive distortion. Engineers who design reflex speakers have long claimed that these speakers can reproduce bass with less distortion than closed-box speakers. Perhaps one reason for this claim was that they measured distortion by feeding controlled signals to the speakers, signals free from infrasonic noise. But when these same speakers went into a home music system and a typical phonograph was included, the rumble from the record player—even though inaudible—often drove the speaker into such wild vibration that the higher tones produced by the speaker were modulated by the excessive cone motion at the rumble frequency. This was one of many instances where what looked good on paper and in the laboratory failed to live up to its promise in the living room.

In addition to producing more distortion, the tendency of the woofer cone in an unprotected reflex to go into orbit could be expensive for the owner. As amplifiers have become ever more powerful, woofer damage has increased. Some speaker manufacturers consider the power rating of a speaker in a reflex to be about one-half of that of the same speaker in a closed box. Reflexes have some valid advantages, and because of the advantages they are now making a comeback. But to realize their inherent advantages without problems, they should be protected by an infrasonic filter such as the one incorporated in the equalizer project described in this chapter.

SPECIAL-DESIGN EQUALIZED REFLEX SPEAKERS

During the last decade much interest and study has been spent on the reflex speaker. Beginning with the work of A.N. Thiele of Australia, along with further work of Richard Small, D. B. Keele, Jr., and others, the reflex speaker is now much better understood than in the days of its initial popularity. Theile published a paper almost 20 years ago, which gained recognition in this country a decade later, that established a firm guide to future reflex design. He showed how to design ordinary reflex speakers for flat bass response. But he also went a step beyond that, giving specifications for speaker systems that would have a predictable bump or dip in their bass response curves.

To generalize on reflex theory, a properly designed bass reflex system will have a flat frequency response. If the box is tuned to a frequency that is too high, a hump in the response curve will appear above cutoff. If the box is tuned to a frequency that is too low, the speaker will have a weak bass because of a drooping response curve (Fig. 3-2).

Thiele showed that one can put a speaker in a box that is too small, by typical design procedure, and tune the box to a lower than normal frequency to produce a system with a smoothly tapered bass response. But when a bass equalizer is added, the bass is brought up to a level even with the midrange, and the range of flat response is extended well below that of a larger, properly tuned box with no equalizer (Fig. 3-3). By this method of tuning the equalizer to the speaker system, one can achieve an extra half octave of firm bass response in a box that is significantly smaller than that specified for an unequalized speaker.

Note that the equalizer should be specifically tuned to the speaker system. With a reflex speaker this means that the peak bass boost frequency of the equalizer should be just above the tuned frequency of the reflex port. The beauty of this arrangement is that the equalizer is applying bass boost to the speaker in the frequency region where the cone is best controlled by the port. Such a design produces a bass response that is solid as the Rock of Gibraltar. There is no cone surging because a properly tuned equalizer cuts off the bass just below the port frequency.

Here are some of the advantages of an equalizer:

- It reduces cone motion and distortion.
- It saves amplifier power by removing infrasonic noise.
- It extends the undistorted bass range of the whole system.

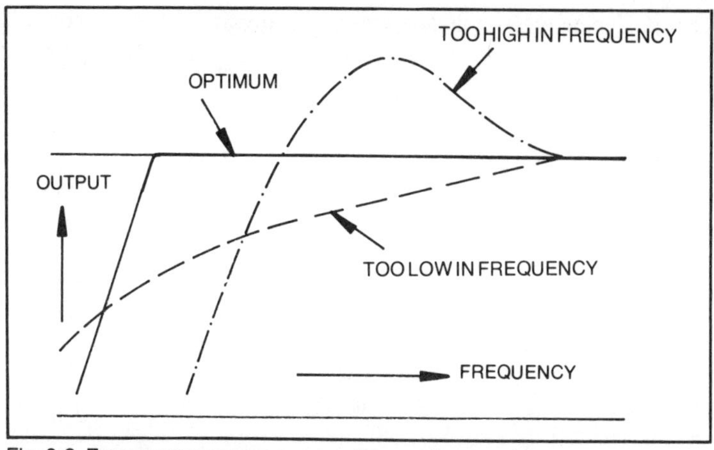

Fig. 3-2. Frequency response curves of bass reflex speaker with various kinds of enclosure tuning.

To design a complete equalized reflex system from scratch requires attention to several kinds of speaker characteristics. If you are interested in doing this, you can find the details in TAB book No. 1064, *How To Design, Build & Test Complete Speaker Systems*.

HOW TO USE AN EQUALIZER WITH A REFLEX

As mentioned earlier, a reflex designed for use with an equalizer should be tuned to a lower frequency than an unequalized

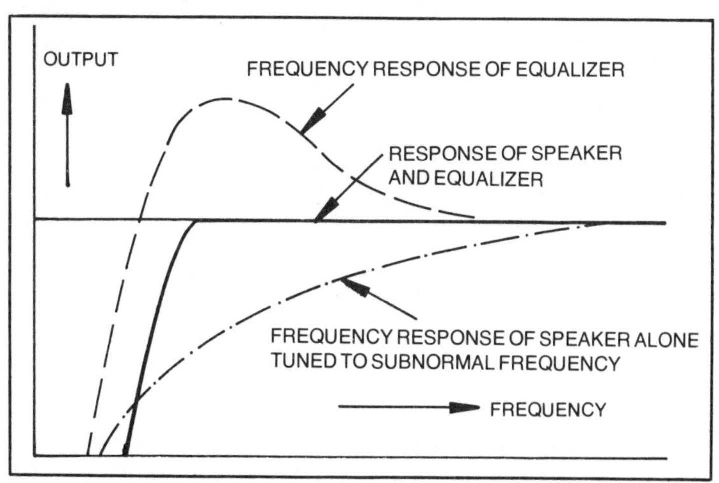

Fig. 3-3. Frequency response curves for a bass reflex speaker before and after adding an equalizer.

reflex. Typically, it should be tuned a half octave below the normal frequency, and the equalizer frequency should be set about 7 percent above the box frequency.

Here is an example. Suppose you have a speaker that should be put in a reflex tuned to 50 Hz. To use an equalizer in such a system, you would tune the box down to 37.5 Hz and set the equalizer frequency at 7 percent above that, or 40 Hz.

If you don't know how to tune either a box or an equalizer at this point, don't despair. Choosing a capacitor value to set the equalizer frequency is a simple matter, as you will see. And you can experiment with box tuning even if you have no test equipment.

Ported boxes are tuned by varying the size of the port—if the port is a simple hole—or by changing the length of the tube—if the port is a duct (Fig. 3-4). Virtually all recent reflexes have ducts for ports. Consequently, to lower the box frequency, you simply make the duct longer.

Here is the procedure if you have no test equipment of any kind. Assuming that your reflex speaker system is working properly, simply lower the tuning by lengthening the duct until the bass begins to sound thin. Then connect the equalizer and run the peak height up. If you hear no change in the sound, the equalizer is tuned too low. You can raise the tuning frequency by substituting

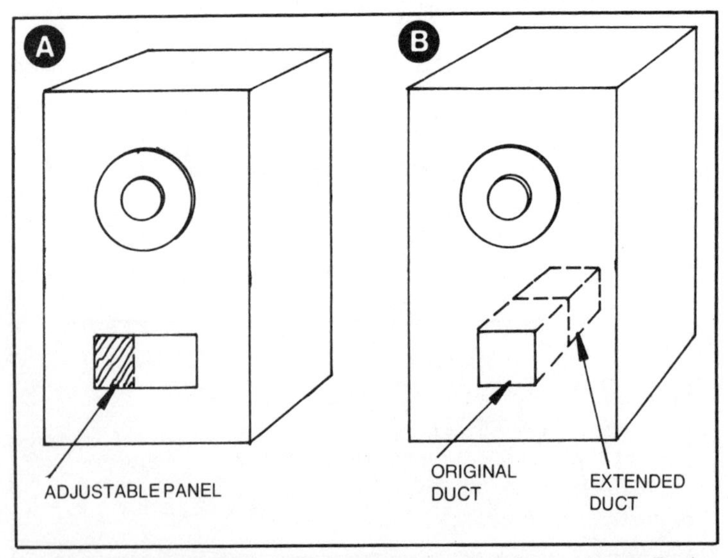

Fig. 3-4. How to lower the frequency of tuning of a bass reflex enclosure. For the simple port at A, reduce the port area. For the duct port at B, increase the length of the duct.

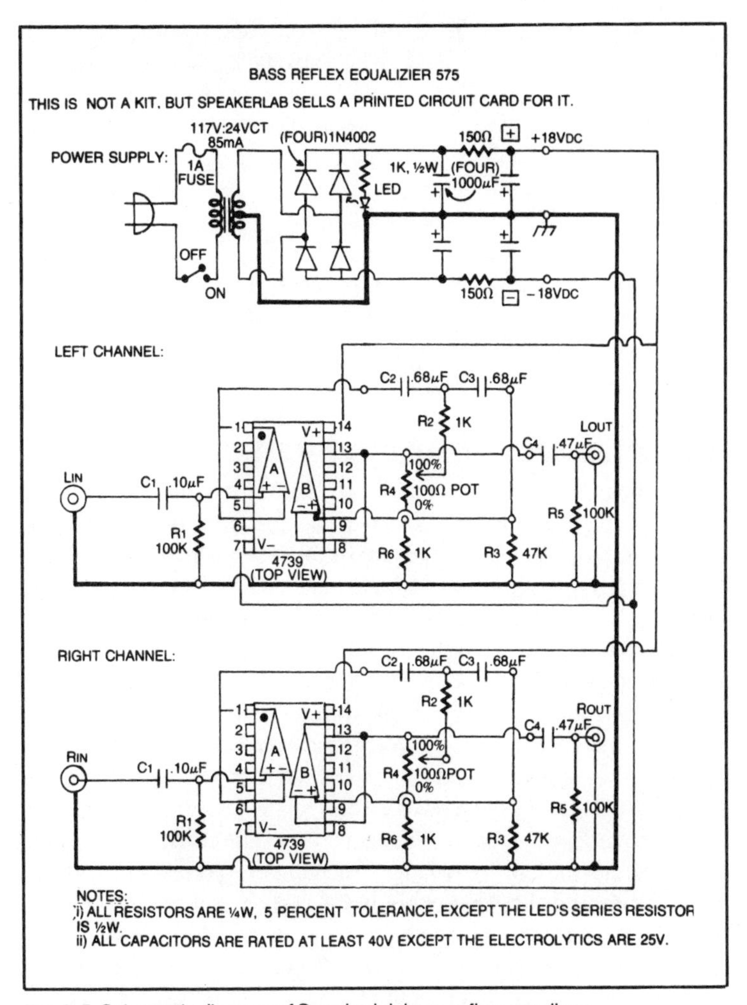

Fig. 3-5. Schematic diagram of Speakerlab bass reflex equalizer.

small capacitors for the ones used at C2 and C3 in the schematic diagram of Fig. 3-5. But if the equalizer changes the sound without improving it, you probably have it tuned too high and need to increase the capacitance at C2 and C3. When it is tuned right, the bass will have a solid feel that indicates extended lows. If the bass is too prominent, simply roll back the peak height by adjusting the pot controls counterclockwise. You will notice that the equalizer reduces cone motion no matter how you set the controls.

You can do the job in less time, but not necessarily any better, if you have some kind of calibrated variable audio frequency source, such as an audio generator or even a test record. To locate the tuned frequency of the box, feed a strong signal to the speaker system and watch the woofer as you reduce the frequency. When you get to the port frequency, the woofer cone will move very little or even seem to stand still. Below that frequency, it will vibrate with increasing amplitude. When you have found the box frequency, change the port until you have lowered the frequency to about 75 percent of the original frequency. Then try to peak the equalizer at 7 percent above that point. For example, if the box frequency is 40 Hz, you would tune the box to 30 Hz. and the equalizer to 32 Hz. You can get good performance even if the equalizer isn't tuned to that precise value. In this case, 35 Hz would probably be close enough.

USING AN EQUALIZER WITH A CLOSED BOX SPEAKER

Although the equalizer is ideally suited for use with reflex speakers, it also can be a valuable aid for any kind of speaker system. In most cases, the equalizer should be tuned to peak below the frequency of maximum bass output, where the response is declining. Used in this way, the equalizer will extend the frequency response of the system without producing excessive boom. If the speaker system is weak in the lower bass, the equalizer can produce a striking improvement in system performance.

In some cases, even better performance can be obtained by altering the speaker system before adding the equalizer. A correctly designed closed box system can be detuned by enlarging the box to reduce system Q, or resonance magnification, and lower the frequency of bass resonance. Such an altered system may be overdamped and show a drooping bass response at the lower end, much like a reflex that is tuned too low. Enter the equalizer. By tuning the equalizer to the new lower frequency of resonance, the system response can be brought back up to a level equal to that of the midrange. The result is a flatter, more extended bass response. Because enlarging a box is usually impossible, this means building new speaker cabinets. If you try this, aim at enlarging the cubic volume behind the woofer by at least 50 percent but by no more than 100 percent. You should also be aware that a larger box leaves the woofer cone free to move farther because of reduced air pressure on the cone. Because of this effect, the power handling ability of the woofer will be decreased unless you use the equalizer with its sharp infrasonic cutoff.

Another way to take advantage of the bass equalizer is to apply extra mechanical damping to a cheap woofer to lower its Q, as shown in Project No. 21, and then use the equalizer to restore any missing bass. This kind of combination can make the low-priced woofer sound like a much more expensive model. But you don't get something for nothing! Such a system will have much lower efficiency than a high-quality woofer with a larger magnet. This means that you must have adequate amplifier power to bring the acoustical output back up to your normal listening level.

PROJECT 10: BASS EQUALIZER

Here is a practical equalizer that was designed by Pat Snyder of *Speakerlab*. You can order a circuit board from Speakerlab that will greatly facilitate construction, but the unit is not available as a kit. After building a couple of these and finding that they both worked perfectly at first try, I can recommend the project as a reliable one, especially if you get the circuit board. You can write to Department BRE at Speakerlab, 735 Northlake Way, Seattle, WA 98103, for the latest catalog which will list the current price for the low-cost booklet on the equalizer. When you get the booklet, you will have the latest update on equalizer parts availability, cost of the circuit board and other details.

Construction

Before starting to work on the circuit board, you should get all the parts (Table 3-1) so you can drill the printed circuit card to match the parts you get. Note that some parts require a different size or placement of holes from that on the original card (Fig. 3-6). All of the parts should be readily available in a medium-sized to large-sizedcity. If you live in a small town, however, you may have to order at least some of the parts, specifically the transformers.

Table 3-1. Parts List for the Equalizer.

- 1—Circuit Card (optional)
- 1—117 V Primary, 24 or 25 VCT Secondary, 85mA (or more) Transformer.
- .4—1N4002 or other 100 V. 1A(or greater) Rectifying Diodes.
- 4—1000 μf, 25 V or Greater Electrolytic Capacitors.
- 1—LED (Light Emitting Diode).
- 2—Raytheon 4739 Integrated Circuit Operational Amplifiers
- 2—14-Pin Dual-in-Line IC Sockets (Optional).
- 2—100Ω Trim Pots for Mounting Directly on Printed Circuit Board
- 1— 1K, 5%. ½W carbon comp. resistor goes next to LED
- 2—150Ω, 5%, ¼W Carbon Film Resistor (Color code: Brown-Green-Brown- Gold)
- 4—1K, 5%, ½W Carbon Composition Resistor (Brown-Black-Red-Gold)
- 2—47 K, 5%, ¼ W Carbon Film Resistor (Yellow-Purple-Orange-Gold).
- 4—100 K, 5%, ¼ W Carbon Film Resistor (Brown-Black-Yellow-Gold).
- 2—.10μf, 25 V or Greater Capacitor
- 2—.47 μf, 25 V or Greater Capacitor.
- 4—Phono Jacks for Inputs and Outputs.
- 1—On-Off Switch.
- 1—Fuseholder and 1 A Fuse.
- 1—Line Cord and Plug.

"Plus capacitors for C_2 & C_3. These to be selected to match speaker system. See text."

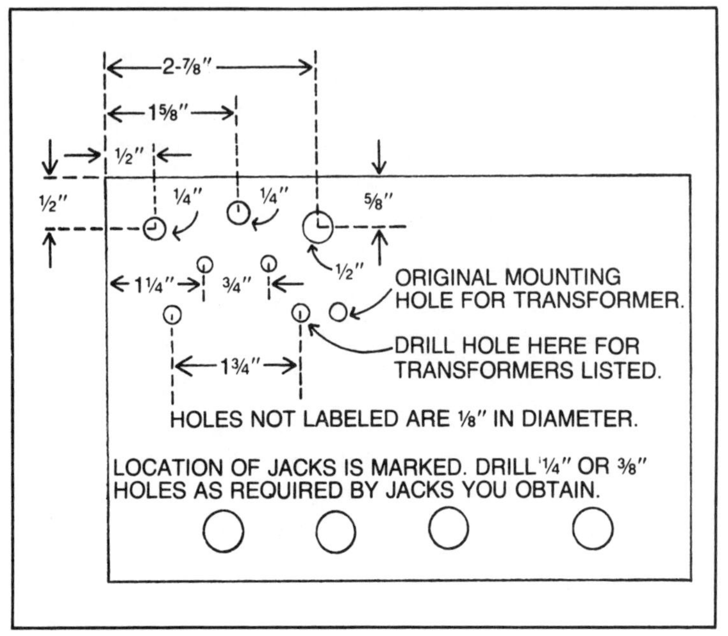

Fig. 3-6. Drilling pattern in circuit card.

While you are drilling new holes for the transformers, the switch, fuse holder and other components, make sure you check the diameter of the printed circuit jacks. Some of these will fit ¼" holes; others require ⅜" holes. Drill the proper size of hole in the marked position for the four holes on your card. To make sure you get the holes in the marked location, use a small drill bit to locate the hole and then the full size. While drilling the circuit card, hold it firmly against a block of wood to prevent damage. Even better, use a drill press. Either way, use great care to avoid damage to the card.

The holes provided for transformer mounting are spaced too far apart for the Stancor P-8394 or Triad F-115X transformers, which match the required specifications. Note that you can use the prepared hole near the edge of the card and drill a new ⅛" hole spaced 1-¾" from it for the other mounting hole.

When you complete the drilling process, install a leg at each corner of the card. One easy way to provide legs is to use ⅛" x 1-1½" bolts, running a nut up each bolt to tighten the card to the bolt. These legs will prevent the weight of the equalizer from disturbing the under-chassis wiring before you solder the connections.

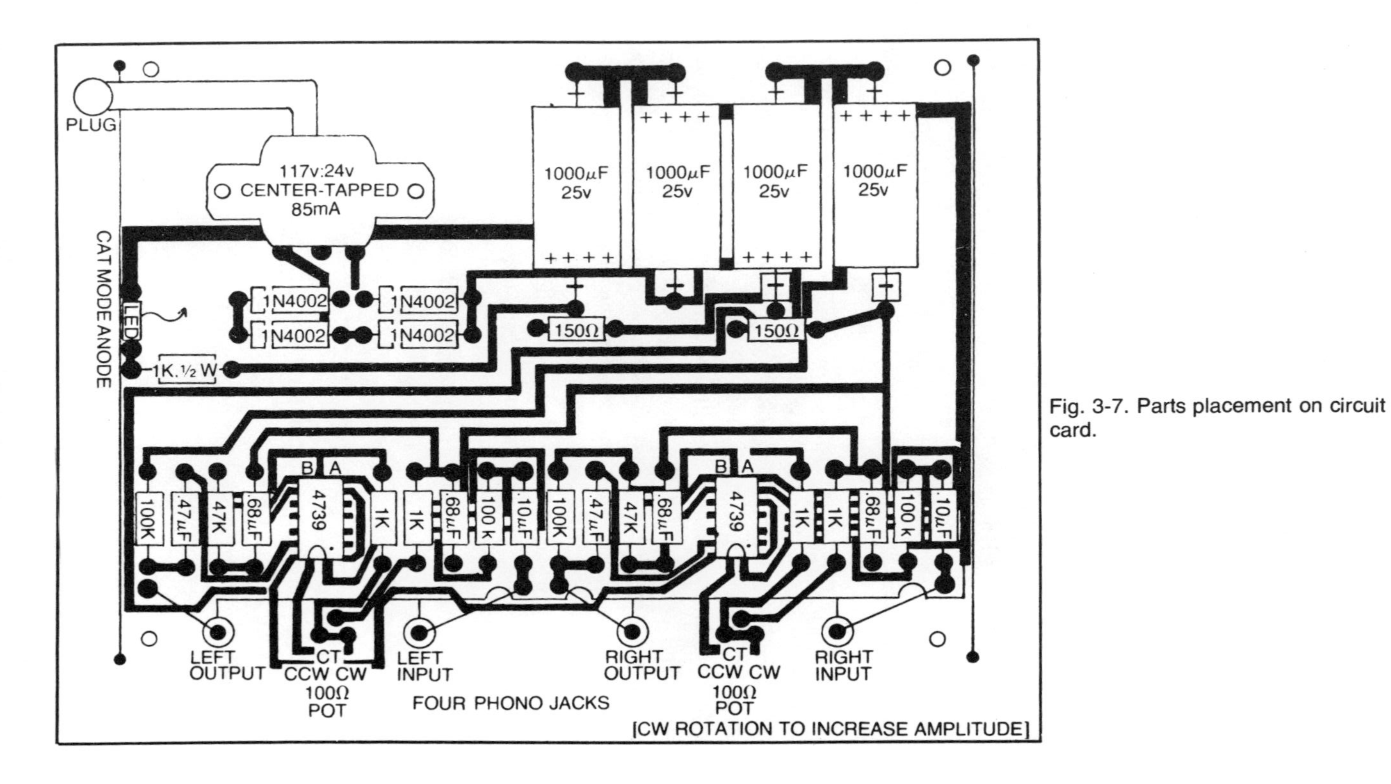

Fig. 3-7. Parts placement on circuit card.

Begin wiring by installing the four jacks. Run short lengths of wire from each jack to the nearest point on the bottom of the card marked "in" or "out." Run insulated wire from the ground lug on the left output jack to the ground lug on the left input jack. Run a short length of wire from there to the ground lug on the right output. Then run another length of wire from the right output to the ground lug on the right input. Finally, run a short length from the right input ground to the point on the board marked "gnd." Make all of these runs as short and straight as possible. Check them to see that all are insulated from the hot terminals of each jack. Solder the connections with rosin core solder.

Parts Installation

Install the IC sockets. Note that pin 1 of these sockets must be placed at the end nearest the jacks; this is the end with the half-moon indent. Use an soldering iron with a small tip to solder the connections under the card. And use solder sparingly for the closely spaced pins. Solder every pin. Inspect your work carefully with a magnifier to make sure you have no shorts between pins.

Install the resistors next. They are small and so pose no problem of access to other parts that will be installed later. Push the leads through the proper holes in the card. Then bend the lead on the bottom of the card to lock the resistor in place until you solder it. After you have installed one-half dozen or so, stop and solder the leads. Clip off the excess wire. Check your solder joints to see that they are bright and smooth. A rough-looking joint means a cold joint, the greatest cause of trouble in amateur soldering. Resolder any joints that do not look right.

Install the capacitors, diodes and the light-emitting diode (LED). The value of the capacitors you use at C2 and C3 in each channel will depend on the speaker system you intend to use with the equalizer. If you know in advance what frequency you want to tune the equalizer to, you can calculate the values of C2 and C3 by this formula:

$$C2 = C3 = \frac{23.2}{f}$$

You will often have to place two capacitors in parallel to get the right value. For example, if the frequency you want is 30 Hz, the value of the capacitors would be:

$$C2 = C3 = \frac{23.2}{30}$$

$$= 0.77\mu F$$

This value of capacitor is not available, and the next smaller value that you can obtain easily is 0.68 μF. This would tune the equalizer to 34 Hz. If you want to tune the equalizer to 30 Hz, with the calculated value of 0.77 μF, you could use a 0.68 μF and a 0.1 μF capacitor in parallel to provide an indicated total capacitance of 0.78 μF.

If you don't know what frequency is best for your equalizer, you can start with a 0.5 μF capacitor for each channel. This value would tune the circuit to about 46 Hz. If that frequency is too high, you can either add a small capacitor in parallel or change to the 0.68 μF value.

When you are installing the electrolytic capacitors, give special attention to their polarity (Fig.3-7). These capacitors are plainly marked with a (+) at one end and a (–) at the other. Polarity must also be observed with the diodes and the LED. The diodes have a band at one end. The LED may have an indent on the body or one lead shorter; check the information on the package before you discard it.

Set the transformer temporarily in place with the three secondary leads facing the side of the chassis with the jacks. Check to see how long to make these leads; then cut them off. Remove ¼" of insulation from each lead and install the transformer by feeding the secondary leads through the three holes on their side of the transformer and the two primary leads through the holes drilled on the other side. Note that the center tap of the secondary—the wire with the different color coding from the other two—goes through the middle hole. This center tap is *yellow* on a Stancor transformer and *yellow striped* on a Triad. Solder the secondary leads and clip off any excess wire.

Install the fuse holder and the switch. Run one of the primary leads from the transformer to the middle lug on the switch. Run the other primary lead to one of the lugs on the fuse. Solder all connections. Connect one AC cord lead to the remaining lug on the fuse and the other lead to one of the outside lugs on the switch. Install a 1A fuse in the fuse holder.

Use a magnifier to inspect your work for possible shorts. If everything looks good, you can test the equalizer. Advance each of the two controls to full clockwise position for maximum bass boost. Plug the equalizer into the tape monitor circuit (Fig.3-8) of your receiver and turn on both the receiver and equalizer. Try your receiver at low volume with the tape monitor circuit "out." When you switch it to "in," you should hear a difference in the sound. If

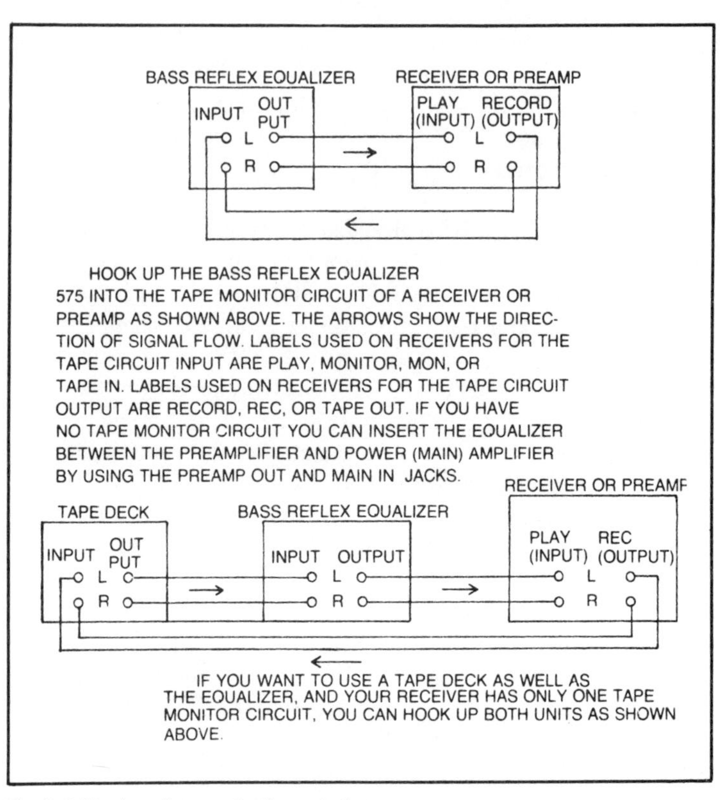

Fig. 3-8. Hookup diagram for the equalizer.

you hear no change, the equalizer is probably tuned too low for your speakers or it isn't working. If you are using two capacitors to get the right value for C2 and C3, try removing the one with the less value.

If you try various values of capacitors and still hear no change, you must do some troubleshooting. Speakerlab's Compendium No. 3 offers some steps to take if you don't get a signal through the equalizer. The chances are great that you made a wiring mistake, so first go back and double check your wiring and soldering. Even better, get a friend to help you check it. It is easy to overlook a mistake that you have made, but another person can usually spot it.

Using the Equalizer

You can tuck the equalizer away and out of sight in any recess near your amplifier or receiver, using the legs made by the bolts installed during construction to support it. If you have no place to

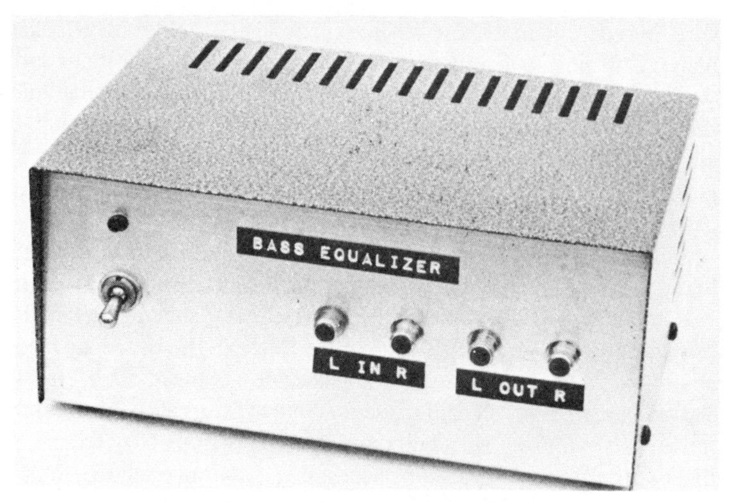

Fig. 3-9. Equalizer installed in a metal cabinet (Radio Shack #270-269).

hide it and want a more professional looking job, mount it in a metal cabinet (Fig.3-9). For one that fits well, go to your local *Radio Shack* store and ask for stock number 270-269.

Here is how Bill Wiede of *IBM* adapted this cabinet to his equalizer. He installed the on-off switch, the LED and the phone

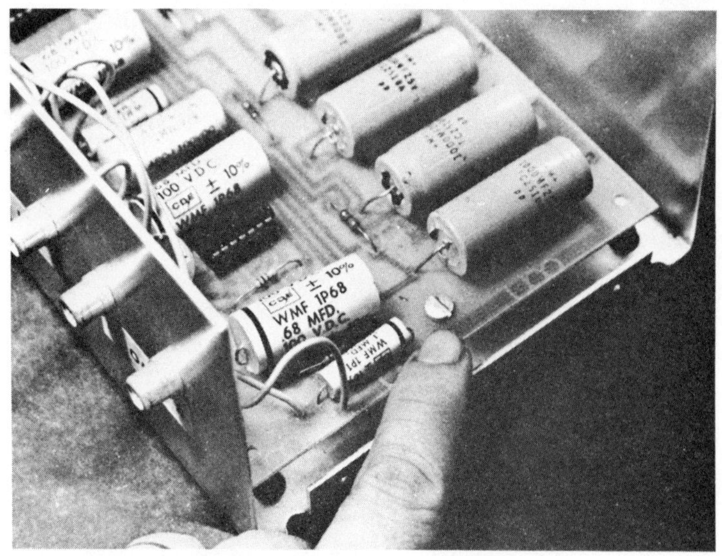
Fig. 3-10. To install the circuit card in the cabinet, you must drill two mounting holes located at each end of the card (photo by Bill Weide).

jacks in the front panel, and the line cord and fuse holder in the back panel. For more convenience in using mating cables for input and output, Wiede placed the input jacks for each channel together and the output jacks together (Fig. 3-9). He removed the jacks from the circuit board and installed them in the panel, but you can use extra jacks and run wire from the jacks in the circuit board to those in the panel.

To install the circuit board in the box, Wiede couldn't use the four mounting holes at the corners of the board because they align with the holes in the cabinet where rubber feet should be installed. His solution was to drill new mounting holes in the board, spaced about midway between the front and back edges and about ½" from the side edge. You can drill matching holes in the cabinet base and install the equalizer with bolts and spacers (Fig. 3-10). To make the two potentiometers more accessible for future adjustments, bend them back slightly.

Note that this equalizer can improve your stereo system in several ways. Its basic purpose is to boost the bottom bass, but the sharp cutoff feature, effective just below the frequency of maximum boost, is even more valuable if your amplifier has no infrasonic filter. Not only does it save the speaker from subsonic garbage, it removes the burden of useless low-frequency signals from the amplifier too. When you add an equalizer such as this to your stereo system, the overall reduction in distortion can be dramatic.

4
Crossover Networks

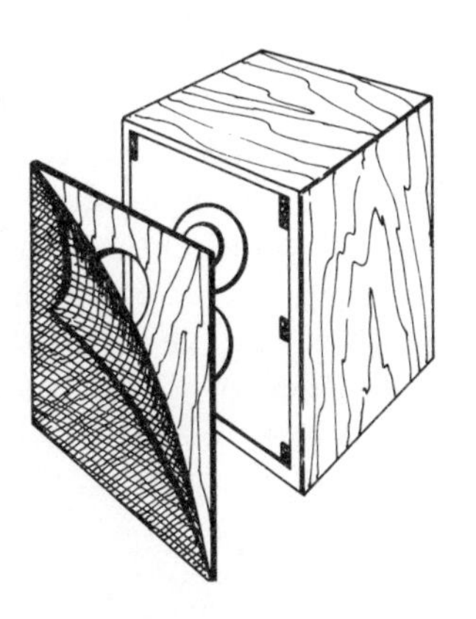

Any speaker system that has one speaker for bass and another for treble must have a crossover network to divide the electrical signal. A full crossover network feeds low-frequency energy to the woofer and high-frequency energy to the tweeter. If the speaker system is a three-way set-up, it must also include a band-pass filter, one that sends the middle frequencies to the midrange driver.

Crossover networks vary tremendously in their complexity. The simplest networks have few components and provide a gradual slope for rolling off the undesired energy from each passband. Other networks, with more components, are more effective at blocking unwanted energy but are not necessarily better. The best choice for any set of speakers depends much on the characteristics of the speakers. Some networks shape the total response curve to make imperfect speakers give better performance.

KINDS OF CROSSOVER NETWORKS

The simplest kind of network is a high-pass filter used in series with a tweeter to block the bass (Fig. 4-1A). Because a capacitor has more *reactance* to low frequencies than to highs, it reduces the low-frequency energy that gets through it to the tweeter; the lower the frequency is, the more effective is the filter. This kind of filter is a necessary minimum for every speaker system that includes a tweeter. When fed low-frequency signals, tweeters not only produce distortion, but they can be damaged by such misuse. The value of the capacitor chosen usually offers an impedance to the signal that is equal to the impedance of the tweeter at the desired crossover frequency.

The next step in network development is to place a coil in series with the woofer (Fig. 4-1B). A coil of wire possesses *inductance*, which has a frequency characteristic opposite to that of a capacitor. A coil therefore passes low-frequency energy but blocking high-frequency signals. Its reactance to an alternating current varies directly with the frequency of the current, so by knowing the value of inductance we can predict just how it will react to any given frequency. To choose the right value of inductance for a crossover network, the same rule applies as for capacitance: Choose a coil that offers the same reactance at the crossover frequency as the impedance of the speaker at that frequency. Such coils are often called choke coils, or just chokes, because they choke off the rapidly alternating current of the highs.

The crossover networks just described are the most popular kinds. Because they use no more than a single element in each leg of the speaker system, they produce a gentle slope instead of a sharp cutoff to undesired frequencies. This slope is called a 6 dB slope because when graphed it shows a 6 dB decline for each octave beyond the crossover frequency. Another characteristic of the crossover networks shown here is that they are called *parallel networks*. Although the components of the network are in series with the speakers, the speakers themselves are in parallel with each other. Parallel networks put each speaker in a separate leg of the total circuit, so the crossover element for each speaker can be altered from the theoretical value to better match the requirements of good sound. For example, a woofer may have an increase in efficiency near the crossover frequency, so it would be desirable to use a larger choke than the value suggested by calculations or obtained from charts. This larger choke would reduce the energy to the woofer more near the crossover point, flattening its response curve. In other cases, the value of the choke may be reduced below the indicated figure because of inductance in the speaker voice coil itself, which has the same effect on frequency response as inductance in the crossover network. With the parallel network circuit, the reactors can be selected for the special requirements of the driver in one branch of the circuit only.

Another way to hook up the crossover components is to put them in series with each other and to wire the speakers in series

(Fig. 4-1C). The choke in this circuit appears as a shunt across the tweeter, giving a low-impedance path for the lows, while the capacitor shunts the woofer. Notice that each speaker is in series with a parallel circuit that consists of a reactor and the other speaker.

The choice of component values for the series network is made much like that for a parallel network, but here the speakers should be equal in impedance, at least in the region of the crossover frequency. For the traditional series crossover, the reactance of the two elements at the crossover frequency is made equal to the impedance of the speakers, but other choices are possible. For example, the reactance of each element can be made just one-half that of the speakers to obtain a slightly sharper cutoff at crossover. Even in such special cases, it is customary to make the reactance of the coil equal to the reactance of the capacitor at the crossover frequency.

All the networks described above have no more than a single crossover element per speaker. The next step up in complexity is a network that has a choke *plus* a capacitor for each speaker (Fig. 4-1D). Note that this kind of crossover network places a coil in series with the woofer and a capacitor in series with the tweeter, but in addition to these components, a capacitor is across the woofer and a choke is across the tweeter. This kind of circuit provides a double action in each leg, making it a 12 dB per octave network.

If the speaker system includes a midrange speaker, the crossover network must include a band-pass filter to feed it (Fig. 4-1E). The simple band-pass filter for the 6 dB network contains only a choke and a capacitor with the values selected to block the low bass and the upper highs. Because of interaction between components, some calculations should theoretically be done to get the exact values. In all cases, the values one gets from design charts or by calculations are not necessarily those that will give the best sound. For optimum values, one should experiment with a range of component values and choose those that produce the best performance.

A three-way 12 dB-per-octave network is more complex, having eight filter elements in the circuit (Fig. 4-1F). If you draw an imaginary vertical line just to the right of the second choke, this circuit can be simplified by considering it in two parts. Each part works like a two-way 12 dB per octave network. Many 12 dB three-way systems have a simplified midrange section that

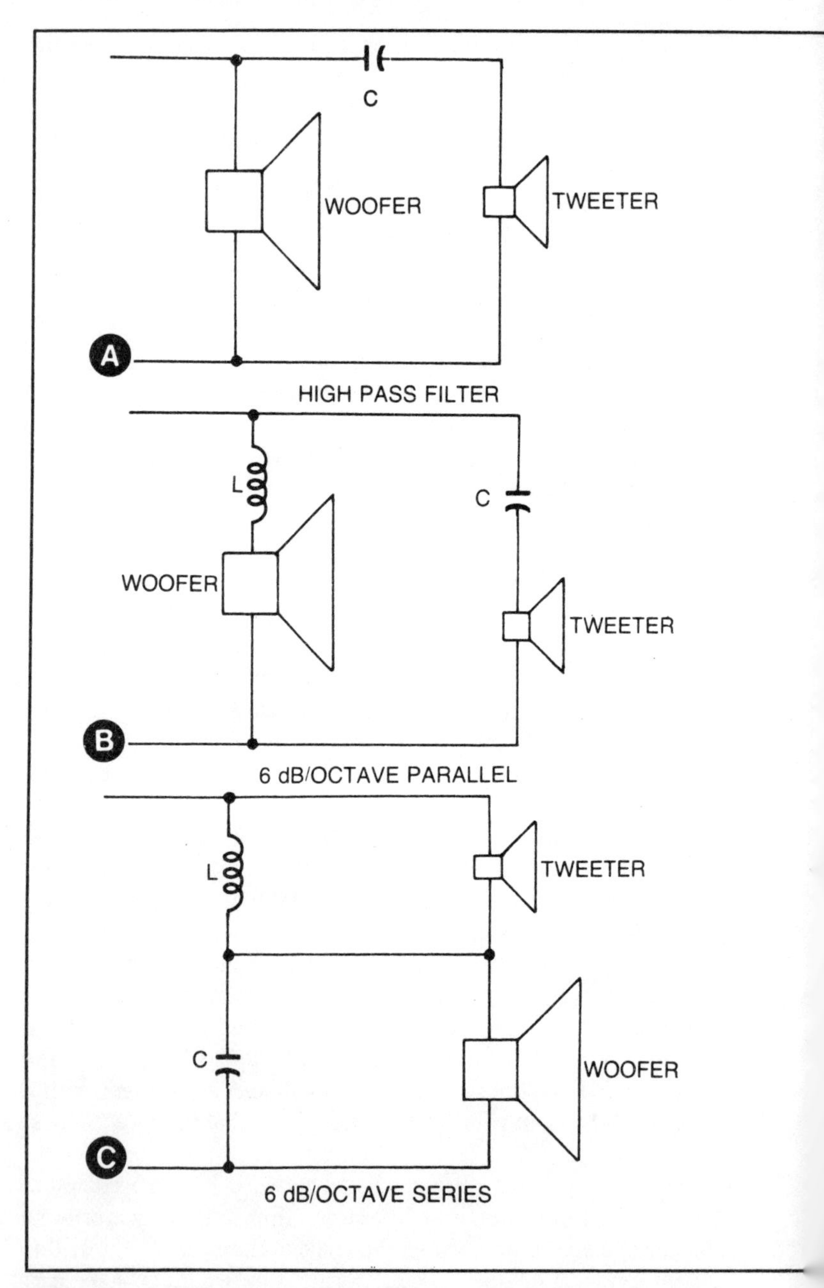

Fig. 4-1. Various frequency-dividing networks for two-way and three-way speaker systems. A high-pass is at A, a 6 dB-per-octave parallel network is at B,

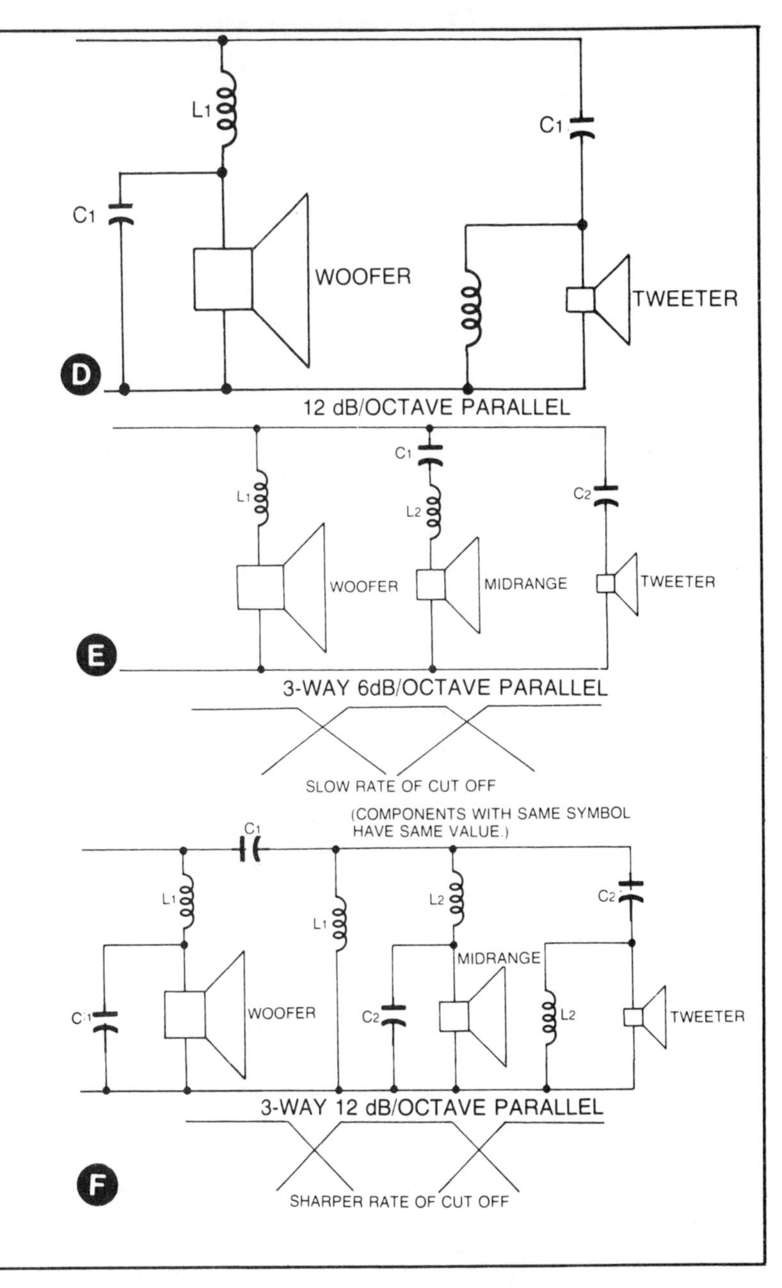

a 6 dB-per-octave series network is at C, a 12 db-per-octave parallel network is at D, a three-way 6 dB-per-octave parallel network is at E, and a three-way 12dB-per-octave parallel network is at F.

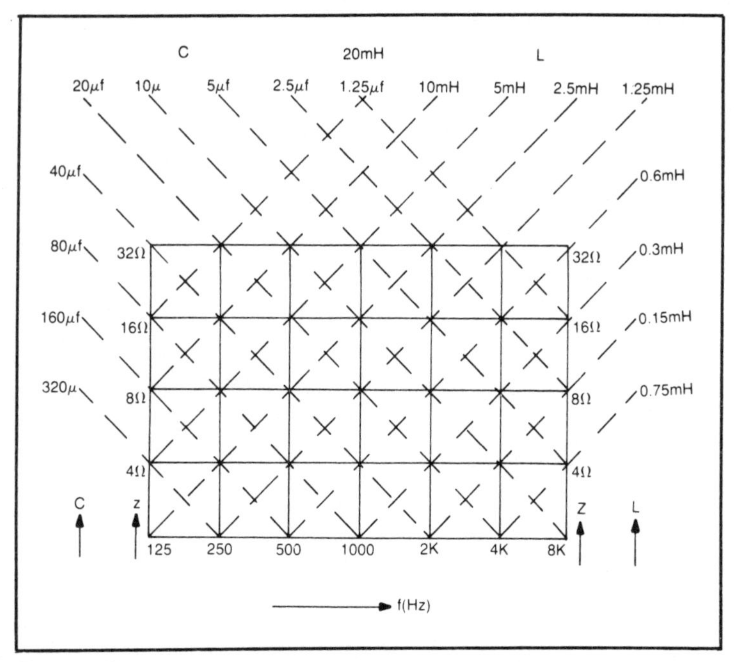

Fig. 4-2. Crossover network design chart. Approximate values of inductance and capacitance for speakers with ratings from 4 ohms to 32 ohms.

contains only a single choke and a capacitor, just like a 6 dB network, saving the cost of a choke and a capacitor.

The 6-dB-per-octave three-way series network shown in Fig. 4-3 has one advantage over the corresponding 6 dB three-way network; it offers a double action blocking of lows to the tweeter because of the position of the two coils that appear in cascade across the tweeter. This extra protection is especially important for small tweeters, the kind now in favor for their superior dispersion at ultrahigh audio frequencies. But 6 dB parallel networks can be altered to provide better tweeter protection as shown in Project No. 15.

There is considerable argument about the different degrees of phase shift caused by various kinds of crossover networks and whether such phase changes are audible. In the simplest case, that of the first-order 6 dB network, the capacitor shifts the phase in one direction and the coil in other, making a net displacement of 90 degrees between the two legs of the circuit. With a 12 dB-per-octave (second-order) network, the shift is increased to 180 degrees, putting the signals to the woofer and tweeter out of phase at the crossover frequency. To avoid a hole in the frequency response at this point, the conventional wiring practice is to connect the woofer and tweeter out of phase with each other in second-order systems. For a three-way second-order system, the midrange driver can be connected out of phase with the woofer and tweeter. Note that this kind of hookup applies to the 12 dB-per-octave circuit. Some critics of second-order, or 12 dB, circuits say that this does not adequately solve the problem; because of complex phase problems with all sharper cutoff filters, networks should be limited to the gentle slope of first-order networks that contain just one element per speaker. Still other network engineers advocate third-order circuits that produce 18 dB-per-octave rolloff to adequately split the signal. Considering the advantages claimed for various kinds of networks, there is one undeniable virtue of the 6 dB-per-octave circuits: They cost less and are simpler to wire.

HOW TO USE THE CROSSOVER NETWORK DESIGN CHART

Although you may never want to design a complete crossover network, it is useful to be able to choose the correct value component for add-on tweeters and other speaker improvement projects. The design chart, Fig. 4-2, enables you to find these values for many situations quickly and accurately without resorting to calculations. For crossover frequencies not listed, you can usually interpolate between the values given and get satisfactory performance. Or you can calculate the proper value from the formulas in the next section of this chapter.

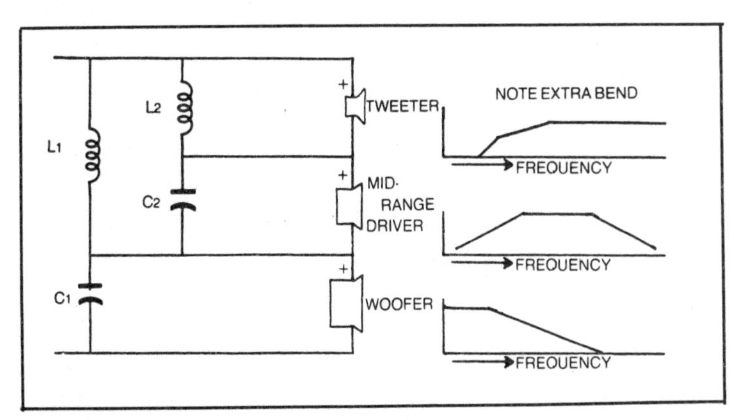

Fig. 4-3. A three-way series crossover network with 6 dB-per-octave rolloff but slightly better tweeter protection because of the double-choke action.

Figure 4-2 shows the correct values of filter components to use with speakers of various impedances for any crossover frequency from 125 to 8000 Hz. To use the chart, follow the vertical line from the crossover frequency up to the horizontal line that corresponds to the impedance of the speaker. Then follow the dashed diagonal lines upward to the left for capacitor values and upward to the right for chokes. The capacitor values are listed in microfarads (μF). Choke inductance values are shown in millihenries (mH).

Here's an example. A woofer and a tweeter have impedances of 8 ohms, and you want to find the correct values of a 6 dB-per-octave network with the crossover point at 1000 Hz. Moving up from the middle vertical line, for 1000 Hz, to the middle horizontal line, for 8-ohm speakers, you then follow the dashed lines to find that you will need a 20 μF capacitor for the tweeter and a 1.25 mH choke for the woofer. Your crossover network design is completed by this simple chart reading, except for any minor changes you might want to make after trying the components.

CREATIVE CROSSOVER NETWORK DESIGN

You can sometimes tamper with the theoretically correct values shown by the chart and come up with a network that helps your speakers to perform better. In the example just given, with a crossover point at 1000 Hz, it is likely that most speakers, except for large woofers, will be more efficient there than at other ranges. To offset a potential peak in the response curve, you can stagger the crossover points for the woofer and tweeter. This is permissible with a 6 dB network because the gentle slope of the network will leave a shallow dip instead of a hole in the response curve when that is done.

To try such an approach, you might choose values that fall between the values shown for 500 and 1000 Hz for the woofer and between 1000 and 2000 Hz for the tweeter. The average of the values shown would produce figures of 1.88 mH for the choke and 15 μF for the capacitor. Such a choice would make a slightly asymmetrical placement of crossover points, putting them at just under 700 Hz for the woofer and just over 1300 Hz for the tweeter; however, it would be a useful way to start your experiments. The idea in this kind of fiddling with crossover components is to make a system that *sounds* right rather than one that looks right on paper. It is the kind of practical approach that can put the final touch to an otherwise unexceptional set of speakers.

COMPONENT VALUES FOR MIXED DRIVER IMPEDANCES OR SECOND-ORDER NETWORKS

You might sometimes want to mix speakers of different impedances in the same system. For example, I recently noticed a high-quality dome tweeter at a bargain price because the tweeter happened to have an impedance of 4 ohms. Such a tweeter could be added in two ways, either by wiring two tweeters in series and hooking up the usual 8-ohm network, or by adjusting the network to 4 ohms in the tweeter branch. Because two tweeters would have increased the price to about the same as for a similar 8-ohm driver, the logical approach is to change the network. If you wanted to add this tweeter to an 8-ohm woofer or full-range speaker, you would use the same value of inductance found in the chart for an 8-ohm woofer, but the capacitor should be selected for the 4-ohm impedance. In the earlier example with the crossover point at 1000 Hz, the choke value should again be 1.25 mH, but the value of the capacitor obtained from the 4-ohm line on the chart would be 40 μF, instead of the 20 μF shown for an 8-ohm tweeter.

If you had the opposite situation, an 8-ohm tweeter and a 4-ohm woofer, the values would be 20 μf and 0.6 mH. In systems with mixed impedances, such as these examples, the parallel network would be the logical choice: The drivers are in different branches of the total circuit, permitting you to adjust the value of one circuit element without affecting the response of the driver in the other branch.

If you want to build a second-order crossover network, such as the ones shown in Fig. 4-1D and 4-1E, you must multiply the values obtained from the chart by a factor. The factor for the chokes is 1.41; for the capacitors, it is 0.707. For a 12 dB two-way network with a 1000-Hz crossover point, the chokes should be 1.25 mH x 1.41, or 1.76 mH. The capacitors for this same network would be 20 μF x 0.707, or 14 μF.

HOW TO CALCULATE VALUES NOT SHOWN ON THE CHART

In some cases, you might want to get exact values of components for crossover frequencies between those listed on the chart. These values can be obtained from two simple formulas and then altered as shown in the previous section, if necessary, to fit into 12 dB networks. The formula for inductance is:

$$L = \frac{Z}{2\pi f}$$

where L = inductance in henries, $\pi = 3.14$, Z = impedance of speaker in ohms, and f = frequency in hertz.

And for capacitance:

$$C = \frac{1}{2\pi fZ}$$

where C = capacitance in farads.

For example, suppose you want to find the exact value of a choke and a capacitor for a 6 dB two-way crossover for an 8-ohm tweeter and an 8-ohm woofer with the crossover point set at 1600 Hz.

$$L = \frac{8}{(6.28)(1600)}$$

$$= .0008\ \text{H} = 0.8\ \text{mH}$$

$$C = \frac{1}{(6.28)(1600)(8)}$$

$$= .000012\ \text{F} = 12\ \mu\text{F}.$$

Notice that you must convert henries to millihenries by moving the decimal point three places to the right, and farads to microfarads by moving the decimal six places to the right.

CHOOSING VALUES FOR A BAND-PASS FILTER

One way to improve the typical two-way system is to add a midrange driver. If there is space in your cabinets for a 5″ midrange driver, you can rework the crossover network to lower the woofer cutoff and raise the tweeter crossover frequency to make a middle band for the added member.

The easiest way to add a midrange driver is to use a 6 dB three-way crossover network as shown in Fig. 4-1E. You can easily obtain the values for L1 and C2 from the chart, but you should do some simple calculations for the components used in the band-pass filter. Except for some interaction between crossover elements, you would look up the value of C1 for the lower crossover frequency and the value of L2 to match the upper frequency. To compensate for the shift, you should subtract the lower frequency from the upper frequency to get the proper frequency to use in calculations for the inductance of the coil.

For example, suppose you want to add an 8-ohm midrange driver to a system and make the crossover frequencies at 500 and

4000 Hz. Doing that, you get 3500 Hz as the adjusted frequency, so:

$$L = \frac{8}{(6.28)(3500)}$$

$$= 0.00036\ \text{H} = 0.36\ \text{mH}$$

And for the correct value of capacitance, another adjustment must be made to use a frequency somewhat higher than the desired crossover frequency in the calculations. The new frequency for calculating the correct capacitance is:

$$f' = \frac{1}{\dfrac{1}{f_L} - \dfrac{1}{f_H}}$$

For the network outlined in this example:

$$f' = \frac{1}{\dfrac{1}{500} - \dfrac{1}{4000}}$$

$$= 571\ \text{Hz}$$

And:

$$C = \frac{1}{(6.28)(571)(8)}$$

$$= 0.0000348\ \text{F} = 35\ \mu\text{F}$$

So you would put together a network consisting of a 0.36 mH choke and a 35 μF capacitor instead of a 0.3 mH choke or a 40 μF capacitor, as found on the design chart. From a practical standpoint, the change in capacitance would probably be more important to observe because the inductance of the midrange driver's voice coil might easily have the effect of increasing the apparent inductance in the midrange circuit.

CAPACITORS FOR CROSSOVER NETWORKS

Don't use polarized, electrolytic capacitors in crossover networks! Nonpolarized electrolytics are available in a wide range of values and work fairly well. These are the only suitable capacitors that you are likely to find that have large enough values to use in woofer circuits. Nonpolarized electrolytics can also be used as high-pass filters for tweeters, but mylar capacitors are preferred.

Recent studies of capacitor dielectric absorption have suggested that superior performance can be obtained by substituting polypropylene or polycarbonate film capacitors for mylar or especially nonpolarized electrolytic capacitors. Whether this move is advisable for crossover networks is far from certain because the preferred capacitors can cost up to 40 or more times as much as the commonly used kinds. And some experienced speaker builders have concluded that the more expensive capacitors offer no audible advantage over selected capacitors of the common types.

Remember that you can increase capacitance by wiring capacitors in parallel. The net capacitance of parallel-wired capacitors is the sum of the individual values.

HOMEMADE CHOKES

Some electronics dealers offer ready-made chokes, but you may have to wind your own coils to get the precise value of inductance you need for specific crossover frequencies. To wind a coil, get some magnet wire and make a coil form to the proper dimensions for the choke you need. Magnet wire is copper wire with a transparent plastic or lacquer insulation. It varies greatly in diameter, according to an assigned gauge number. One of the most useful gauge numbers for woofer circuits is No. 18, with a diameter of about 0.04″. For small chokes, such as those used in midrange band-pass circuits, you can use smaller gauge wire, such as No. 20 or even smaller gauge. A typical goal in designing a choke for speaker circuits is to limit the total resistance in the choke to about one-tenth that of the speaker impedance. Resistance varies directly with wire length. Therefore, the larger the value of inductance needed, the more wire you must use. This makes a heavier gauge more appropriate.

To select the right wire gauge and the best coil dimensions for a particular circuit, you must first find the value of inductance you need for either the crossover network schematic diagram or, if you are designing your own network, from the chart in Fig. 4-2. Then go to Table 5-1 in Chapter 5 to find the specifications for each value of inductance you need. Make up a coil form to the suggested dimensions. Then wind the coil on that form, using wire of the specified gauge. You can either make a single, temporary, come-apart form and use it for more than one coil of the same value, or you can make a permanent coil form for each choke (Fig. 4-3). If you use a permanent form, make sure there are no iron screws, nails, or other iron parts in its construction. Such material would increase the inductance unpredictably by changing the coil from an air core model to an iron core choke. An easy way to make a permanent form is to glue hardboard sides to a wood dowel core of right dimensions.

To make a temporary form, use a hollow core, such as a cardboard or plastic tube of right diameter. Pieces of plastic pipe or plastic pill bottles are good choices. If you cannot find any 1″ or 1-½″ tubes around home, you can go to hardware store and get a 1″ overflow tube for toilets or a 1-½″ drain extension tube. These plumbing parts are available in plastic pipe with outer diameters that are just right for coil cores. You can saw off sections of pipe with a hack saw to fit the coil specifications and then leave the core in the coil when you take the form apart.

You will need only a couple of form sides for each coil size because the sides are reusable. The sides should be made of thin material, such as ⅛″ or ¼″ hardboard (Fig. 4-4). Drill holes in the center of each side that are large enough to accept a ¼″ bolt. Set the core in the center of one side and mark an outline of it on the side. Then drill an exit hole for the wire at the start of the coil winding operation, located just outside the marked position of the core. Drill this hole with a small bit. Bolt the forms together, using a couple of washers, such as a ¼″ and a ½″ washer outside each piece of hardboard. Use a bolt long enough to reach through the core, accept the washers and two nuts, and have enough extra

Fig. 4-4. Materials for hand winding a choke.

Fig. 4-5. Use two nuts when winding with an electric drill. The tape on the left makes counting easier.

length to be gripped by a drill chuck. You can wind the coil by hand, but an electric drill makes the work go much faster. A variable speed drill is useful; you can control its speed so that the winding doesn't go too fast to control.

Note that two nuts were specified. If you use only a single nut, you may find that the coil form will loosen as you wind, depending on the direction of winding. To avoid this nuisance, install two nuts. Tighten the first one against the hardboard side and then hold it with a wrench while you cinch the outer nut against the first one. The coil specifications in this book list both the correct length of wire and the approximate number of turns to make a coil with a specific inductance. You can use either method of wire measurement, depending on which is more convenient for you. If you plan to count the turns, place a narrow strip of tape on the edge of one form side to help you keep track of each rotation of the form (Fig. 4-5).

To wind the coil, thread a few inches of wire through the center exit hole from inside to outside. Note that you should add enough to the specified wire length to include this extra wire lead plus the one at the outside of the coil. Begin winding by making sure that the first loop is placed tightly against the side of the form with the exit hole. Try to make each turn lie flat against the previous one. If you stop winding for any reason, wrap the part of the coil already wound with plastic tape to hold the wire in place until you start winding again. Another way to lock the wire temporarily in place is to make an angled cut in one form side with a fine saw such as a hack saw and draw the wire into it, where friction will hold it until you are ready to wind again (Fig. 4-6).

When you have made the specified number of turns or used the length of wire required, tape the coil around the exposed surface of copper wire to hold it (Fig. 4-7). Cut off the excess wire and, if you are using a temporary form, remove the nuts and pull off the coil sides. Tape the coil thoroughly by winding tape through the hole in the "doughnut" and over the outer circumference. Go all the way around the coil with this taping, making it as tight as possible so the wire cannot shift position.

Do not be intimidated by the coil winding process. If you are reasonably careful in making the forms to proper dimensions and in winding the coil, you will be successful. Even a scramble-wound air core coil will perform in a speaker system better than the average iron core commercial choke. In fact, there is nothing wrong with scramble winding unless it is extremely sloppy. If you have never tried your hand at coil winding, now is a good time to start.

L-PADS

Tweeters and midrange drivers are usually more efficient than large woofers, so there must be some method of controlling the volume of sound from the smaller speakers. In the early days of hi-fi, experimenters often used simple 50-ohm potentiometers to control tweeters, but these reflect a different impedance to the crossover network at each different setting of the control. Now

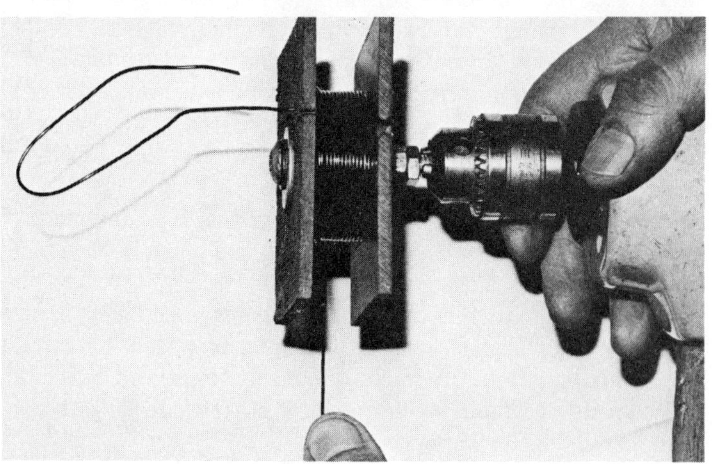

Fig. 4-6. Winding a coil with a drill. Notches in the coil form are handy to bind wire so it will not loosen after stopping.

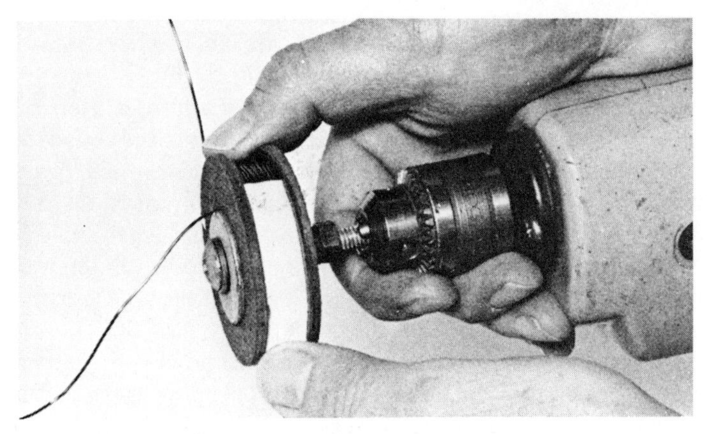

Fig. 4-7. When winding is completed, tape the coil.

L-pads are used almost exclusively because an 8-ohm L-pad will maintain the same impedance no matter what the setting. This keeps the crossover frequency constant at the frequency for which the network was designed.

POWER RATINGS OF CROSSOVER NETWORKS

You can estimate the power rating of crossover capacitors by checking the voltage rating of the capacitor. As a rule of thumb, don't use capacitors with voltage ratings of less than 25 volts in a crossover network. If you have an amplifier capable of putting out 50 watts or more of power per channel, you should get capacitors with voltage ratings of at least 50 volts, preferably higher.

L-pads come in at least three different sizes for low- , medium- and high-power speaker systems. For use with a powerful amplifier, get the largest L-pads—the kind rated for high powered systems. But if you don't have enough of the heavy-duty L-pads for both midrange and tweeter circuits, use the heavy ones in your midrange circuit. In typical musical programs, the percentage of power above 2500 Hz is about 5 percent, and it's only 10 percent above one-half of that frequency.

In addition to the crossover elements mentioned in this chapter, other components are sometimes added to correct impedance variation or to smooth the response curve. These extra components and their use in practical circuits are described in some of the Chapter 7 projects.

5
Crossover Network Projects

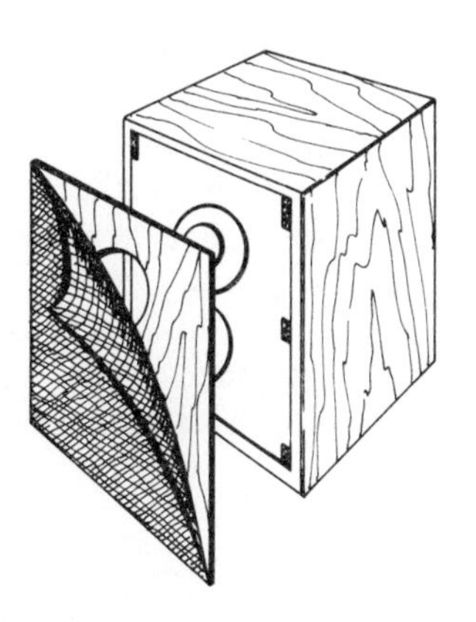

No multidriver speaker system is any better than its crossover network. While some speaker combinations can work well with the minimum of crossover components, most systems can be improved by adding a full crossover network to replace a high-pass filter or other halfway measures. Commercial crossovers vary in quality from inadequate to superlative, but the ones that fit the latter category can cost as much as many complete speaker systems.

Should a stereo fan build his own crossover network? The answer to that depends on the quality of the original network supplied with the speakers. If your speakers are either nameless or carry a department store house brand name, you can almost certainly improve the sound of your stereo rig by reworking the crossover networks. There are many ways that manufacturers cut corners on bargain speakers, such as using iron core chokes. Iron core chokes can give satisfactory performance with careful design, but many of these add distortion. You can eliminate the possibility by replacing the iron core chokes in your speaker systems with homemade air core inductors. And if your speaker is one without chokes, the chances are even greater that you can improve it by adding suitable ones.

Another substitution in many bargain systems is the use of an inexpensive potentiometer, instead of an L-pad to control the tweeter. This is a subtle problem, but the potentiometer will change the impedance seen by the crossover network, causing the crossover frequency to move up or down the audio band with changes in the control setting. This situation can be dangerous to the tweeter if you drive your speakers hard because the tweeter may be getting energy at a lower frequency than it can handle. Even if the tweeter can take the punishment, your ears may object to the distortion it produces.

Again, some manufacturers use an even more effective cost cutting trick: They just omit the controls on the midrange driver and tweeter. In a few cases, these controls are omitted because of careful design and they aren't really needed. If this is true of your speakers, you will know it by their relatively high cost. Inexpensive speakers usually consist of a collection of drivers whose efficiencies match well only by chance. This leads us to the first crossover project of adding a control, or controls, to your speaker system.

PROJECT 11: L-PAD ADD-ON

Tweeters are nearly always more efficient than woofers, but many low-priced speaker systems have no balance control. Salesmen for such speakers sometimes say that you can control the tone with the treble cut on your receiver, but that is a poor way to do it because it usually kills the upper highs. To see why, look at the graph in Fig. 5-1. Notice that when you use the amplifier tone control, the highest frequencies are depressed much more than the lower highs. With a typical tweeter, those ultrahighs are the very frequencies that are not well reproduced anyway. To make matters worse, furniture, carpets and draperies absorb these delicate top notes more effectively than they do lower frequencies. This situation means that you must choose between using your tone control to boost all the highs—risking a harsh sound because of the efficiency of the tweeter in its lower ranges—or a muffled reproduction. At the least you can't expect to hear the shimmering sound of those airy, upper treble frequencies.

To analyze your speakers to see if they would profit by adding L-pads, begin by checking the cabinet exterior for any control knobs or shafts. If you find none, you can be pretty sure there are no controls on the midrange or tweeters. The lack of a control knob is not 100 percent proof that no level balancing was attempted. When you break into the cabinet, as you must to make sure, look for any resistors in the tweeter branch of the circuit. If resistors were used to control the tweeter output level, you will probably find two; one will be in series with the tweeter, and the other will be in parallel.

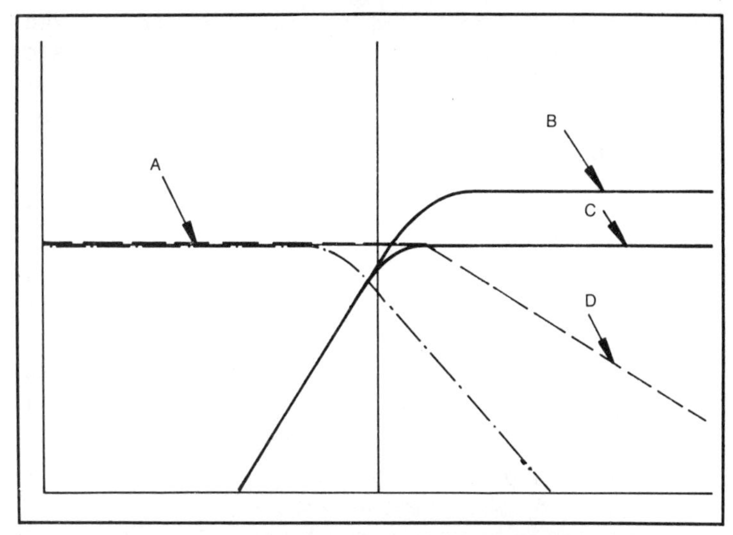

Fig. 5-1. How the action of a tweeter control on the speaker system differs from the treble tone control on the amplifier. The woofer response curve is at A, the tweeter response curve with no control is at B, the tweeter response with L-pad control in use is at C, and the high-frequency response curve when the amplifier tone control is used to balance the tweeter to the woofer is at D.

The presence of such resistors shows that some design care went into balancing, but you may still want to add an L-pad. Such fixed balancing resistors do not take into account the variation in acoustic environments that can occur from one room to another. Even if fixed balance resistors are used, you will find that an L-pad offers more versatility and so better balance in many situations.

Is the control that you find in your speaker an L-pad or an inexpensive potentiometer? An L-pad is made so that it varies a resistance that is parallel to the load as the series resistance is changed, maintaining a constant impedance to the crossover network and amplifier at all settings. To see if your controls are true L-pads, look for identification markings on the metal cover at the back of the control. An L-pad will usually be labeled by a short description such as "8-ohm L-Pad." Potentiometers usually have the value of the pot listed in numbers, such as a stock number. When used in speaker circuits, such potentiometers usually have maximum resistance values that can range from about 15 ohms to about 50 ohms. To positively identify the control, disconnect it and use an ohmmeter to measure the resistance between lugs at various control settings. Another clue is that L-pads are usually larger than potentiometers, measuring from about 1-½" in diameter up to about 2-½″ or more in diameter for those designed to be used in high-powered systems. Potentiometers used in speaker circuits usually measure about an inch in diameter.

Construction

Your first task is to get into the cabinet to see just what kind of network you have. See if the back of the speaker cabinet is attached with screws. If it isn't, check the front panel. If the grille cloth is installed permanently with trim, or is glued on firmly, go to the back again and enter the box from that panel. Many modern enclosures have the front and rear panels installed in dadoed grooves in the walls of the box, and the panels are held together by glue. If you have such a sealed box, you must cut into the cabinet and make an alternate provision for replacing the removed panel. You can cut out the back panel with either a sabre saw or any taper-bladed hand saw, such as a keyhole saw. To start the cutting, use either a drill or the sabre saw held at an angle to the back, as shown in Chapter 6.

You will probably want to hide the control and yet have access to it for future adjustments, so a good location is in the back panel. You can sometimes put it on the speaker board, but in that case the grille cloth must be removable. If the back panel is not over ½″ thick, you can probably install the control directly in the panel (Fig. 5-4). If it is thicker than that, you will probably have to install a control board in the panel because most L-pads do not have a mounting shaft long enough to reach through such a panel. Inspect the L-pads you get to see how thick a panel they will fit into without a special board.

For thick back panels, you can cut a window in the panel and glue a piece of ¼″ tempered hardboard over the window on the inside surface of the panel. Before gluing the control board onto the panel, drill a hole in it to accept the L-pad body and install the pad with the hardware provided, as shown in Fig. 5-2. The right diameter of the mounting hole is usually ⅜″, but check your L-pad to make sure. Where an L-pad is mounted on a board with the box terminals, the typical size of the window in the panel can be 3″ x 4″; it can be smaller for an L-pad alone. Before cutting the window, make sure there is no wiring in the way and temporarily remove any damping material from the panel.

You will likely find a single capacitor high-pass filter wired between the woofer and tweeter. This capacitor will nearly always run directly from one solder lug terminal of the woofer to a lug on

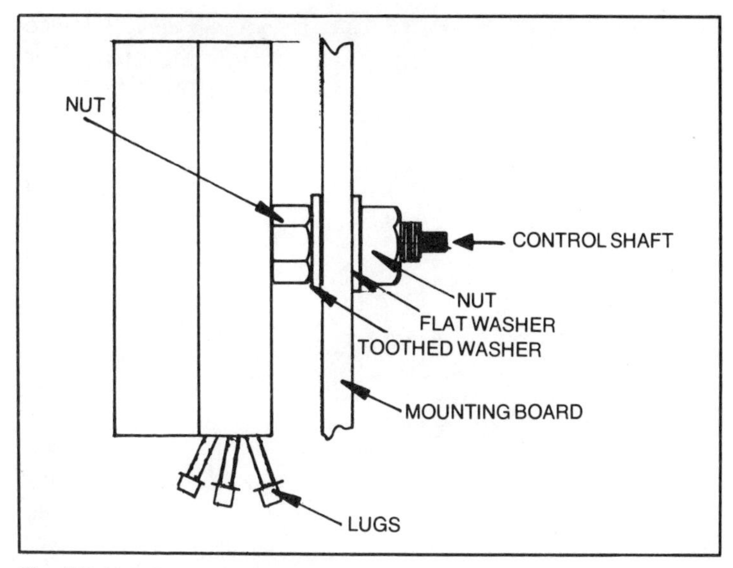

Fig. 5-2. How to use the hardware provided with an L-pad to install it on a mounting board or enclosure panel.

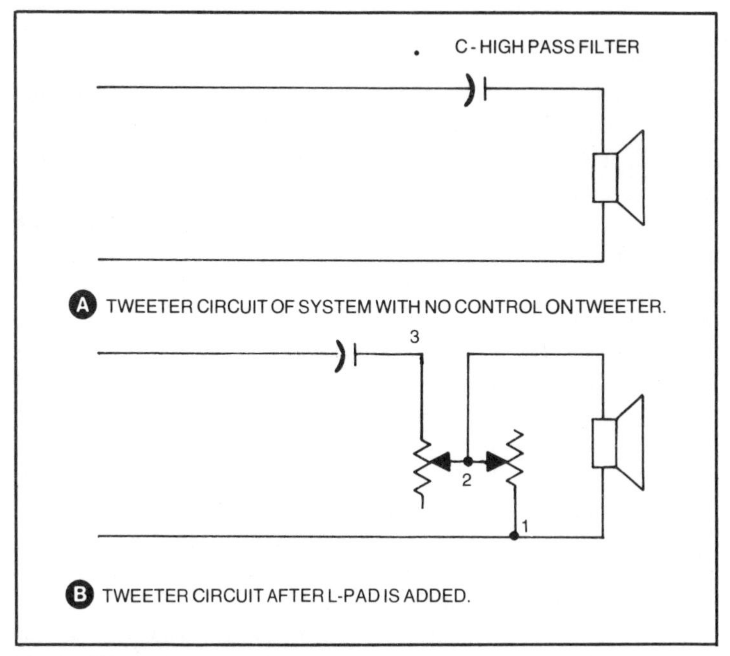

Fig. 5-3. Wiring diagram for Project 11, L-pad add-on. No control is shown at A, and the L-pad add-on is shown at B.

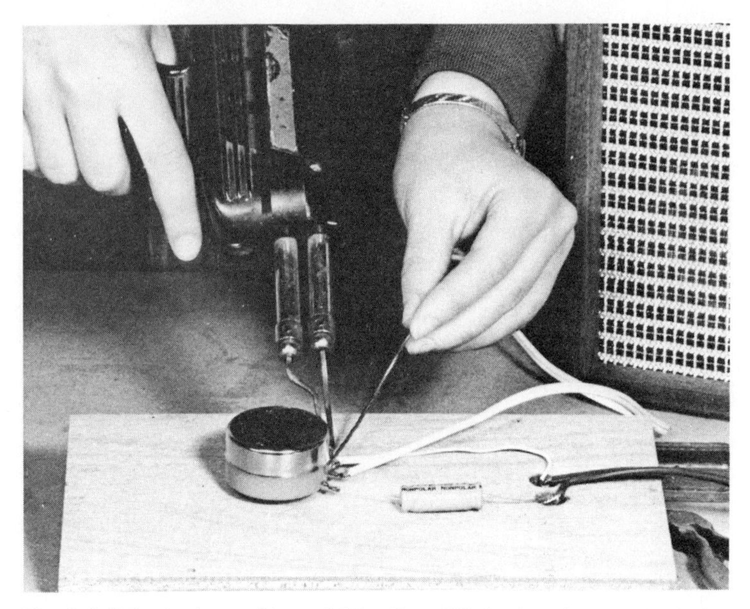

Fig. 5-4. If the back panel is no thicker than ½″, the L-pad can probably be installed directly in the panel.

the tweeter. To get the wiring right, remove this capacitor and place it on the inside of the back panel so it can be connected between one speaker box terminal and pin No. 3 of the L-pad (Fig. 5-4). Connect a wire from pin No. 1 of the L-pad to the other box terminal. Run leads from pin No. 2 and pin No. 1 of the L-pad to the tweeter, feeding the lead from pin No. 2 to the positive tweeter lug. Connect leads from the box terminals to the woofer; make sure that the lead from the terminal to which the capacitor from the L-pad is connected goes to the positive terminal on the woofer.

That completes the wiring. Try your speaker at low volume to see if the circuit is working right. You should hear only highs from the tweeter, and the tweeter level should be adjustable by rotating the control shaft. See if a clockwise rotation increases the tweeter level, as it should. If all is working right, solder the connections with rosin core solder. Cover the back panel with damping material and install the back in the cabinet. If you had to saw out the back, you must install cleats around the top, bottom and sides of the box to receive screws to hold on the back. These cleats must be installed to fit closely so they will permit no air leaks when the back is screwed in. If you suspect air leaks, use caulking material to eliminate them.

Place the speaker in the room where you will use it. Rotate the L-pad shaft fully counterclockwise and connect the speaker to your stereo set. Turn up the volume to your normal listening level and set the tone controls at 0, or a *flat* setting. Rotate the control shaft clockwise very slowly until the sound from the tweeter blends with that from the woofer. Leave it in this position. You will find that you can listen with less fatigue to a balanced system than to one in which the tweeter overpowers the woofer. Most systems that have no controls on the tweeter have too much tweeter output, even if it isn't obvious at first hearing.

PROJECT 12: WOOFER CHOKE ADD-ON

If you have never gone into the cabinets of your speakers, you won't know for sure whether chokes were used in the woofer circuits, but you can make a guess by careful listening. Get a frequency test record and listen with your ear directly in front of the woofer while the record plays. It may help to cover the tweeter with a piece of thick felt or other damping material. Then listen to the tweeter while you play the record again. Write down the frequencies where each driver has an obvious drop in response. Particularly note whether you can hear any increase in woofer output at frequencies above the point where its response begins to roll off. Such increases indicate spikes in the response curve that should be eliminated, if possible. They are no proof that there isn't a choke in the circuit, but they suggest that you can improve the system with some tinkering. And even if the response of the woofer seems to fall off in an orderly fashion above the crossover point, there might be no choke, especially if the woofer is large. A woofer designed for a compact, closed box usually has little response above its normal passband. It is difficult to keep small woofers from acting as full-range speakers without chokes, so a small woofer that acts right probably has a choke in its branch of the circuit.

Common sense will tell you whether a given speaker is likely to have a full crossover network or something less. Such clues as the original price of the speaker, the care with which it was manufactured, and whether level controls are installed give an indication of how complete is the design. But if you tire of such guessing games, there is just one way to make sure: open the cabinet and look.

To get into the box, use the methods suggested in Project No. 11, plus any special methods that might occur to you after an inspection of how the box is put together. For example, if the speakers are front-mounted, you can sometimes get into the box well enough to check the wiring by just removing the woofer. This assumes that you can get the grille cloth off without damaging the cabinet.

When you have laid bare the crossover circuit, you will probably find a single capacitor wired between one woofer terminal and one tweeter terminal (Fig. 5-5). You might be tempted to leave this capacitor in place, but if you do that, it will pass very little energy on to the tweeter after you add a choke, because the choke will roll off the highs to the woofer. To get the circuit right, you must remove this capacitor—or at least remove its connection to the woofer—so you can wire the woofer and tweeter in two separate branches of the circuit (Fig. 5-6). The easiest way to correct the circuit is to make a crossover board, as described in later projects, and place all crossover components such as choke, capacitor and L-pad on this board. To connect the speakers, you can run lamp cord lead from the appropriate lugs on the board to the

Fig. 5-5. If a high-pass capacitor is mounted between the woofer and tweeter lugs, it must be remounted so it connects to the side of the choke opposite from the woofer.

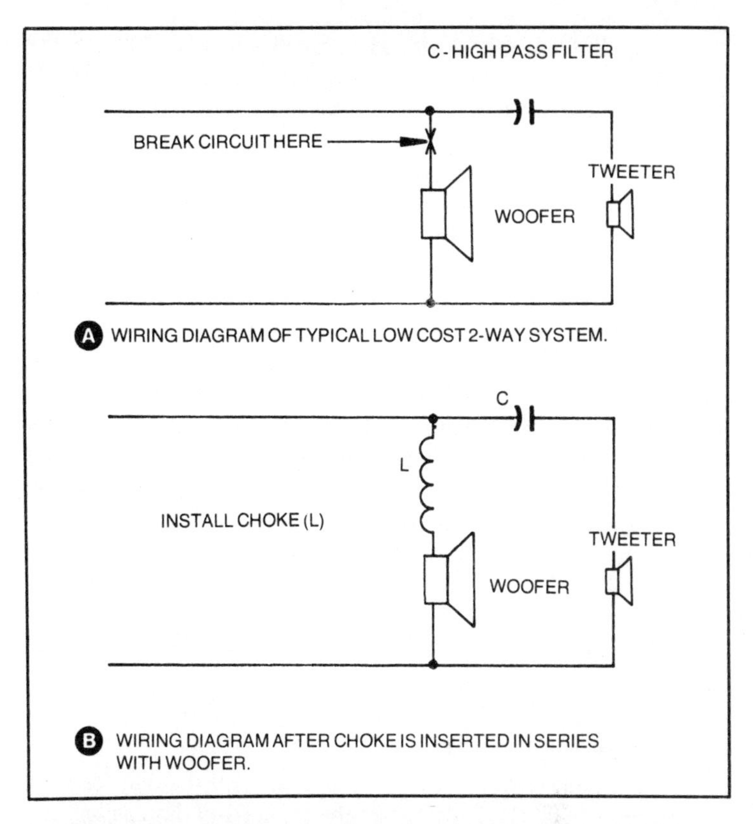

Fig. 5-6. Speaker circuit diagrams before (A) and after (B) doing Project 12.

woofer or tweeter. You can install this board on the inside of the back panel where there is usually more than enough room.

Choosing the Right Value of Inductance

While inspecting the crossover network originally installed in your speaker, jot down the value of the capacitor used in the tweeter circuit and check the back of the tweeter for any indication of an impedance rating. When you find these facts, locate the crossover frequency in Fig. 4-2.

Here is an example. The tweeter has an impedance of 8 ohms, as indicated on the back of the magnet, and the value of the capacitor in series with the tweeter is 10 μF. Locating the dashed diagonal line for 10 μF in Fig. 4-2, follow it down to the point where it intersects the 8-ohm horizontal line, then downward to the base of the chart where you find 2K, or 2000 Hz. This is the crossover frequency. If the woofer is also an 8-ohm driver, move back up the vertical line from 2000 Hz until it intersects the 8-ohm horizontal line, then upward to the right on the dashed line to find the value of choke to use. It is 0.6 mH in this case.

A few speakers have no impedance values printed on them. For those you can make a simple measurement with an ohmmeter. Disconnect the speaker from its circuit so you will be measuring only the resistance in its voice coil, not its resistance in series or in parallel with any other component. Use a simple DC ohmmeter set on its lowest scale to measure the DC resistance of the voice coil. This will not give you the exact figure for impedance because impedance is a measure of all kinds of reactance in the voice coil to an alternating current. If the meter indicates about 5 ohms or 6 ohms resistance, it is an 8-ohm speaker. If the resistance is about 3 ohms, it is a 4-ohm speaker. For a resistance of about 12 ohms, you would use the 16-ohm impedance line on the chart in Fig. 4-2.

As you can see, finding the correct value of choke for your woofer is a simple matter. Remember that it does not have to be a perfect value every time. You can even make the choke to a higher value of inductance than that shown in the chart. Then gradually unwind it until the sound pleases you.

Construction

When you have found the theoretically ideal value for your choke, refer to Table 5-1 to locate the proper coil dimensions and the correct wire gauge. The wire length, listed in Table 5-1 and in the various graphs in Appendix A that show exact coil data for a wide range of coils, is useful in purchasing wire. But you should expect to use at least an extra foot of wire for coil leads. To get the approximate weight of wire you will need for a coil, refer to Table 5-2, which lists the number of feet of wire per pound for various wire gauges.

The schematic diagram of a two-way speaker system with no choke in the crossover network appears in Fig. 5-6A. After you have made the choke according to the data found in Table 5-1 and using the instructions in Chapter 4 to guide you, insert the choke in the woofer circuit, as shown in Fig. 5-6B. Most magnet wire has tough insulation, so you must use a piece of sandpaper, a grinder or a sharp knife to remove a strip of it at least ½" long before soldering the coil wire into the woofer circuit. This is important because any added resistance in the woofer circuit can alter the damping of the woofer by the amplifier. Such a change can produce boomy bass.

Table 5-1. Wire and Form Data for Some Frequently Used Coils for Crossover Networks.

Frequency (8 ohm speaker)	L (mH)	Wire Length (ft)	No. Turns	Wire Gauge	Form Dimensions
10000 7000 5000 4000	0.127 0.18 0.25 0.32	17.5 20.8 24.5 27.7	89 106 125 141	#24	A = ¼″ B = ½″ C = 1″
3500 3000 2500	0.36 0.42 0.51	36 39 43	122 132 145	#22	A = ⅜″ B = ¾″ C = 1-½″
2000 1500 1000	0.64 0.85 1.27	56 64 78	141 162 200	#20	A = ½″ B = 1″ C = 2″
900 700 500 300	1.41 1.82 2.55 4.24	101 115 136 175	171 194 203 296	#18	A = ¾″ B = 1-½″ C = 3″ to 4″

The coil should be permanently installed to the interior of the back or other wall of the speaker cabinet so it won't move around and cause future wiring problems. A dab of silicone rubber sealer can hold it in place. Simply pull aside the cabinet damping material, glue the coil to the cabinet and push the damping material back over the coil. Don't use screws, nails or other iron pieces to install the coil and don't install the coil near any iron.

Test the speaker thoroughly before sealing the box. Hook it to your receiver and listen at low volume to see if the lows are coming

Table 5-2. Length of Magnet Wire per Pound.

WIRE GAUGE	APPROX. FT. PER LB.
14	80
16	125
18	200
20	315
22	500
24	800

from your woofer and the highs from the tweeter. If the sound appears wrong, recheck your wiring.

An added choke has several advantages. It removes the dual presence you sometimes get when two speakers are reproducing the same tonal range. Also it saves power by eliminating the need for the amplifier to deliver high frequencies to a speaker that can do little more in that range than radiate heat from the voice coil.

PROJECT 13: QUASI-SECOND-ORDER TWO-WAY CROSSOVER NETWORK

In Project No. 12, a choke was added to a typical low-cost speaker system to make a full 6 dB-per-octave, or first-order, parallel network. First-order networks have fewer parts than higher order networks, which means lower cost. And they have an even more important advantage: lower phase distortion. Various methods have been used to avoid the phase problems of networks with sharper cutoff rates, but most engineers agree that the simple networks are the most foolproof and, except for one disadvantage, are the best choice. That one disadvantage is more evident in some systems than others. It is the gentle cutoff slope which permits drivers to be heard over a range of about four octaves beyond the crossover frequency.

Speaker designers would like to be able to combine the simplicity of the first-order network with a sharper cutoff rate. Several years ago, Dr. Allan Kaminsky of the University of Colorado suggested that if a series network could be designed without demanding that it have constant input impedance, a quasi-second-order network could be made with the same number of components as a first-order network. By quasi-second order, he meant that it works like a second-order network at the crossover point, cutting response sharply there, although the ultimate slope is only 6 dB per octave. This kind of network has several advantages. As mentioned, the cutoff is sharper than with a conventional network, the network has few parts, and any variation in component values affects both the woofer and tweeter in a complementary way, so that a mistake means simply a shift in crossover frequency rather than gap or overlap between the output bands of the two drivers.

Considering these advantages, you are probably wondering what kind of *disadvantages* exist. This is a series network, so the two drivers should have the same impedance because they do not appear in a separate branch of the circuit and the component values cannot be adjusted to fit a different impedance for each one. Speaker impedance is quite variable, though, and no two speakers will have exactly the same impedance over a very wide band of frequencies. Minor impedance variations will not cause significantly affect performance. But for predictable results, use this network with 8-ohm tweeters and 8-ohm woofers.

To choose the components for your crossover, you simply consider that you are making a crossover network for 4-ohm speakers even though they really have an impedance of 8 ohms. The network diagrammed in Fig. 5-7 is designed for a 1000-Hz crossover frequency to match your drivers.

For a 1000-Hz crossover, refer to Fig. 4-2 for determining the correct component values. Starting on the 1000-Mz line at the base of the chart, go upward to the second horizontal line, the one for 4-ohm speakers. Moving upward to the left, the capacitor should be 40 μF, twice that for an 8-ohm speaker. This value of capacitor is not common, but any two more commonly available capacitors with values that add up to 40 μF can be wired in parallel. For example, a 24 μF capacitor can be used in parallel with a 16 μF capacitor, or a 36 μF capacitor can be used in parallel with a 4 μF capacitor, and so on.

Going back to the chart for the choke value at 1000 for a 4-ohm speaker, this turns out to be 0.6 mH, instead of the 1.25 mH of a more conventional network for 8-ohm woofers. To get the ideal choke winding data, refer to Table 5-1. But note that this table is for 8-ohm drivers, and the woofer here is being treated as a 4-ohm

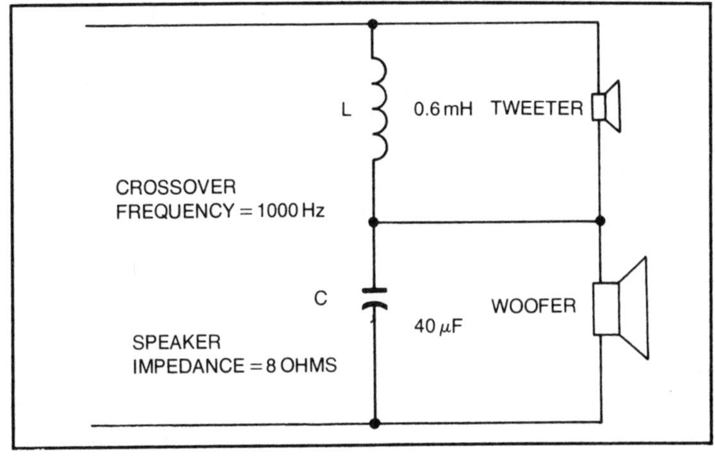

Fig. 5-7. Wiring diagram for quasi-second-order crossover network.

speaker. Because an 8-ohm speaker will require the same value of inductance for a crossover point at 2000 Hz as a 4-ohm speaker does at 1000 Hz, you must go to the 2000-Hz data. Just remember to double the frequency to get 4-ohm values from Table 5-1. Here, the calculations have been carried out to an extra place beyond the decimal point—to 0.64 mH. Also, 56 feet of No. 20 wire will be needed to wind a 141-turn coil on a 1″ core. In cutting the correct wire length, allow *at least* an extra foot for two 6″ leads from the coil. You can always cut off any extra length.

This completes the decision. You must have a 40 μF capacitor, a 0.64 mH choke and an 8-ohm L-pad to control the tweeter. With materials for these items assembled, the crossover can be constructed.

Construction

The crossover network can be easily built on a 6″ x 8″ piece of ¼″ tempered hardboard. To install this board in the speaker enclosure, you can cut a 3″ x 4″ hole in the back of the box. Then, after completing the network, simply glue the board into the back, from inside the cabinet, with silicone rubber cement.

Drill a ⅜″ hole in the center of the board for the tweeter control. Drill two ¼″ holes to match the spacing on the screw-type terminal strip that you use for box terminals. These holes should be spaced about 2″ from the hole for the L-pad.

Glue the screw-type terminal strip to the exterior side of the hardboard. You can use carpet tacks to hold the terminal strip in position while the glue sets. Install the L-pad with the hardware provided with it. Wind the coil, according to the specifications shown earlier, and mount it just above the L-pad with silicone rubber cement. When the cement has set, you can complete the wiring. Cut short lengths of wire to connect the positive box terminal to pin No. 3 of the L-pad. Connect the capacitor, or capacitors, from the negative box terminal to pin No. 1 of the L-pad (Fig. 5-8). Connect one lead from the coil to pin No. 3, and the other lead to pin No. 1 of the L-pad. Prepare two lengths of lamp cord, one for the woofer and one for the tweeter. Connect the leads for the woofer to pin No. 1 of the L-pad and to the negative box terminal. Connect the tweeter cord leads to pin No. 2 and pin No. 1 of the L-pad. Note that lamp cord has one lead with a ridge on the insulation. Use this coding to make sure that the positive terminal of the woofer is connected to pin No. 1 of the L-pad, and that the positive terminal of the tweeter goes to pin No. 2 on the L-pad.

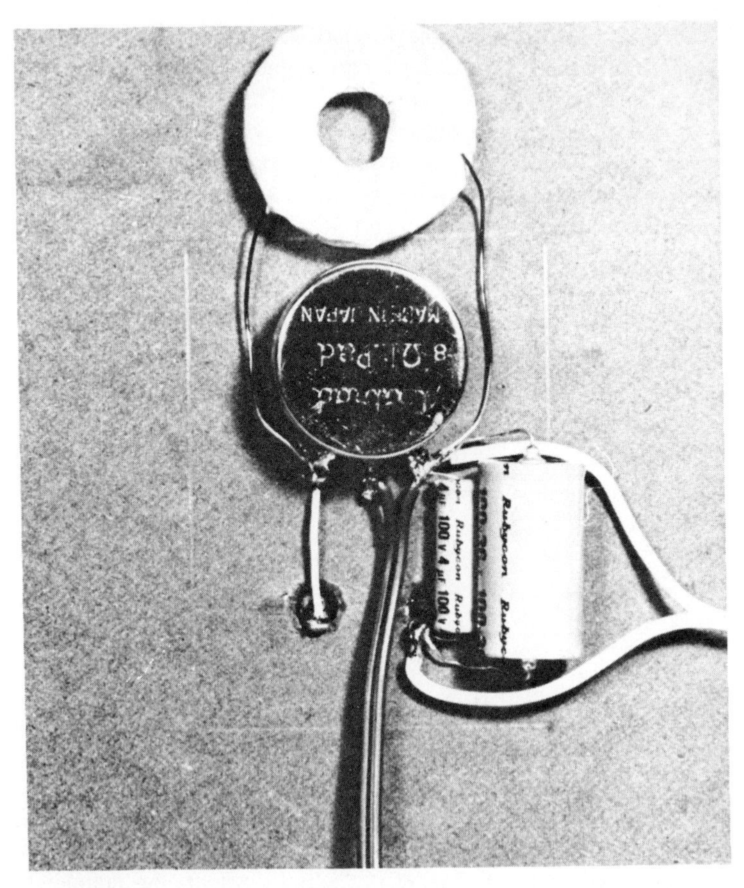

Fig. 5-8. Project 13 built on a piece of ¼″ tempered hardboard. White leads go to the woofer, and brown go to the tweeter.

Connect the speaker to your receiver and try the system at low volume. Check to see if the bass is coming from the woofer and the treble from the tweeter. Rotate the control and see if a clockwise rotation increases the treble and a counterclockwise rotation decreases it. If everything works right, turn off your receiver and complete the wiring by soldering every connection with rosin core solder.

After all the components are sealed in the speaker box, place it in its permanent location and turn the tweeter control fully counterclockwise. Turn on your receiver and gradually bring up the treble until it just blends with the sound from the woofer. When the combination sounds like a single speaker, you have it right.

You can adapt this project to any compatible woofer-tweeter pair. Don't forget to consider 8-ohm speakers as 4-ohm speakers when choosing crossover components. This same series circuit can also be used in a more conventional way by selecting 8-ohm components, but it will not have the sharper cutoff at crossover. Such a combination would be just another first-order network. If you like to experiment, you can try both and choose the one you like best.

PROJECT 14: A FIRST-ORDER THREE-WAY NETWORK

If you are considering a new speaker system, or just improving an old one, here is a three-way network for 8-ohm speakers. The crossover points are set at 500 and 4000 Hz. Using the procedure outlined in the preceding chapter, we find for this 6 dB-per-octave network, as diagrammed in Fig. 5-9, the following components:

$$L1 = 2.5\,\text{mH}$$
$$L2 = 0.36\,\text{mH}$$
$$C1 = 35\,\mu\text{F}$$
$$C2 = 5\,\mu\text{F}$$

Using Table 5-1, we find the value listed there for L1 is slightly different from the value obtained from Fig. 4-2 but only because the calculations were carried out to two places farther to get 2.546 mH and then rounded off. Note that if we had rounded them off to one place to the right of the decimal, it would still be 2.5 mH. But the important data to get from Table 5-1 are the wire gauge, No. 18; the length of the wire, 136 feet; the number of turns, 230; and the coil dimensions. So what we need here is a coil made with No. 18 wire, wound with 230 turns on a 1-½″ core. But start with 137 feet or more wire; you'll need leads from the coil of about 6″ each to reach their connecting points.

Referring again to Table 5-1 for L2, note that we must search for the correct value at 3500 Hz instead of 4000, as explained in the preceding chapter. Here the suggested wire gauge is No. 22 and the core diameter ¾″. We will need 36 feet of wire, which will give about 122 turns. In measuring length, though, don't forget to add an extra foot so that you will have leads from each end of the coil that are at least 6″ long.

For the capacitors, a value slightly different from the theoretical value might have to be accepted. You are likely to find a 4.7 μF capacitor more easily than the indicated 5 μF value, but the

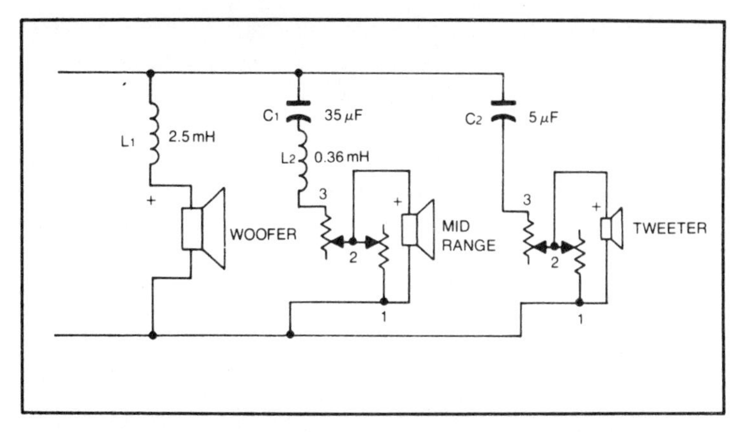

Fig. 5-9. Wiring diagram for first-order three-way crossover network, Project 14.

difference won't be significant. Also you probably won't find any 35 μF capacitors, but smaller values can be used in parallel to add up to about 35 μF. For example, you could use a 24 μF capacitor to make 36 μF. Or, if you are more particular, you could get a 24 μF, an 8 μF, and a 3 μF capacitor and put the three in parallel. This would involve the cost of an extra capacitor, and you almost certainly would not hear the difference in sound.

With this data, and the materials assembled, you are ready to build the crossover network. We will include two L-pads to control the output level of the midrange driver and the tweeter.

Construction

An immediate problem in planning the construction of a crossover network is where to install the network. One solution, which works in almost all cabinets, is to mount the components on a piece of ¼″ tempered hardboard and then glue the hardboard base to the interior of the cabinet back. It is useful to include the speaker terminals and any midrange or tweeter controls on this same piece of hardboard. Access to the controls and the terminals can be provided by cutting a window in the back and then gluing the crossover board over the inside of the window. Typically, there will be plenty of space on the back of your speaker cabinet to allow for a 3″ x 4″ window, about the smallest possible opening that will give access to the controls as well as the box terminals. The crossover components can be mounted on one side of the crossover board with just the control shafts and the terminals extending to the other, exterior side.

Begin construction by winding the coils. It makes no difference whether you use temporary or permanent coil forms, but coils removed from temporary forms should be thoroughly taped so they will be stable.

Cut an 8″ x 8″ crossover board and drill it to the pattern shown in Fig. 5-10. Note that you must locate the box terminal holes to match the distance between the screws on the specific terminals you use. This distance varies considerably with different terminal strips, so you should have the terminals on hand before drilling the board.

Install the hardware first: the screw-type terminal strip, the controls and an ordinary six-lug wiring terminal strip. The screw-type terminal strip should be sealed to the board with silicone rubber or latex caulking compound so that there will be no air leaks around the terminals. Sealer should also be used around the terminal lugs inside the board, but only after all wiring and soldering to those lugs is completed.

Install the capacitors. Mount the coils with silicone rubber cement and leave the board undisturbed until the cement sets.

When you wire the coils into the circuit, make sure to remove all the tough plastic insulation from about ½″ of the coil wire before connecting it to the proper terminal. If you fail to get off all the insulation, the woofer could give intermittent action, working only when the clean part of the wire makes contact.

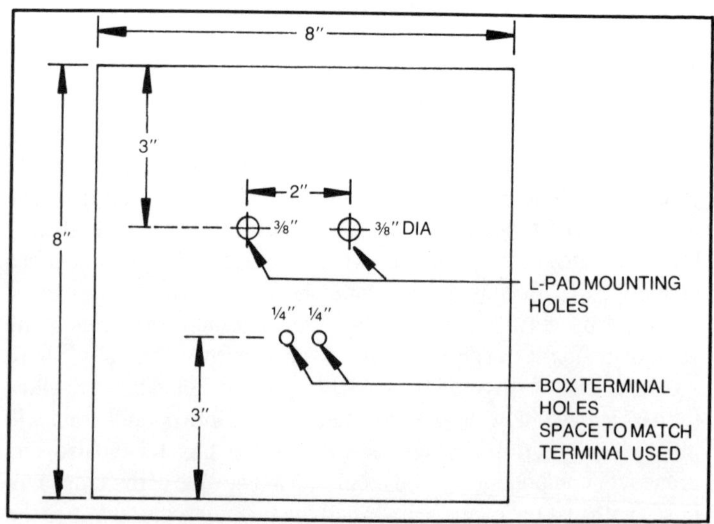

Fig. 5-10. Drilling pattern for crossover board, Project 14.

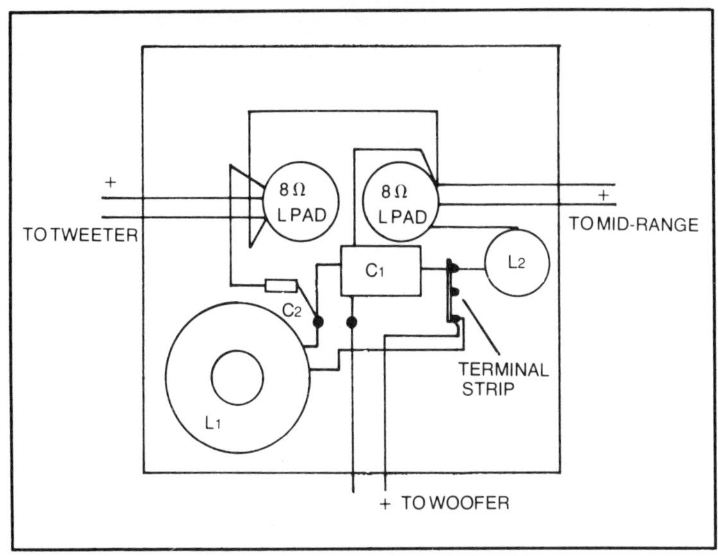

Fig. 5-11. Suggested layout for crossover wiring, Project 14.

Cut short lengths of wire to make the connections between controls and terminal strip lugs and from box terminals to terminal strip lugs (Figs. 5-11 and 5-12). Use ordinary lamp cord for woofer, midrange and tweeter leads. Note that one side of the lamp cord has a ridge on the insulation. Use this ridge to make sure that each speaker is wired in phase. The phase of the speaker terminals will

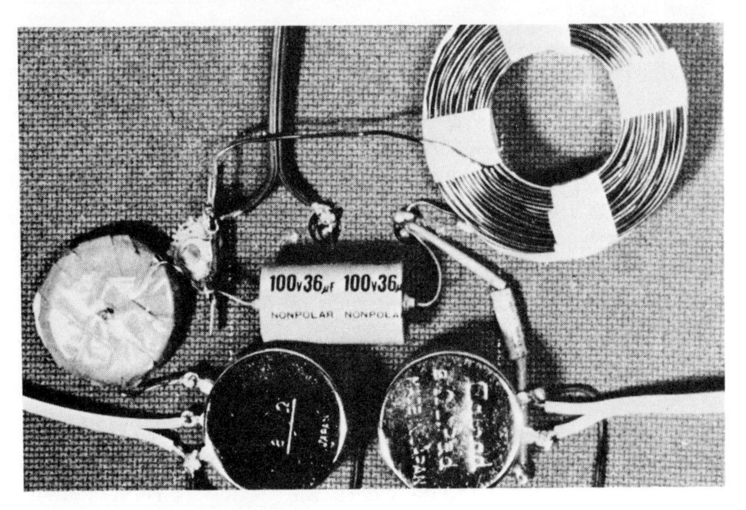

Fig. 5-12. Project 14 completed.

be identified by a red dot or a + at the positive terminal. Follow the guide in the crossover schematic (Fig. 5-9) to see that the proper side of the line goes to the positive terminal in each speaker.

After making the connections, temporarily hook up the speaker to your receiver and test it at low volume. Check first to see that the lows are coming from the woofer, the middle tones from the midrange and the highs from the tweeter. Then rotate the controls to see if they work with the proper driver and if they increase the level of that driver with clockwise rotation and decrease it with counterclockwise rotation. If all appears to be working right, solder every connection with rosin-core solder.

After all components are installed in the cabinet, test the speaker in the location where it will be used. Turn the midrange and tweeter controls fully counterclockwise. Connect the speaker to your amplifier or receiver and adjust the control on the midrange driver first. Bring the midrange volume up until it appears to blend with the sound from the woofer and the two sound like a single speaker. Then bring up the tweeter level until it blends with the midrange. Listen at those positions for a while to determine if the sound is balanced and pleasing.

One of the big advantages of a three-way system is that you have better control over the total balance of bass, midrange and treble. By fine tuning the controls, you can get the kind of balance that best fits your speakers and your room. Used properly, these controls permit you to significantly improve the pleasure you get from your stereo system.

PROJECT 15: SHARP CUTOFF TWEETER NETWORK

As mentioned earlier, first-order crossover networks are simple to design and construct and offer few problems. Their only disadvantage is that they permit such a gentle rolloff that the response of the driver is extended several octaves beyond the desired crossover point. This is of no importance if the drivers have no serious peaks beyond cutoff and if they can handle the power available to them beyond their normal cutoff point. Here is where some tweeters face danger.

Each year, many tweeters are blown out, some by carelessness such as wiring mistakes. Low-power amplifiers cause some of the destruction by adding so much harmonic distortion at dynamic peaks that frequencies appear at the amplifier output that were not present at the input. Even where the wiring is right and the amplifier is good, many tweeters give up when fed signals of

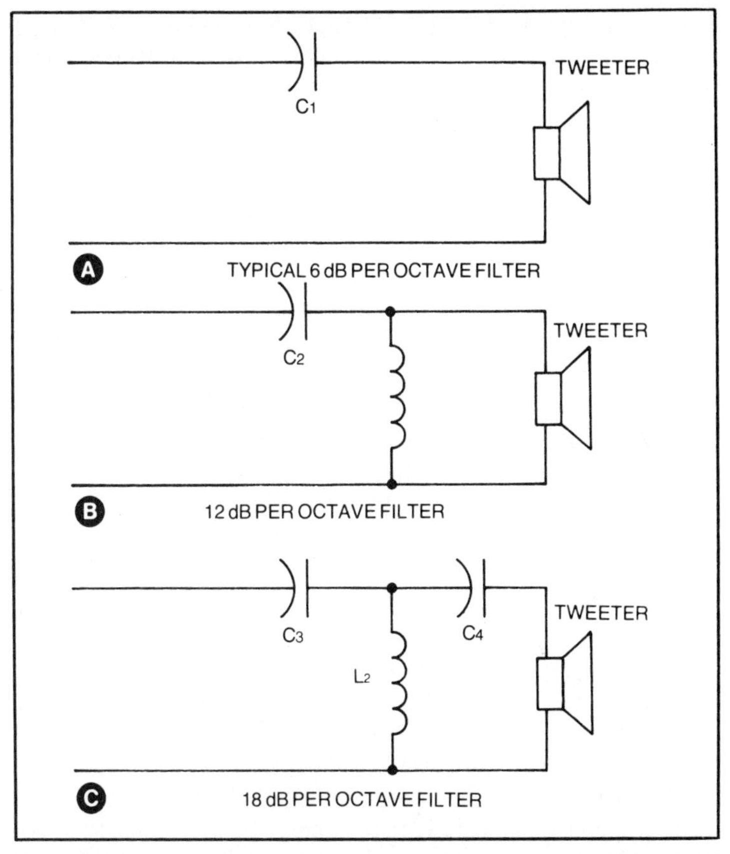

Fig. 5-13. Circuits for sharp cutoff tweeter networks, with 6 dB, 12 dB and 18 dB shown at A, B and C. Refer to Tables 5-3 and 5-4 for various component values.

significant power below their optimum crossover frequency. A first-order 6 dB network doesn't do much to prevent this last kind of damage when a small tweeter is used down to the bottom end of its frequency range.

One solution is to rework just the tweeter leg of the crossover network so that the tweeter sees a much faster rate of rolloff at the lower end of its frequency spectrum. This produces an asymmetrical crossover network, one that cuts off the lows from the tweeter at a much steeper rate than it does the highs from the woofer. That is acceptable because a gentle slope at the upper end of the passband of the woofer makes it invulnerable to such damage.

The typical tweeter circuit has only a single capacitor in series with the tweeter, as shown in Fig. 5-13A. The solution is to change the 6 dB-per-octave filter to a 12 dB-per-octave network or even an 18 dB-per-octave network. The schematic for a 12 dB-per-octave filter appears in Fig. 5-13B for an 18 dB-per-octave filter in Fig. 5-13C.

To get the right values, you must decide what crossover frequency was chosen by the speaker designer. In some cases, you can lower this frequency as you go to a sharper filter, but there is probably no need for this change. The original capacitor must be changed. Table 5-3 shows various crossover frequencies with corresponding original capacitor values. It also shows the new value for a capacitor to be used with a choke in a 12 dB filter, along with the value of choke that you would add as a shunt across the tweeter.

To apply an even sharper 18 dB-per-octave filter, go to Table 5-4. Again this table shows the crossover frequency as implied by the value of the original capacitor and the value of the new components that you would choose to make up a sharp cutoff filter that would operate at the original frequency.

Construction

If the original capacitor is installed with other crossover components, you can install the new components near its site. But on many systems you will find the original capacitor wired directly into the tweeter, sometimes connected between an extra terminal on the tweeter frame to a voice coil terminal. In some cases, you might have to remove the tweeter from the box to get at the wiring. In such installations, I prefer to run a dual conductor lead, such as a piece of lamp cord, from the tweeter to the inside of the back panel where the crossover components can be mounted. This permits you to make future network changes without disturbing the tweeter mounting.

Table 5-3. Values of New Capacitors and Added Inductor for Typical Crossover Frequencies.

Crossover Frequency	Original Capacitor Value (μF)	Capacitor Value To: (μF)	add L with a value of: (mH)
5000	4	2.8	0.36
6000	3.3	2.3	0.3
7000	2.8	2	0.26
8000	2.5	1.8	0.22
9000	2.2	1.6	0.2
10000	2	1.4	0.18

Table 5-4. Component Values for 18 dB-per-Octave Tweeter Network.

C1 (μF) (Original)	f (Hz)	C3 (μF)	C4 (μF)	L2 (mH)
20	1000	13	40	0.95
13	1500	8.8	27	0.64
10	2000	6.5	20	0.48
6.6	3000	4.4	13	0.32
5	4000	3.3	10	0.24
4	5000	2.7	8	0.19
3.3	6000	2.2	6.6	0.16
2.8	7000	1.9	5.7	0.14
2.4	8000	1.7	5	0.12
2.2	9000	1.5	4.4	0.11
2	10,000	1.33	4	0.10

As in other crossover network projects, an ideal way to install the components is on a piece of ¼" hardboard. Because these sharp cutoff networks have more parts, it's a good idea to make a full-size drawing of the crossover board on a sheet of paper before drilling the board. Set the components in place on the paper, trying different locations for each until you find the combination that permits short connecting leads. There is no one way to build a crossover network, but don't stack the choke coils one above the other; the fields will interact, and the inductance of each will be unpredictable. Avoid *floaters*, hooking two components together with no support for the connecting wires. Install terminal strips with solder lugs where appropriate. Connect each part mechanically; then when all are connected, solder the connections with rosin-core solder, but not before testing the filter.

If you follow the precedure suggested above, you will find that even a sharp cutoff filter isn't critical in construction. When you have completed the wiring, connect the speaker to your receiver and listen at low volume to see if the highs come from the tweeter. Listen for the upper highs to make sure they are present.

If you have small tweeters in your system, this project is worth its cost. Small tweeters are the best for reproduction and dispersion of the upper highs, but they are more easily damaged by inadequate crossovers. One of these sharper networks will give more protection than most other measures, such as fusing. A fuse will protect the system from an overdose, but the tweeter can be damaged while being fed power *within* its normal limits if that energy is outside its proper passband.

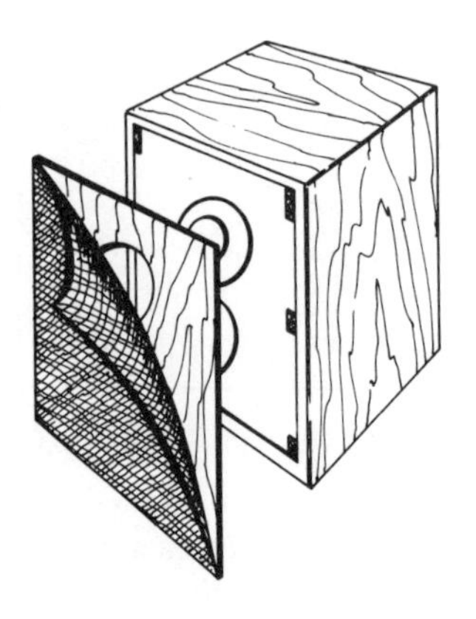

6 Speaker Enclosure Projects

The crossover network projects in the preceding chapter offer a way of controlling the signal that goes to each driver so it can do the job it does best. In addition to feeding the right band of frequencies to each driver, there are many ways to improve a speaker system. The speaker projects in this chapter require no wiring changes; some don't even require that you gain access to the interior of the speaker cabinet. One project, for example, can nullify an external force that sometimes puts a lid on the sound level in your living room. Another offers a way to improve dispersion and apparent high-frequency response without any alteration of cabinet or drivers.

The speaker enclosure is one of the most important components in any stereo system. A good enclosure is much more than a piece of furniture; it is an acoustical accomplice for the speakers. Because many low-priced commercial speakers have enclosures that violate the rules of good design, this is a good place to start in improving any speaker system. The first project of this group is one that can take many forms, depending on the specific kind of speaker system you own.

PROJECT 16: IMPROVING A LOW-COST ENCLOSURE

Someone recently asked me, "Why would anyone want to build his own speakers when department store speakers are so cheap?"

To anyone who has bought any plywood or particle board recently, the question makes you wonder, why indeed? But if you look beyond the smooth-looking surface of these plastic-covered boxes, you can see why they are so cheap. Cheap speakers have been throughout the high-fidelity and stereo eras, and construction of bargain boxes has gone through several stages of evolution. Early, cheap stereo speakers were often installed in small boxes with slotted hardboard backs. The next wave of flimsy speaker boxes usually had solid backs that were held on by a few small screws. Current manufacturing practice for wood or wood product enclosures is to produce a permanently sealed box by gluing in both the front and back panels which are set in wall grooves. This use of dadoed grooves in the walls, which are then wrapped around the front and back panels, saves material and labor over the earlier practice of installing cleats to receive screws. Each stereo age has seen its share of plastic boxes with highly reflective interior surfaces.

If you have any of these cut-rate models, or find some at a garage sale or flea market, you can make a significant improvement in the enclosure at little or no cost, an improvement you can hear in the sound. These modified boxes can serve as extension speakers or, by making further changes as described in the next chapter, they can perform well enough to meet the requirements for main speakers in many stereo rigs.

CABINET DEFICIENCIES

Your first step in improving a cheap cabinet is to analyze what is wrong with it. Here are some typical flaws:

- Cabinet volume wrong for speaker.
- Air leaks in a "sealed" enclosure.
- Cabinet walls too thin.
- No damping material.
- Loose panels that rattle.
- Poor box shape.
- Mistuned port.
- Obstructions or heavy grille cloth in front of speaker that block highs.
- Open-back tweeters or midrange speakers with no subenclosure.

By nit picking, we could find other faults, but the ones listed above are most common. Such speakers also have deficiencies in driver quality and crossover networks, but these deficiencies are dealt with in other projects. Here we are concerned with the box and nothing else.

To analyze your enclosures, you may first have to do major surgery. This is especially true if you have one of the modern glued-up models. As a first step, you can get a good estimate of cabinet construction by rapping the walls of the box with your knuckles. Start with the middle of the back and rap it sharply. Does it sound like a hollow drum? That's a sure sign the box panels are either too thin or made of low-grade material such as low density particle board. Particle board is a good choice for speaker boxes, but only if the high-density grade is used. Flaky board is weak and easily broken.

Assuming that the cabinet fails the knuckle test, and almost any department store box will, check next to see how you will get into the box to rebuild it. Again, if you have a cabinet made in the current mode of glued-up parts, you must cut out a panel, usually the back. Make a starting hole by either drilling a series of small holes along one edge of the back panel next to a side panel or use your sabre saw. If you choose the sabre saw method of starting the cut, you must use care to avoid breaking the saw blade. Angle the saw so the blade barely touches the panel (Fig. 6-1). Press the saw tightly against the panel so you can control the blade. Then, after switching on the saw, gently allow the blade to gradually make contact with the panel. As the blade cuts into the back, rotate the saw very slowly to permit the blade to go deeper. If you take it easy, you can cut a slit through the back that allows the blade to penetrate the panel without using other tools.

If you have one of the older style cabinets, entry is easy. Remove the screws that hold the back panel on the box. In some cases the front panel is removable, but only after you take off the grille. A careful inspection will show you which way to go.

Cabinet Volume

This is a fundamental flaw that must be considered first. Many cheap speakers consist of a speaker in a box with no design effort made to match box volume to speaker requirements. Traditionally, one often found a mismatch on the side of too little box volume, but sometimes the ratio is reversed, as when a 6″ speaker is mounted in a box more than large enough for a typical 8″ model (Fig. 6-2). The reason for the box that is too small is obvious: It saves material and space. But why a box that is too big? Evidently the box is made to impress the customer, rather than to function properly.

Fig. 6-1. If your speaker cabinet has no removable panels, you can saw out the back panel if you use care.

The correction for a box that is too small may take two paths: either a large homemade box or a different speaker. Driver upgrading is a legitimate way to solve the problem, but that process is fully described in the next chapter. The cure for a box

Fig. 6-2. This box is too big for the speaker and has no damping material.

that is too large is to reduce the interior volume by lining the interior walls with scraps of wood.

To precisely determine optimum box volume requires a series of driver tests, but for closed box speakers—the most common type—you can make a good estimate of proper box volume by considering nothing more than speaker diameter, as shown in Table 6-1. This estimate is based on the average requirements of modern speakers with reasonably high compliance.

Examination of several cheap speaker cabinets showed that about one-half of them were badly mismatched, sometimes by a volume factor of 3 or more. In the case of the enclosure shown in Fig. 6-1, the cubic volume was just over ¾ cubic foot, but the speaker inside was only a 6″ model. This speaker system was a model originally sold by a discount store at a price of $40 for the pair. Customers were impressed by the neat appearance. At low volume, the sound wasn't obnoxious, but there was a slight ringing in the middle and upper base. At high volume, the middle and upper bass became increasingly resonant, and it was evident that the low bass was missing. The overall effect was that of a noticeably muffled high range coupled with the obvious bass resonances. When rapped, the box sounded like a drum.

After discovering the 6″ speaker in a box that was too large, the next decision was whether to change speakers or reduce box volume. The speaker had been advertised as a bass reflex model, which could require more volume than a closed box. For cheap bass reflex speakers, however, the best policy is usually to close the

Table 6-1. Suggested Cubic Volume of Sealed Enclosures for Various Sizes of High-Compliance Speakers.

	Typical Enclosure Volume	
Advertised Speaker Size (In.)	Cu. In.	Cu. Ft.
5	430	0.25
6	625	0.36
8	1,200	0.7
10	2,600	1.5
12	5,000	3
15	15,000	9

Fig. 6-3. Internal volume was reduced and walls were strenghtened by lining them with ¾″ lumber.

port, making a sealed system. Reflex drivers should be highly damped by a heavy magnet, something this speaker didn't have.

Further inspection showed that the foam grille was glued on by a bead of glue run around the perimeter of the speaker board, about an inch back from the edge. The foam would have to be removed if the speaker board were to be cut out for a larger speaker. This could be done by using a sharp knife or wood chisel to cut through the narrow glue bead. Because the job would require considerable time and care—plus a new speaker—that path was abandoned. Instead, the original speaker was retained. This demanded that the box volume should be reduced to one more appropriate for a 6″ speaker. The volume was reduced by lining the side walls with ¾″ material and the top and bottom panels with a double layer (Fig. 6-3). The original height to depth ratio was greater than optimum, at almost 4:1. Most speaker design engineers aim at keeping the ratio of greatest internal dimension to the smallest dimension at or below 3:1. Even with the internal depth unchanged at 5-½″ and the internal height reduced by 3″, the ratio remains above 3:1.

Air Leaks

An important step in rebuilding your speaker boxes is to plug all air leaks. These are most likely to occur where a screwed on back was used with inadequate cleating inside the back. The same kind of leak can also occur around the speaker board. Check the box corners for good glue joints. Small holes or gaps between cleats can be filled with *Mortite* weather stripping compound or a good grade of caulking compound (Fig. 6-4). Latex or silicone rubber caulking compounds are the most useful kinds for these jobs because they don't dry out and crumble. If a gap is greater than ¼″, it should be plugged with a piece of wood glued in place. Where sections of cleats are missing or never installed, cut new cleats to fit the spaces and glue them in. These cleats can be held in place by C-clamps while the glue sets, but use some kind of scrap material on the outside of the cabinet to protect the surface (Fig. 6-5). If you have no C-clamps, you can install the cleats with glue and small screws or nails to hold them until the glue sets.

In addition to cleats to receive screws from the back panel, install similar cleats around the rear edge of the speaker board if necessary. You won't need these on most cabinets of recent manufacture because of the grooving and gluing method of construction mentioned earlier. Don't forget to add glue blocks in the interior of each corner if they are needed for a strength or to plug air leaks.

Potential air leaks exist in some cabinets where the speaker cable enters the box. Plug any air space there with caulking compound, or install screw-type terminals. Another possible point for air movement is around the speaker itself. An ideal way to seal the speaker to the box is to use silicone rubber sealant under the speaker. This method is especially useful where the speakers are front-mounted because most speakers have no gasket behind the frame, and the silicone rubber forms a perfect gasket as well as a glue to hold the speaker on the box.

Stiffening Walls

Most cheap enclosures have walls that are too thin or made of low-grade material that lacks sufficient rigidity, even for small

Fig. 6-4. Gaps in cleats (left) were filled with Mortite (right).

boxes. You can glue almost any building material to the walls to add stiffness. Some materials, such as paperboard or roofing felt, also help to damp the walls so they don't vibrate and absorb sound energy from the speaker. The walls have much greater total area than the speaker cone, so a minor wall movement can produce more sound than that from the cone, accentuating certain frequencies.

Another way to improve wall rigidity is with *braces*. The goal is to make the walls stiff, which will raise their frequency of resonance. At first mention this may seem illogical, but the idea is to reduce the amplitude of wall vibration, and a higher frequency of vibration does just that. It also brings the frequency of resonance up to where its sound can be absorbed by the damping material inside the box.

To gain maximum rigidity, run your wall braces in the direction that will divide each panel into the narrowest portions. That means *lengthwise*. Ideally the brace should be placed off center, so it divides the panel unequally, but any kind of bracing is good.

Damping Material

Most cheap speakers have little or no damping material in the cabinets (Fig. 6-2). Damping material is essential for uncolored sound. If there is no damping material, the sound from the back of the speaker hits the cabinet walls and is reflected about in the box. Some of this reflected sound hits the cone. If the sound hitting the cone is out of phase with that being produced by the cone, there is *cancellation*, producing a dip in the response curve. But if it is in phase, the situation is worse because it reinforces the sound produced at the cone, making a peak in the response of the cone.

The typical result of inadequate damping material is a speaker that sounds loud, even at low volume. This kind of speaker may sound lively at first hearing, but after a short time one has the impulse to lower the volume or even shut off the system because of listening fatigue. If you do nothing else to improve your speakers, add an appropriate amount of damping material, if needed.

This brings up the question of how much material to add. The rule with a typical cheap speaker is *the more, the better.* With some expensive speakers that have large magnets, the system can be overdamped. In nearly all small speaker cabinets, you can get better sound by filling the box. For larger systems, with speakers of 8″ diameter or larger, line the walls and cover the back panel

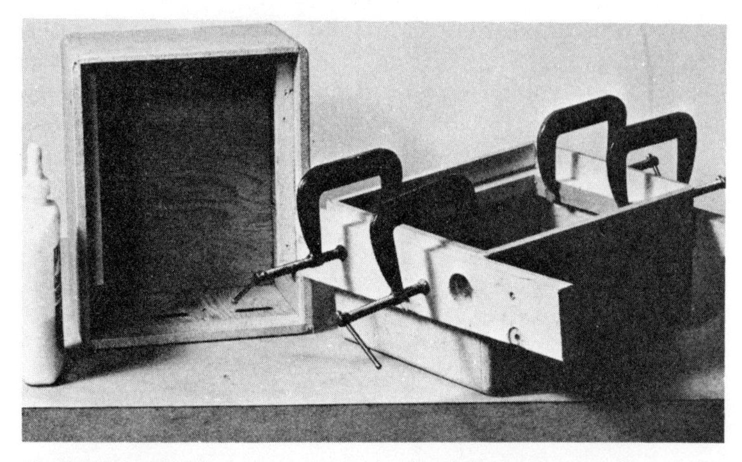

Fig. 6-5. This pair of bargain enclosures had inadequate cleating (left) so cleats were glued in (right).

Fig. 6-6. Cover all internal walls except speaker board with 2″ to 3″ of damping material. This speaker was further improved by adding a tweeter; otherwise the port would have been closed.

with at least an inch or more of damping material. This is a minimum. You can often get more uncolored sound by installing a 2″ or 3″ thick batt on each wall and, especially, on the back (Fig. 6-6).

In choosing the kind of damping material for your bargain speakers, let availability decide this because any of the commonly used materials will work well. Fiberglass is the traditional choice, but polyester batting is more pleasant too handle and works just as well. Sometimes you can get leftover scraps of carpet pads free from furniture stores or carpet layers, and this makes a good base for other fuzzier material. Shredded rags work better than nothing. Just make sure you install some kind of material in your box to absorb sound.

Plastic Enclosures

Plastic boxes are less versatile than wood or wood product boxes because you can't easily add cleats or change internal dimensions. The single most important change you can make with these is to add damping material. Plastic walls are much worse than wood walls in reflecting sound, particularly in the troublesome mid- and high-frequency range where secondary sound waves can impinge on the cone and produce a ragged response.

Use duct tape to seal plastic boxes. This is an aluminum-colored tape that plumbers use to seal heating ducts. It will adhere to plastic well enough to stay in place, and it is impervious to air movement.

Ported Boxes

If your cheap speaker system is ported, you can probably improve the sound by simply closing the ports. In checking several cheap ported systems, I found that they all produced numerous peaks in port output at frequencies far above the useful port range, peaks that added harshness to the music. These peaks can be subdued with proper damping material in the box, if you find the port is desirable. Unfortunately, ported box design is a highly critical process, and the manufacturers of cheap speakers are unlikely to expend the necessary time for proper design and testing. Another reason to avoid a ported box for cheap speakers is that such enclosures demand a highly damped speaker, and cheap speakers nearly always have small magnets.

If you are still determined to maintain a ported box for your "cheapie" driver, you can apply lots of damping to the box and—even better—damp the driver as shown in Project No. 21. Another possibility is to damp the port itself by stuffing it with damping material. This is not advisable in a properly tuned reflex, but theory is unimportant for a cheap reflex. Let your ears be your guide.

Other Changes

While the enclosure improvement techniques described earlier will be adequate for most cheap speaker systems, you will sometimes find a special problem that requires a specific prescription. I found one of these unusual situations recently when I removed the back from a bargain "three-way system." The speakers inside looked to be of far-above-average quality for low-priced systems, but the middle speaker had an open back and no subenclosure (Fig. 6-7). Pressure developed by the larger woofer cone in the sealed box could easily produce distortion in the unprotected midrange driver. The easiest way to correct this kind of problem is to build a box to cover the small speaker and glue it to the rear of the speaker board. The box should be filled with damping material, and provision must be made for bringing the speaker leads out through the box without air leaks. In the case of the speaker mentioned, I used a paperboard shipping carton that measured 5″ x 5″ inside, just right to enclose the midrange speaker without overlapping the frames of the other drives. I reinforced the walls and bottom of the carton with several layers of paperboard and hardboard, glued in place with white wood glue. Then I drilled a ¼″ hole through the center of the bottom of the carton to bring a piece of lamp cord speaker lead out into the main box. Finally I ran a heavy bead of latex caulking material around the rear of the speaker board where the carton would contact it and pressed the carton down on the board (Fig. 6-8). As I installed the carton, I brought the previously soldered speaker cable out through the hole in the bottom and filled the gaps in the hole with latex caulk. This improvised midrange box made a very satisfactory subenclosure at minimal cost.

The ornamental features in some bargain boxes can affect the sound, usually badly. Some enclosures have obstructions in front of the speakers which can block the highs or at least color the sound. Other cabinets have heavy ornamental grille cloth that is too densely woven. To test for these problems, remove a speaker, if full-range models are used, or a tweeter, and listen to the bare speaker at low volume. Make sure you don't remove any crossover

Fig. 6-7. This bargain speaker system had no subenclosure for the open-back midrange speaker.

components in the tweeter circuit while making this test. Then switch to the other channel by using the balance control and compare the sound from the baffled speaker. Keep your receiver switched to the *mono* mode while testing so each channel gets an identical signal. Pay no attention to bass reproduction. The unbaffled speaker will sound thin because of sound cancellation at low frequencies. Listen to the high range. If you hear a significant difference in high end response, or if you notice any coloration, the cabinet is altering the sound of your speakers. In some cases such alteration is done by design, but the more common situation is that it happens by accident because the cabinet stylist did not consider the acoustical effect of his design. If you find that the cabinets filter out the high notes, you might have to rework the entire front panel, or at least the grille frame and cloth.

In choosing a new grille cloth, pick one that you can see through when you hold it up to the light. Lightweight double knit fabrics work well. This kind of material is quite stretchy when you pull on it. It usually comes in solid colors, such as dark brown or black. Burlap is usually the cheapest material available, but it is not as *acoustically transparent* as the lighter double knits. You might find that you like the sound better with a material that dulls the upper register. Choose one you can live with.

Fig. 6-8. Subenclosure for midrange speaker protects it from pressure developed in enclosure by woofer.

Install the grille cloth on a grille frame that is cut out to fit over the speakers, if they are front-mounted. Discard any grille frames that have slotted openings in front of the speakers. At the very least, cut away the material there so the speakers face an open grille.

The last word on improving speaker enclosures is to *remember the damping material*. This treatment will solve more problems with cheap speaker cabinets than any other single modification.

PROJECT 17: FRONT MOUNT YOUR SPEAKERS

If you rebuild a cheap enclosure, as described in the preceding project, make a note of whether the speakers are installed in front or behind the speaker board. In practically all low-priced systems, the speakers are mounted *behind* the board. This was the traditional practice with all speakers until engineers began to realize that it sometimes causes audible problems. First, there is the possibility of air resonance in the chamber formed in front of the speaker. This air resonance can add a peak to the response, the frequency of the peak determined by the cubic volume of the air in the cavity. And the sharp front edge of the speaker cutout can diffract the sound waves as they pass, breaking them up and producing interference, which affects response. These effects occur mostly in the important midrange and treble. Speakers that are rear-mounted must work into a tunnel—a short tunnel, but nevertheless a tunnel. If you can remove the grille cloth on the cabinets, you can change the speaker mounting from rear to front, a step that eliminates the tunnel and improves the clarity of the midrange and highs (Fig. 6-9).

Construction

Remove the speaker and plug the bolt holes or screw holes with caulking compound. Make sure the leads to each speaker are long enough to reach the speaker lugs in the new position. Since you have to remove the back panel to get the speakers out, you can leave it off until the speakers are mounted.

When you install the speakers from the front of the speaker board, you will probably have to enlarge the mounting holes. To find the right hole size, set the speaker face down and check the diameter of the raised part of the speaker frame with a crude caliper made from stiff wire (Fig. 6-10). It's a good idea to make a practice hole in a piece of paperboard first as a precaution against ruining the baffle by making a hole that is too large or having to make two or

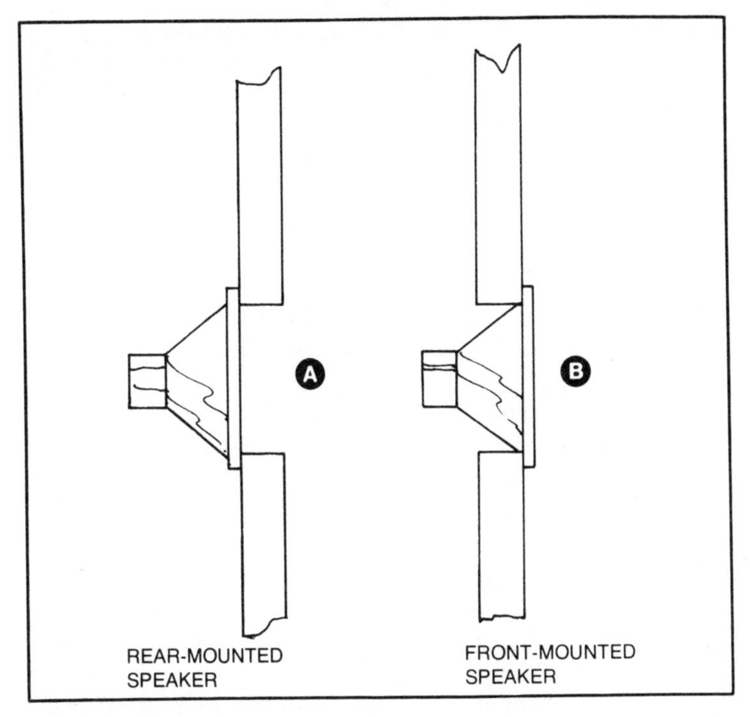

Fig. 6-9. Cavity formed when speaker is rear mounted, as in A, can be eliminated by front mounting, B.

more cuts to get it right. With many speakers you can enlarge the speaker holes enough by using a wood rasp, a *Stanley Surform* tool, or even a coarse sandpaper to round off the front edge (Fig. 6-11).

Almost all speakers are manufactured with a mounting gasket at the front of the frame for back-of-the-board mounting. To mount

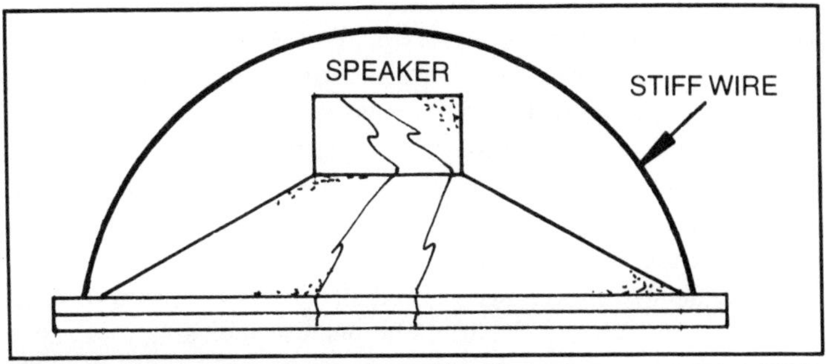

Fig. 6-10. How to make a crude caliper with stiff wire to measure approximate diameter of hole to receive front mounted speaker.

your speakers in front of the board, get the enclosure on its back, and run a bead of silicone rubber sealant around the edge of the mounting hole and then set the speaker into position (Fig. 6-12). Press the speaker down gently and rotate it a few inches to insure that the sealant will be uniformly distributed around the circumference of the hole. Leave the cabinet in this position overnight for the sealant to set. You need not install bolts or screws except for very large woofers, such as 15″ models. If the woofer is extremely heavy, you can drill holes through the speaker board and drive T-nuts into the board from the rear to receive bolts one size smaller than the holes. For example, you would drill ¼″ holes to use 3/16″ bolts, or 5/16″ holes for ¼″ bolts. You can buy T-nuts in 3/16″ or ¼″ sizes at most hardware stores. Use silicone rubber for an air-tight gasket when you install a speaker on the front of the board with bolts and T-nuts.

Grille Frame

When you move the speakers from the rear of the board to the front, the grille cloth will probably have to be discarded and a new grille installed. Most cabinets with rear-mounted speakers have either a stretched cloth grille held by trim or a foam grille glued on the board. Neither of these are likely to work as is with front-mounted speakers. The solution is to install a grille frame, or board, with cutouts to allow it to fit over the speakers and lie flat on the speaker board. The new grille cloth can be stretched over this frame to hide the edges, and the entire assembly can be installed on the cabinet with spots of glue or with brads driven through the cloth and frame into the speaker board. A more refined arrangement is to use pieces of *Velcro* on matching surfaces.

The frame should be cut to fit the front of the box so that its edges are flush with all of the cabinet sides (Fig. 6-13). The edge of the grille board can be cut square or, if of ¾″ or thicker material, tapered at a 45-degree angle. If the board is rather thin, say ⅜″ to ½″ thick, the edges will be too narrow for a significant taper.

After installing the grille cloth, the project is completed. This project won't make the kind of striking difference in sound that some audio projects make, but the difference is that of a more natural, open sound. And if you intend to install a quality high-frequency speaker, such as a small dome tweeter, this project is virtually mandatory. There is no point in investing in a super tweeter that has superior high-frequency range and dispersion, and then hiding it behind a thick speaker board.

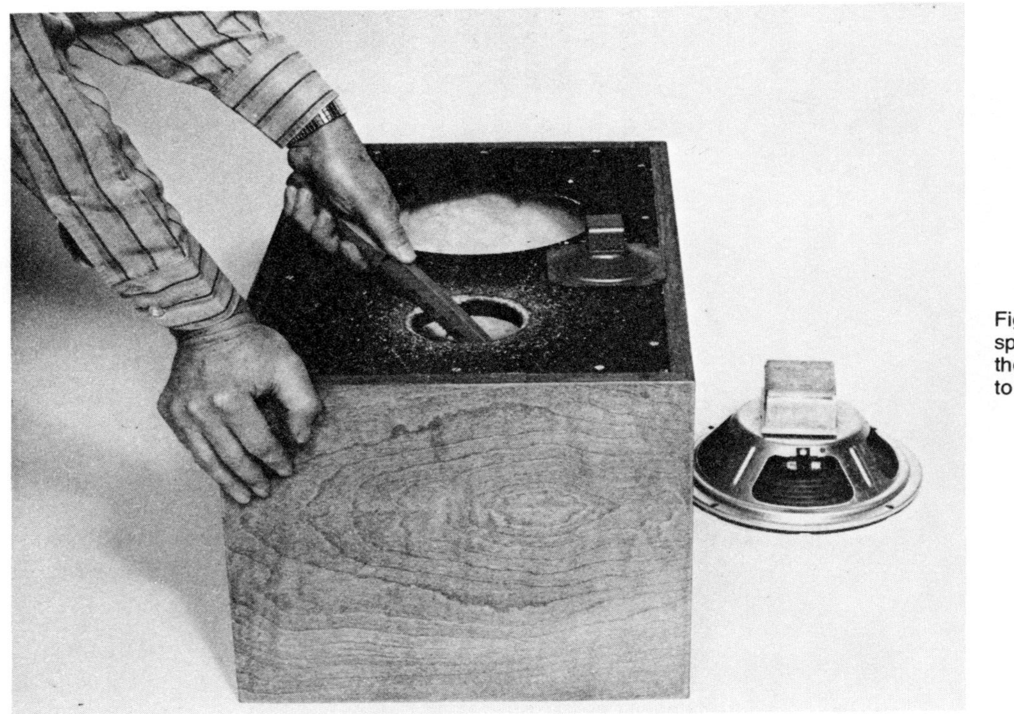

Fig. 6-11. If you keep the same speakers, you can usually rasp out the front edges of the holes enough to front mount them.

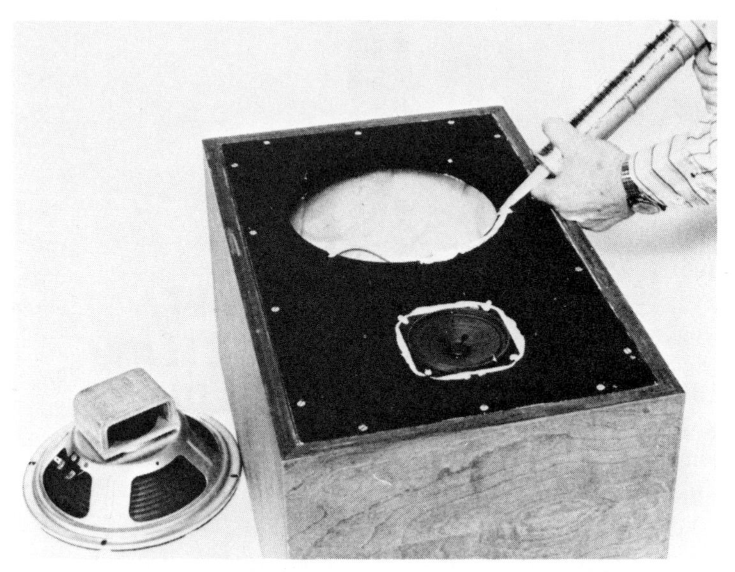

Fig. 6-12. A bead of silicone rubber around speaker hole will make an air tight seal and hold speaker to board.

PROJECT 18: SPEAKER STAND

Although the size of stereo speakers varies from the submini to a few of closet proportions, the vast majority fit into the class known as *compact*. Typically, this most popular of speakers is a bit too large to fit into a bookshelf or onto a table, so the speaker system often goes on the floor. For most speakers, the floor position is bad: The high frequencies are projected across the room at knee height where much of their energy is absorbed by the furniture. With even the most expensive tweeters, dispersion of the upper highs leaves something to be desired when compared to live music. If the tweeter is aimed across the room at a level far below that of the listener's ears, the problem is compounded.

Another problem with floor placement arises in the bass. Room resonances are often set up, particularly if the speaker is placed in a practical location, such as at a wall-floor junction. While this can boost low-frequency sounds, the boost is often selective, or boomy.

As speaker technology has advanced, it has become apparent that the floor position can seriously impair the performance of any good compact speaker. With that insight, some manufacturers have begun to furnish a kind of pedestal or stand for their compact speakers. Here is a stand that you can easily build with hand tools.

It raises the speaker a few inches off the floor and tilts it back so that the highs cover the room at ear level. The upward angle also avoids direct reflections. If you have a compact speaker system, it's a must.

Construction

Almost any kind of ¾″ material will do for this project. For a furniture-quality appearance, birch plywood is ideal. You can use strips of veneer, available from lumber dealers, to cover the edges and stain the stand to match other room pieces. If you use solid wood, such as pine boards, you won't have to cover any edges and the appearance can be attractive, although it won't precisely match that of hardwood. For economy, you can use particle board and either paint it or cover it with an adhesive-backed plastic veneer. Because the stand is close to the floor, its appearance is not as important as that of a taller item or one that stands at eye level.

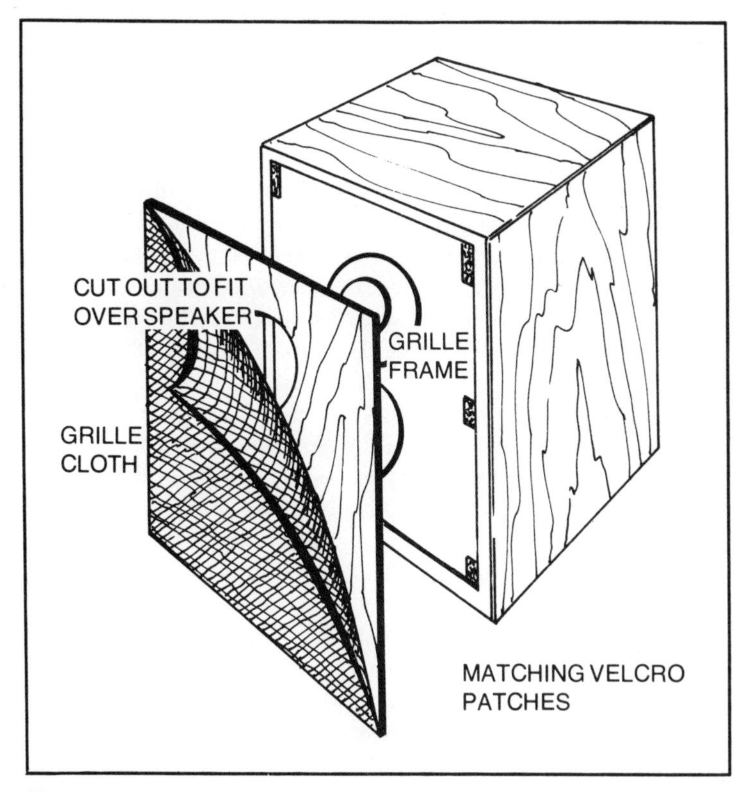

Fig. 6-13. How to add a grille frame after front mounting your speakers.

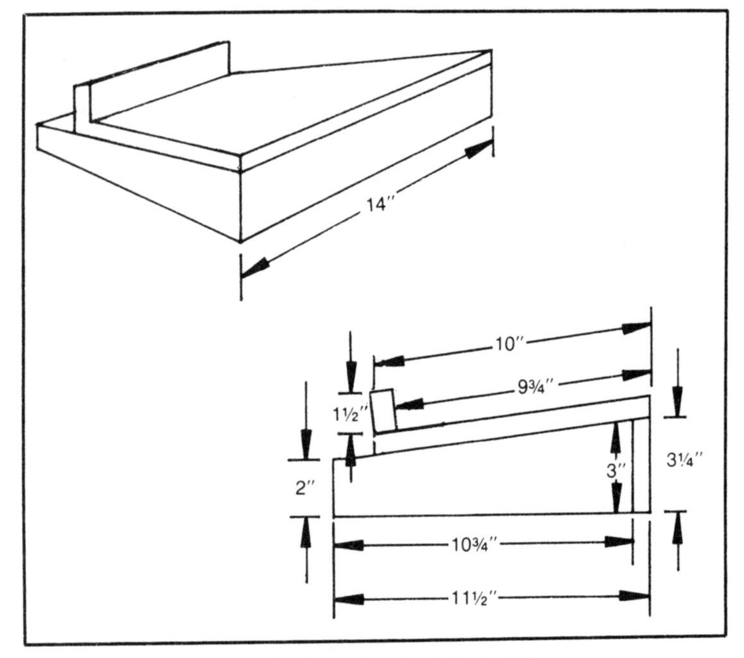

Fig. 6-14. Construction details of speaker stand, Project 18.

As mentioned earlier, you can build the stand with hand tools, but you must make an angled cut along the top edge of the front base piece. If you don't have a power saw, you can either slightly angle your hand saw or use a rasp or sandpaper to taper the cut. If you have a power saw, set it at 7 degrees to rip out the 3-¼″ piece. Note that it measures 3-¼″ wide at the front side, but only 3″ at the rear (Fig. 6-14).

The only other pieces that are not cut squarely in each direction are the base sides. They have square edges but taper from 3″ wide at the front to 2″ wide at the back end (Table 6-2).

Assembly is by wood glue and No. 6 finishing nails. Cut out the parts. Then glue and nail the front base piece to the base sides. Glue and nail the top piece to the back rail by inverting the pieces and driving the nails from the underside of the top into the rail. Then glue and nail the top to the base (Fig. 6-15).

Sand the stand until it is smooth. Then stain it to match your other furniture. Apply a satin varnish if you wish, or just wax it. When the stand is dry, set it in your usual speaker location and place the speaker on it (Fig. 6-16). Note that the base side pieces extend back far enough to reach the wall and prevent the base from

Table 6-2. Parts List for Speaker Stand.

Part	No.	Dimensions
¾" Material:		
Top	1	10" x 14"
Front	1	3-¼" x 14"
Sides	2	10-¾" —tapered from 3" to 2"
Back Rail	1	1-½" x 14"

being shoved too far against the wall. Without this extension, someone might push the stand too close to the wall and topple the speaker forward.

When you fire up your stereo with the speakers on their stands, you will probably be amazed at how much more open they sound. The slight upward tilt disperses the sound as well as if the cabinets were kept level and lifted much higher.

PROJECT 19: THE LEASE KEEPER

The trouble usually begins with a tap on your apartment wall, and you know why it's there. Suddenly the full enjoyment of your stereo rig is spoiled, because even if you disregard the tap, and the knock, and the pounding, you know there will be the inevitable visit from the apartment manager. He will show you a sheaf of complaints and, perhaps, deliver an ultimatum. It's a fact of life if you live in the typical urban or dormitory environment.

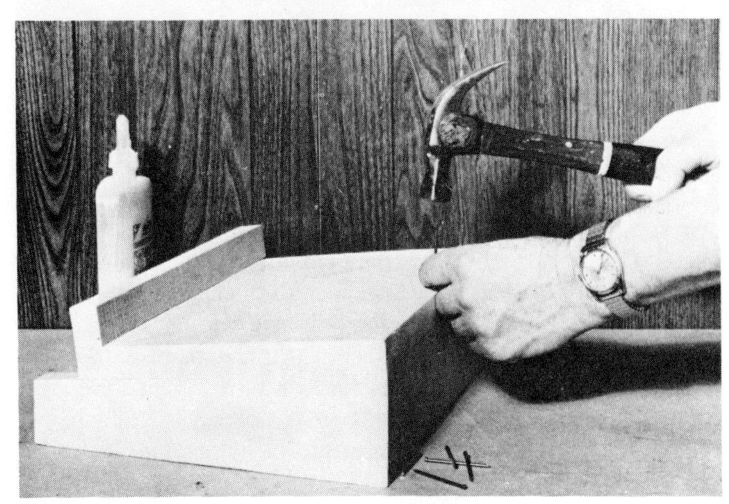

Fig. 6-15. Assemble stand with glue and finishing nails.

Fig. 6-16. Speaker stand in use.

Sometimes the biggest complaint comes from the people who live in the apartment below yours. That's not too surprising when you analyze the building as an agent of energy conductivity. Your speakers probably sit on the floor, or on a stand that sits on the floor. This direct coupling transfers vibration from the cabinet into the beams that support your floor, or the neighbors' ceiling. The low notes are especially effective in bridging the space between apartments, so even if your neighbors hear nothing else, they will notice a boom, boom, boom every time you turn on your stereo set. In some apartment buildings, the booms seem almost louder below than above. In one Los Angeles apartment I visited, the family who lived on the ground floor thought the people upstairs had a Great Dane until one day when the upstairs people took their small poodle for a walk. If you live in such a building, this project will be of interest to you.

Construction

The solution to complaints from neighbors is to acoustically isolate your speakers from the floor or walls of your apartment. The ideal solution is to suspend the speakers in midair, so they touch nothing solid that touches the wall or floor (Fig. 6-17). If you have small speakers, hanging them is a possibility. The speakers can be hung from the ceiling much like hanging lamps. But be prepared for reduced low-frequency response from hanging speakers. To hang your speakers, go to a hardware store and get the kind of hardware sold for hanging lamps (Fig. 6-18). If you can't use the ceiling for anchor hooks, you can erect a support system by

Fig. 6-17. Hanging speakers transfer less sound to neighbors, but bass response may suffer.

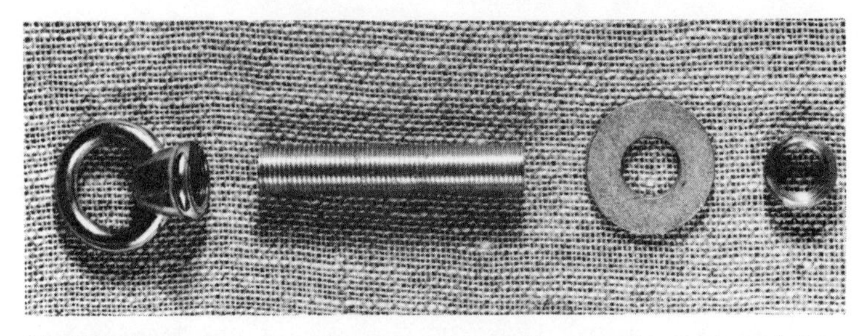

Fig. 6-18. Brass loop, threaded pipe, washer and nut can be purchased at an electrical supplies store and used to hang small speakers.

clamping a horizontal beam to any vertical item in your living room. To connect the chain or nylon rope to your speaker, you can use a section of macrame. Wrap the macrame around the speaker and gather it at the top to connect it to the chain or rope. Or, if you don't

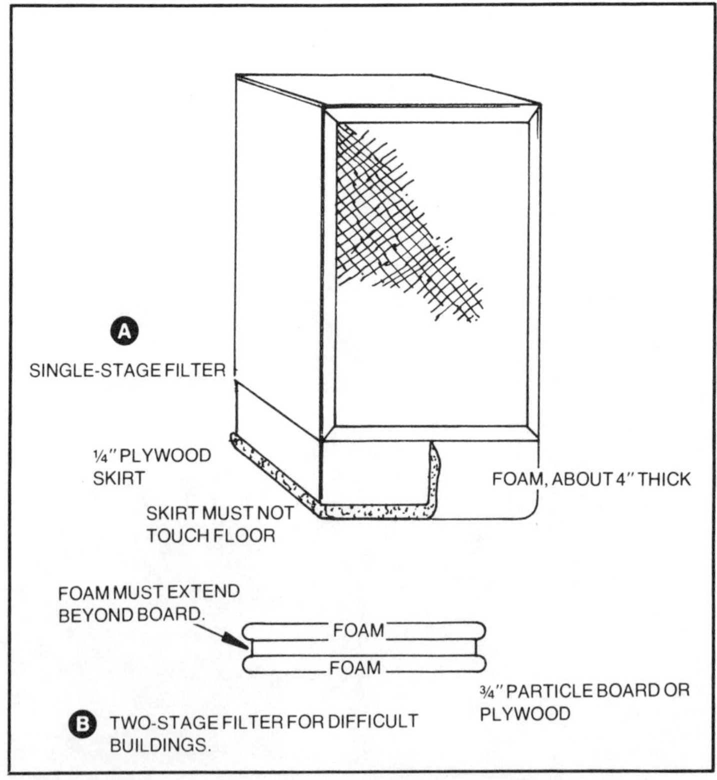

Fig. 6-19. How to acoustically isolate your speakers by a foam filter. A single-stage filter is at A, and a two-stage filter for difficult buildings is at B.

mind marring the appearance of your speakers, install a heavy eyebolt in the center of the top of each speaker cabinet and hang the speakers by that point. To wire your hanging speaker, wind a lamp cord from the amplifier through the chain or around the rope that holds the cabinets.

If the hanging method is impractical for you, or just doesn't seem appealing, you can obtain some degree of decoupling by placing a foam foundation under the speaker cabinets. Go to a surplus store, or any kind of store that carries odd pieces of thick foam cushion material, and buy two pieces of foam 4″ thick, each large enough to fit under one of your speaker boxes. Place the foam cushions under the speakers, making sure the cabinets touch nothing more solid than the foam cushion (Fig. 6-19A).

If you want to hide the foam, you can make a shirt, or plinth, from ¼″ hardwood plywood that extends around the base of the speaker box. To make the shirt, measure the distance from the bottom of the cabinet to the floor surface with the foam in place. Then make the shirt sides slightly shorter than this distance. It is imperative that the shirt *not* touch the floor or any other surface, such as the wall.

For problem apartment houses, such as the one with the "giant" dog upstairs, a multiple stage decoupler can be used. To make the decoupler, lay down successive layers of foam and ¾″ particle board (Fig. 6-19B). If you decide to use the skirt over the sandwich decoupler, cut the particle board pieces so they are slightly narrower and shorter than the foam and then place them between layers of foam so they do not touch the skirt.

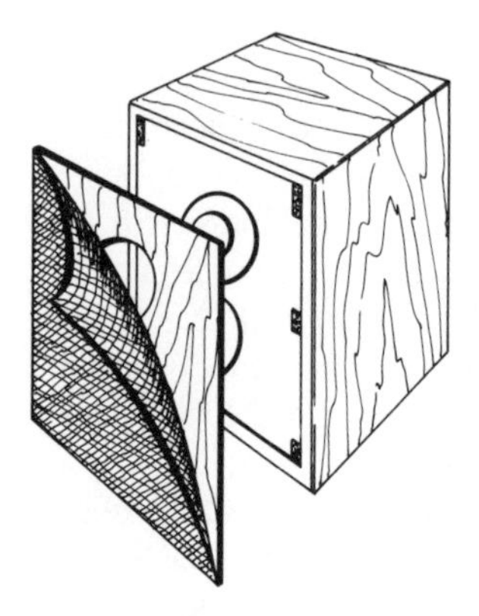

7 Speaker Improvement Projects

Some projects have pizzazz; some are mundane. Often it is the little ordinary project that can put the final touch on a speaker system that changes it from just satisfactory to something really good. Don't downgrade a project because it costs little in cash outlay or time; in fact, the logical way to approach any problem with a speaker system or other component is to try the little, low-cost solutions first. Then, if they don't work, you can go on to more heroic measures such as more exotic speakers. One sure bet, the typical low-cost commercial speaker system has seen little in the way of attention to fine tuning detail.

While most of the projects in this book offer a direct path to better sound, some achieve that goal indirectly by making an essential step in setting up a successful stereo system easier. The *phaser switch* is an example of that kind of project. It's possible that you may try a certain sound improvement project and find that it does nothing to help your particular speakers. There could be several reasons for that, including the inadequacy of your speakers to provide you with quality sound. In that case you can try Project No. 22, "Driver Modification." It is a project that is reserved for really inexpensive speakers. If it doesn't do the job, then you can move on to the next chapter and try the somewhat more ambitious projects there. Meanwhile, give these a try.

PROJECT 20: REWIRE YOUR SPEAKERS

This project involves one of the most mundane parts of your whole stereo system, the speaker cables. If you go to an electronics store and ask for speaker cable, you will probably be offered a lightweight wire with green, translucent, plastic insulation. The typical gauge is No. 24, which adds about one-quarter of an ohm resistance for every 10 feet of wire. When you consider that the total distance of wire in any speaker cable is twice the length of the cable, you can see that the resistance for long runs can get to be a significant fraction of the impedance of a speaker. Suppose you have a 20-foot cable, the resistance of the cable is 1 ohm, well over 10 percent of the impedance of an 8-ohm speaker and 25 percent of that of a 4-ohm speaker. As a rule of thumb it is considered good practice to keep the total resistance in any speaker system hookup and crossover network below 10 percent of the impedance of the speaker. When you consider that all chokes add some resistance, and that connecting hardware often contributes extra resistance, it is clear that small speaker cable should be avoided.

If you wonder why extra resistance causes problems, consider how amplifier manufacturers boast of high damping factors in their products. The *damping factor* is the impedance divided by the internal resistance of the amplifier. The lower the internal resistance of the amplifier is, the better it can damp any spurious speaker cone movements because it acts in a way similar to a short circuit across the voice coil. Any resistance in the hookup will have the same effect on damping as a higher internal resistance at the amplifier. As you can see, it makes no sense for a manufacturer to expend design time and money to produce an amplifier with a high damping factor only to have that feature defeated by high-resistance speaker wire.

Some high-fidelity suppliers offer special cable for speakers, cable that is designed to offer lower inductance than the typical cable. These special cables are also very expensive. All speaker cable has certain electrical characteristics which include series resistance (as mentioned earlier) and inductance, along with a capacitance across the line (Fig. 7-1). The expensive cables offer reduced inductance in the interests of improved high-frequency response, but they have a capacitance that is much greater than ordinary lamp cord wire. Apparently the cable construction that reduces inductance inevitably increases capacitance. Unfortunately, when you connect this kind of cable to any modern wideband amplifier, there is danger of amplifier instability caused by the capacitive load. This problem can be solved by placing a filter across the cable, but it would seem to be much easier just to avoid it by using a less expensive cable.

For most purposes, No. 18 lamp cord, sold by electronics stores, hardware stores and many general merchandise stores, is the most practical choice (Fig. 7-2). It offers low capacitance and fairly low resistance. If your speaker run is long, choose a cord with No. 16 conductors. Table 7-1 shows suggested length limits for each gauge of wire.

The ideal speaker cable would have no resistance, no inductance, and no capacitance. This ideal can be approached only by keeping the speaker cables as short as possible. So if you have extra cable drooping around the back of your speakers, cut it down to size.

Any connecting point can add resistance. While most speaker systems use screw-type box terminals for connections, you can eliminate these possible causes of resistance points by using a single piece cable from inside the enclosure to your receiver terminals. Solder the cable to the proper connecting points in the speaker box and bring it out of the cabinet through a 1/4″ hole in the back. Tie a knot in the cable just inside the box to protect the terminals to which the cable is attached if someone trips or jerks on the cable. Fill the exit hole with caulking compound to prevent air leaks.

The difference in a cable that is too small for the job and one that is adequate is a subtle one. The resonances, and there are always resonances, will be better damped with a husky cable than with one such as many low-priced speakers carry as original equipment. To check the adequacy of your cables, look at the figures in Table 7-1. They show the preferred size of speaker cable for various length.

Regardless of the cable you use with your speakers, don't forget to make the best connections possible at each end of the

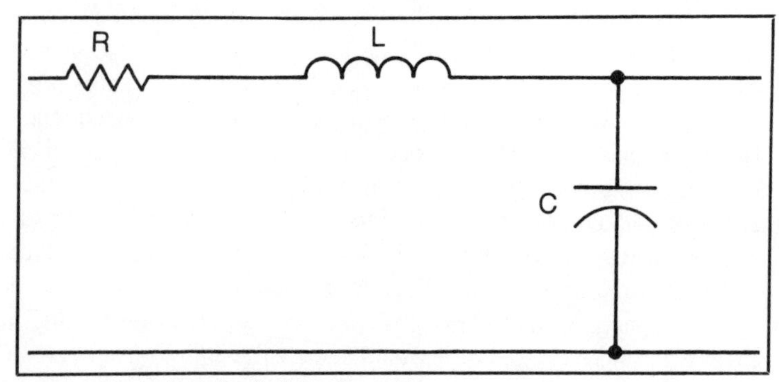

Fig. 7-1. Equivalent circuit of speaker cable.

Table 7-1. Cable Length for Speakers of 4-ohm, 8-ohm and 16-ohm Impedance.

	CABLE LENGTH					
	5% LOAD LENGTH			MAXIMUM LENGTH		
WIRE GAUGE	4 OHM	8 OHM	16 OHM	4 OHM	8 OHM	16 OHM
#14	40 FT.	80 FT.	160 FT.	100 FT.	200 FT.	400 FT.
#16	25 FT.	50 FT.	100 FT.	65 FT.	130 FT.	260 FT.
#18*	15 FT.	30 FT.	60 FT.	40 FT.	80 FT.	160 FT.

*#18 WIRE IS ORDINARY LAMP CORD.

CABLE LENGTH IS LENGTH OF DOUBLE CONDUCTOR CORD THAT RUNS FROM AMPLIFIER TO SPEAKER. ACTUAL WIRE LENGTH IS TWICE THE CABLE LENGTH. THE 5% LOAD LENGTH FIGURES WILL LIMIT WIRE RESISTANCE TO ABOUT 5% OF THE TOTAL LOAD, A CONSERVATIVE FIGURE. IF NECESSARY THE LENGTH CAN BE EXTENDED TO THE FIGURES GIVEN IN THE COLUMNS UNDER "MAXIMUM LENGTH." THESE LENGTHS MAY CAUSE A CHANGE IN SOUND QUALITY WITH SOME SPEAKERS. CABLE WITH CONDUCTORS SMALLER THAN #18—SUCH AS #20 TO #24—SHOULD NOT BE USED UNLESS CABLE LENGTH IS VERY SHORT. SMALL CONDUCTOR WIRE IS OFTEN SOLD AS "SPEAKER CABLE."

cable. The connectors should be inspected regularly for looseness or oxidation of the contacting conductors. Although these connecting points may show low resistance when measured with an ohmmeter, they can offer enough resistance to greatly change the distortion level in even the best of amplifiers.

PROJECT 21: WOOFER DAMPER

The preceding project was designed to improve speaker damping by removing some of the electrical resistance associated with inadequate speaker cables. If you try that and your woofer is still boomy, here is a way to cure the boom without going to a more expensive speaker.

Bass performance is largely determined by the degree of damping on the woofer cone. Too much damping makes for a thin bass response. With too little damping, the bass is boomy and the whole lower range can be muddy. One symptom of too little damping is a one-note bass which emerges from the speaker as a monotonous boom, boom, boom. While some kinds of music have a certain amount of this kind of bass, an underdamped speaker converts all musical passages with bass in them to the same single-tone sound.

Listeners who are accustomed to one-note bass don't miss the real thing until they are exposed to a good speaker and a chance to listen carefully for a while. A woofer with the proper damping reproduces bass notes with great clarity. The bass moves up and down the scale, if that is the kind of bass written into the music, and there is a lack of *hangover,* the tendency of a woofer cone to continue moving after the signal ends. Once you have been exposed to that kind of musical bass, it's "goodbye" to the one-note thumper.

Woofers are damped by various forces. When the cone moves, the suspension system pulls it back toward the rest position, so it offers some mechanical damping. At the fundamental resonance of the speaker, however, the mass of the cone vibrates on the suspension with increased energy, and there is little mechanical damping. At this frequency, the magnetic field offers damping because the moving voice coil must cut through lines of force set up by the field, an action that sets up a back electromotive force (emf) in the voice coil. This back emf is a voltage developed by the coil movement through the field, like a generator. The speaker itself is, in a sense, a motor, with electrical driving force supplied by the amplifier. The back emf generated by the voice coil acts in opposition to the driving force on the coil, so it damps excessive coil movement at the frequency of resonance. This is why a speaker with a large magnet, which has a strong magnetic field, is better damped than one with a tiny magnet.

Speaker Q

The kind of enclosure the speaker is put into also affects damping. To consider how this factor works, one must consider the Q of the speaker. The term, Q, represents resonance magnification, so the higher the Q of a speaker, the greater the effect of resonance and the boomier it will sound. Typical Q values range from a low of about 0.2 for highly damped speakers up to 2 or 3 for speakers with small magnets and stiff suspensions. Speakers designed for use in bass reflex enclosures usually have a low Q, from 0.2 to about 0.6. The Q of a speaker designed for closed box operation is usually somewhat higher than that of the typical reflex speaker.

When a speaker is installed on a flat board, the Q doesn't change very much. But if you put the speaker into a closed box, the Q is higher than when the speaker was hanging in midair or mounted on a flat baffle. It has long been known that to get the maximum bass range from a woofer in a closed box, commensurate with adequate damping, a Q of about 1 is a good choice. This offers a satisfactory compromise between the requirements of good damping and a full bass response. Somewhat flatter response can be obtained with a lower Q, but the bass range will be more limited.

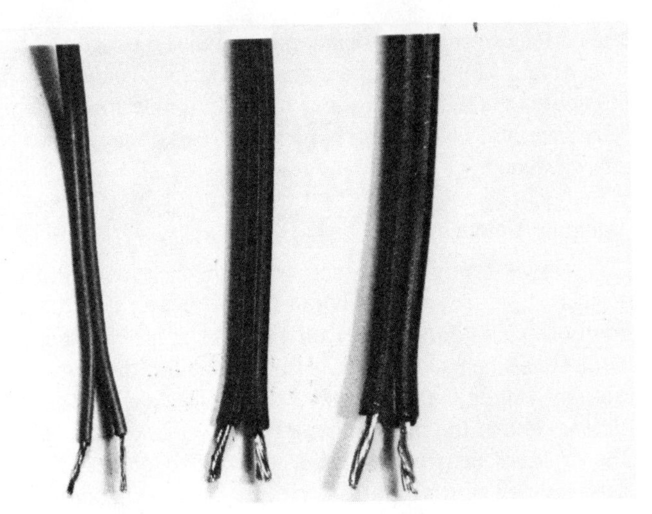

Fig. 7-2. Three grades of speaker cable: No. 24 gauge, No. 18 gauge lamp cord and No. 16 gauge lamp cord.

A *Q* of 1 permits a slight rise in output at resonance, but the rise is small enough to be tolerated by most critical listeners. A higher *Q* means some degree of boom. Rock music fans seem to tolerate a higher *Q* than people who like classical music, probably because of the difference in bass composition of the two kinds of music.

As mentioned above, the *Q* rises when the speaker is placed in a closed box. The smaller the box is, the higher is this rise in *Q*. In some products, the woofer is installed in a box that is smaller than optimum size for two reasons, the desirability of making a compact speaker and the cost of more magnet to make the speaker match that box size. Where little attention is given to speaker damping, the opposite effect can occur if a speaker is installed in a box that is acoustically too large. This kind of mismatch is not as common.

While mismatching often occurs in closed box speakers, the tonal effect is rarely disastrous. This kind of system can tolerate a certain amount of careless combination of components and still give fairly good sound. The situation with a bass reflex, or ported, enclosure is different. Here, even a minor miscalculation can produce a boom box. A properly matched reflex enclosure can improve damping because the air in the port will control the cone at the frequencies where control is needed, making for a flatter, more extended bass response. But even here, box volume is an important parameter of the entire system. To properly design a reflex enclosure for a woofer, the designer must consider the Q of the woofer along with the compliance of its suspension when he computes the right box volume for optimum performance. With the many variables involved, it's no surprise that many reflex speakers are mistuned.

Damping Methods

Down through the years audio fans have tried many kinds of prescriptions to remove boom from a poor reflex system. One method has been adding acoustical resistance to the port. This sometimes helps, but it is a treatment for the *symptons* of the ailment rather than the *cause,* the *underdamped speaker.* And it defeats one of the main purposes of the reflex: its ability to damp the driver at the frequencies around the resonance of the port. It also reduces port output, which can give a substantial reinforcement to low bass. What is wanted from a reflex is a port with a high Q, so air can flow freely in and out, changing the pressure in the box at the right instants to damp the cone.

Another solution that has been tried is filling the box with damping material. This makes the box is changed from *adiabatic* to *isothermal*. Under isothermal conditions, the temperature in the box doesn't change, but heat flows into and out of the damping material. The velocity of sound in the box is decreased, making the box appear to be larger to low-frequency sound. It is an attractive solution, but it often doesn't work as well in practice as in theory. For one thing, only a 20 percent increase in effective volume is usually achieved, although the theoretical increase is about 40 percent. And this kind of filling usually overdamps the whole system, reducing the effectiveness of the reflex as a reflex. It often reduces the Q of the speaker by making the box appear larger, but when it also reduces the Q of the reflex port, there may be a net loss in bass range and level. It can produce a system that is more acceptable than a boom box, so there are occasions when such measures *are* worth trying.

One might compare the methods just listed as similar to an ailment in the human body, say the thyroid gland, which is treated by a massive dose of medicine taken by mouth. If only the thyroid is affected, some scientists say, it would be better to concentrate the treatment at that point, a smaller dose where it was needed instead of flooding every body organ with medicine that can cause serious side effects. Because the speaker is the component that is underdamped in a boomy system, the best place to put damping material is at the speaker (Fig. 7-3). That is the aim of this project. But first, you must decide if your woofer is underdamped.

Damping Tests

The easiest test to apply is careful listening. Does male speech appear to be cavernous, as if the announcer were talking in a barrel? This kind of reproduction suggests that either box design or the damping on the woofer is wrong.

Another, more precise test is to connect a flashlight battery to your speaker and listen to the sound when the circuit is completed and broken. To better match operating conditions with a typical

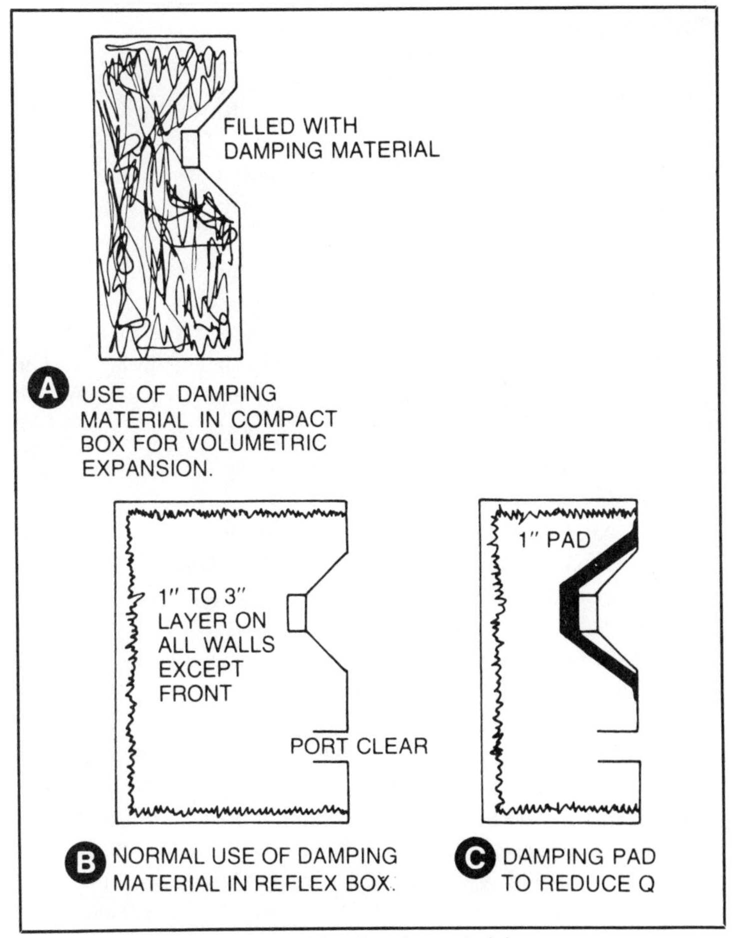

Fig. 7-3. Various methods of applying damping material.

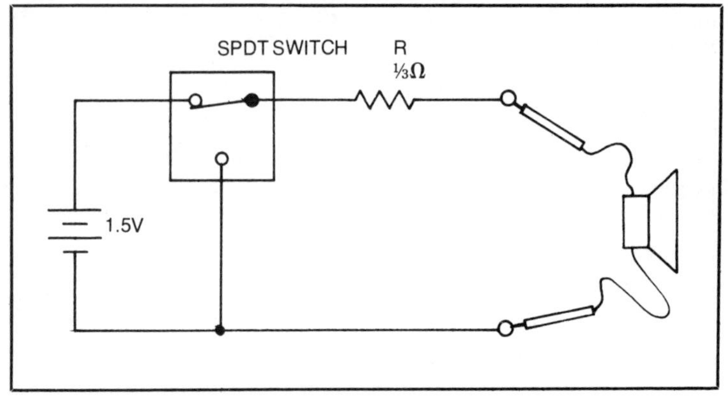

Fig. 7-4. Wiring diagram for damper tester.

amplifier, place a 1/3-ohm to 1/2-ohm resistor in series with the speaker, as shown in the circuit diagram of Fig. 7-4. A useful gadget can be quickly assembled in an empty tuna can that makes for more convenience in making such tests (Fig. 7-5). Construction for this is described later.

When you complete the circuit, you should hear a click in the speaker. If you hear a bong or boom, the speaker is underdamped. Break the circuit and listen to the sound made at the break. Some speakers that pass the test on the *make* action fail on the *break*. The object is to damp the speaker just enough that it passes on both tests.

Construction

To make the damper tester, prepare an empty tuna can by drilling holes for two banana jacks, a miniature single-pole—double-throw (SPDT) switch and a battery holder for a size C flashlight battery. See Table 7-2 for a list of parts.

Fit the components into the can according to the layout shown in Fig. 7-5. Prepare some leads by installing banana plugs at one end and alligator clips at the other. Note that the leads are color-coded according to battery polarity so you can use the tester to check the polarity of speakers. Just connect the tester to any speaker that has a visible cone and flip the switch. If the speaker cone moves forward on the *make* circuit, mark the terminal served by the red lead with a plus mark. If it moves to the rear, place the plus mark on the other terminal. Do this for all your speakers and observe the marks when connecting speakers in stereo.

Table 7-2. Parts List for Damper Tester.

1	SPDT Mini Toggle Switch	Radio Shack #275-326 or eq.
1 pkg.	Banana Plugs	Radio Shack #274-721 or eq.
1	0.33 Ω Resistor	
1 piece	Test Probe Wire, Red	
1 piece	Test Probe Wire, Black	
1 pkg.	Insulated Alligator Clips	Radio Shack #270-378 or eq.
1	Battery Holder	To fit size C-1.5 Volt Battery
1	Small Chassis (Can use tuna can or small pineapple can.)	

To increase the damping on your woofer, staple a 1″ thick layer of acoustical fiberglass or polyester batting over the back of the woofer (Fig. 7-3C). Stretch the material so it won't flap back and forth when the woofer cone moves. Then test the speaker again for proper damping. Use more damping material over the speaker if necessary.

You must have access to the back of the speaker to make this modification, but with some enclosures it is the front panel instead

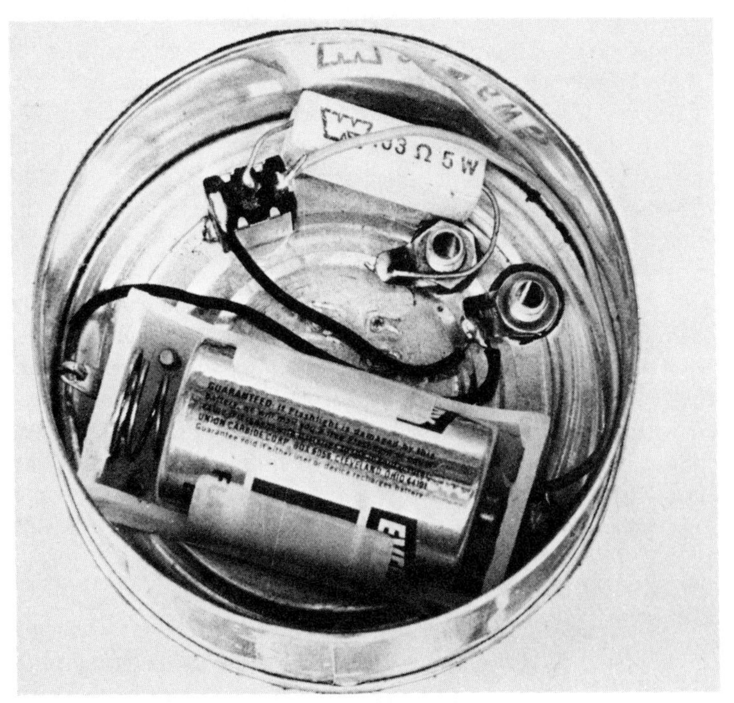

Fig. 7-5. Damper tester installed in a tuna can.

of the back that is removable. Regardless of which way you go to gain access to the speaker, you must install the panel again before testing. To eliminate numerous bouts of panel removal and replacement, install the woofer on the outside of the box, facing inward (Fig. 7-6). This permits you to conduct the tests, choose the optimum amount of damping, and then install the speaker and the damping material in the box. There will be a reduction in effective box volume when the speaker is installed inside the box, and this reduction will affect damping. This change is usually not significant enough to audibly effect the performance of the speaker, however. What is significant is the time saved by making the tests outside the box.

In addition to applying damping for the woofer, this position for damping material offers an extra benefit. Reflected sound inside the box can impinge on the cone, producing dips and peaks in the response curve of the speaker. By placing the damping material over the speaker itself, sound must travel through the blanket of material twice to get back to the cone. Thus the material is twice as effective in this position as in most others.

When you get the damping right, the bass will be much cleaner. You can then roll up the bass control without producing the annoying boom that was ever present in the underdamped speaker. What you have done is substituted *mechanical* damping with the resistive air filter in a speaker that lacked sufficient *magnetic* damping for the box in which it was placed. The mechanical damping is obviously less expensive, but it requires much more time to get right. You have substituted your time for the money that a better woofer costs.

Don't feel that you can get something for nothing by using this trick. In addition to the time required to damp the woofer, you lost a bit of something else, but you may never miss it. The woofer will now be slightly less efficient than before. This means you may have to reset the balance controls, and you may have to use a bit more power from your amplifier to get the room-filling volume you want. But that's a small price to pay for cleaner sound. And most amplifiers have more power available than you really need anyway.

PROJECT 22: DRIVER MODIFICATION

By *driver modification*, we mean making *mechanical* changes in the cones of your speakers. Obviously you wouldn't try this on an expensive speaker, but there are several ways to make an inexpensive replacement speaker sound more expensive. Such

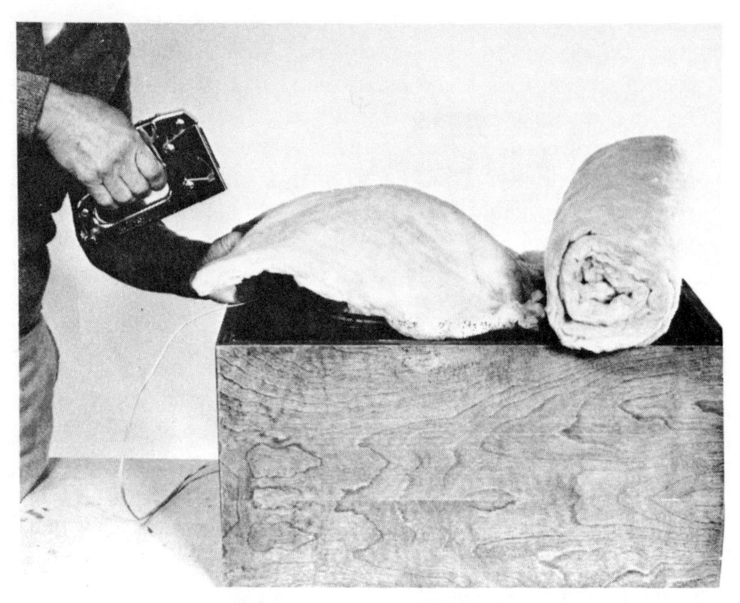

Fig. 7-6. By inverting woofer so it is outside cabinet you can test various thicknesses of damping pads without getting into cabinet for each test.

speakers can make good extension speakers or serve as main speakers for small stereo sets.

The best cure for any cheap driver is to upgrade by getting a high-compliance speaker with an adequate magnet, but that is another project. It also requires an outlay of funds you might not want to invest for whatever use you plan for the inexpensive speakers. Speakers of the quality usually slated for replacement duty in radio or television sets have several likely shortcomings other than a magnet that is smaller than desirable for the mass of the cone and voice coil. The two most common faults are a suspension that is too stiff and uneven frequency response. The stiff suspension limits bass range, especially when the speaker is placed in a compact closed box. This problem often surfaces after modifying an enclosure that originally had an open or slotted back by sealing the box. The trapped air in the cabinet then raises the frequency of resonance by an unacceptable degree. We will attack this problem now; a later project shows how to smooth the frequency response of such speakers.

Increasing Compliance

There are several ways to increase the compliance of the suspension of a speaker, but the easiest and most reliable method is to slot it. Before applying this treatment, check the suspension to see if it appears to be the problem. There are about four main types of suspensions found on speakers of current manufacture. The least expensive, and one that you can easily improve, is a single wrinkle carried around the periphery of the cone to add some degree of flexibility. The next step is a series of two or more wrinkles, the wrinkled section treated with a gooey material that adds to the compliance and damps the cone. Beyond that kind, you will find some speakers that have a cloth surround the cone, with the cloth usually treated with a sealant. And, most compliant, is the *roll suspension.* Typically, the roll is composed of a foam material, but sometimes butyl rubber is used. These differences are easily identified, as you can see in Fig. 7-7. We recommend that you make no suspension changes in any speaker that has any of the last three types of suspensions. Their existence shows that the designer gave some attention to getting optimum compliance for the particular speaker in question. But for speakers with no more suspension than an extension of the cone material in a single stiff wrinkle, it's open season.

To slot the suspension, use a razor or a sharp knife to cut radial slots through the suspension at regular intervals around the cone. Make sure that the cuts do not encroach on the smooth section of the cone. The right technique is to start the cut at the edge of the piston area and draw the blade toward the edge of the speaker frame (Fig. 7-8). Begin with two cuts, spaced at opposite points on the compass. Or, considering the cone as a clock face, make the first cuts at 6 o'clock and at 12 o'clock. Then make cuts at 3 o'clock and 9 o'clock. Next, make four more cuts, spaced equally between the ones already made. Continue this process with eight more evenly spaced cuts, making 16 in all. If you are working with a speaker larger than 6″ in diameter, you might want to make 32 cuts.

After slotting the cone, run your thumb around the suspension several times to make it still more compliant. Use care so you won't damage the cone or the suspension. You can easily lower the frequency of resonance by 20 percent or more in this way. A lower frequency of resonance means lower bass distortion, unless the speaker is overdriven. And it invariably extends the bass range and improves damping. The modified speaker will produce a more musical sound. However, it may not sound as loud and it probably will not handle quite as much power as before the alteration, just as high-fidelity woofers are more susceptible to damage than musical instrument speakers, which have stiffer suspensions.

Fig. 7-7. Three kinds of cone suspensions: untreated paper, treated accordian and foam roll.

There will be some air leakage through the slots in the suspension, but the net effect is better performance. You can seal the slots with a light dab of silicone rubber sealant at each cut, but this treatment can undo some of the good accomplished by slotting, so go easy.

Whizzer Damper

One way that manufacturers of full-range speakers widen the frequency response of their speakers is to add a *whizzer*, which is a flared piece of paper at the center of the cone to reproduce the

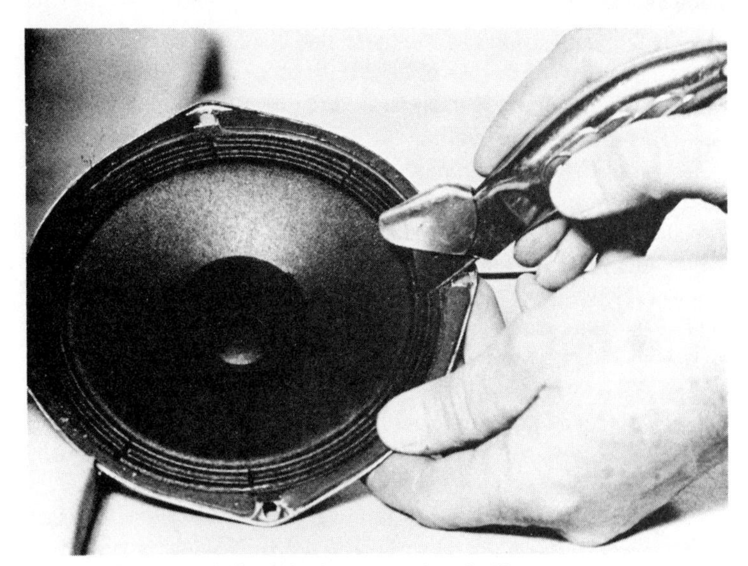

Fig. 7-8. Use a sharp knife to slot the suspension of stiff cones.

highs more efficiently (Fig. 7-9). Whizzers are effective in increasing the frequency range of a speaker, but they sometimes add an unwanted sizzle to the sound because of the undamped edges of the flare. If you have an inexpensive speaker that sounds a bit peaky in the upper ranges, try applying a damper to the whizzer. As in other driver modification jobs, however, don't try to improve the whizzers on name-brand speakers.

To damp the whizzer, simply cut several short pieces of adhesive-backed foam weather-stripping material and stick them around the edge of the whizzer at the mouth (Fig. 7-9). The weather-stripping material should be light; the 1/4" x 3/8" size is about right. Don't try to make one piece go very far around the mouth of the whizzer or you will wrinkle it. Cut about eight pieces. Make the cuts at a slight angle, so the individual pieces of weather-stripping are trapezoidal, or shaped a little like truncated pieces of a pie.

This treatment sometimes makes a speaker with a whizzer sound much smoother, perhaps because it damps the free edge and also may prevent reflections back through the whizzer, which could cause interference with waves traveling out from the apex. Another use is to damp peaks which may override a crossover network when the speaker is relegated to woofer or midrange duty.

Cone Sealer

This modification has the most risk of all because it requires great care to avoid danger of fire. With proper precautions, it can be done with impunity. But if you are the impatient type, forget it.

The purpose of sealing the cone is to avoid some of the disadvantages of inexpensive paper cones. Paper has been criticized by some engineers because it is hygroscopic; that is, it can absorb water from the atmosphere. The characteristics of cone paper might vary from batch to batch, and that of a single cone might vary with changes of temperature and humidity. And, after a while, a cone can dry out and become more brittle. These variations impair performance. The paper cones used in more expensive speakers have usually been treated to prevent the worst effects of changing conditions, but inexpensive speakers have cones that are more variable. One way to improve and stabilize these cones is to seal them with wax.

Place a small piece of paraffin in an empty can and dip the can into hot water. *Don't use direct heat on the can, and make sure there are no open flames near your work.* Hold the can there until the wax

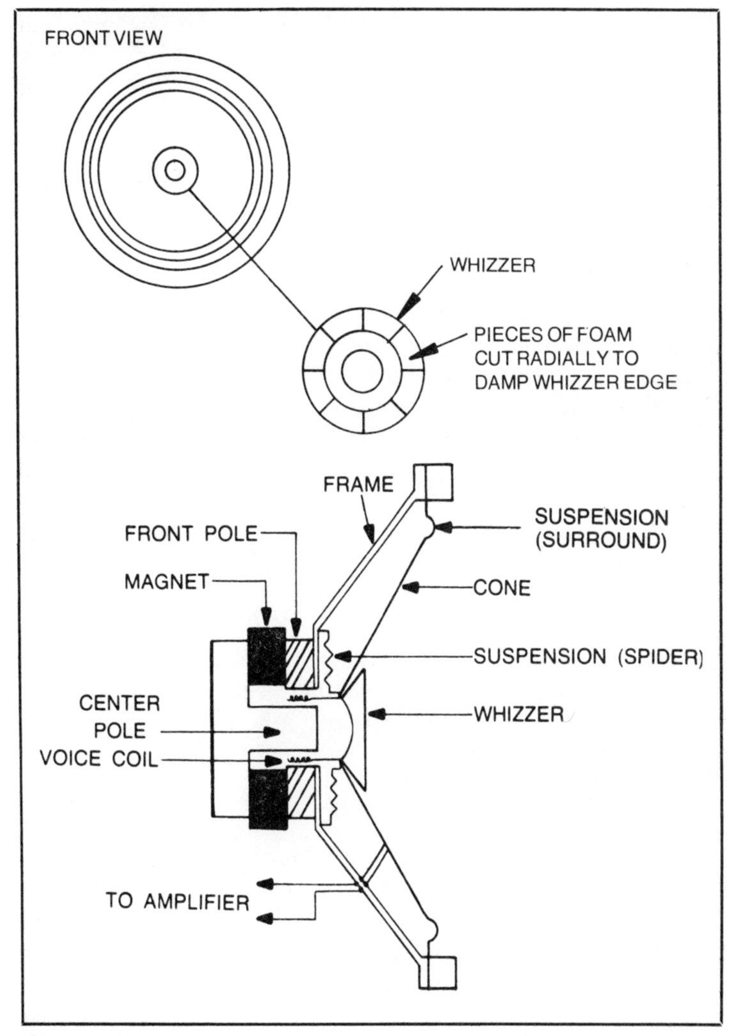

Fig. 7-9. How to add foam damping to a whizzer cone, Project 22.

in the can is a liquid. You will probably have to use gloves and a pair of pliers to avoid discomfort while doing this. When the wax has become liquified, use a small brush or a piece of rag tied to a stick to lightly brush the wax onto the speaker cone. After you have covered the surface of the cone, hold the speaker in front of an electric heater until the wax quits bubbling. This will drive the wax through the cone and eliminate any water in the cone. The wax displaces the water and helps to damp cone resonances.

This method of damping the cone appears to work better with some inexpensive speakers than with others. It won't ruin your speaker, but it may not appear to have done much good. Since the condition of the cone changes in an inexpensive speaker, the degree of immediate improvement will depend somewhat on the atmospheric conditions when you apply the wax. Again, use great care in applying this treatment.

PROJECT 23: SPEAKER PROTECTOR

This project is an appropriate one for anyone who tries to get the last dB of sound from a set of speakers. It won't protect the speakers in every conceivable situation: Some low-power amplifiers blow tweeters regularly by producing harmonic distortion so profusely that they add high-frequency energy not found in any ordinary program material. With that kind of aberration, a system fuse, such as described here, may not help because the total energy to the speaker system may be below the power rating of the speaker. Such ratings are based on normal frequency distribution, not special cases.

Another special case is a reflex speaker used with an amplifier that has no infrasonic filter. This kind of problem was discussed in Chapter 3, and the cure was given in Project No. 10. But for any ordinary situation, the speaker protector here will help preserve the voice coils.

Construction

The easiest place to add a protector is on the outside back panel of the speaker cabinet. The advantage here is that you have instant access to the fuse if it blows. Get a slip-in fuse holder with solder lugs at each end. Mount the holder near the speaker box terminals with a small screw through the center mounting hole. Connect a power resistor from the lug at one end of the fuse holder to the lug at the other end, as shown in Fig. 7-10. The value of this resistor should be somewhat greater than the impedance of the speaker. If it were made the same as the impedance, it would dissipate one-half the power after the fuse has blown. Its purpose is to maintain a load on the amplifier when the fuse blows. A 50-ohm, 10-watt resistor will usually be adequate for this job. If the resistor is subjected to too much current, it will get hot and its resistance will rise, cutting the current flow. You will not be likely to leave the speaker unfused for any length of time so a 10-watt resistor is probably adequate.

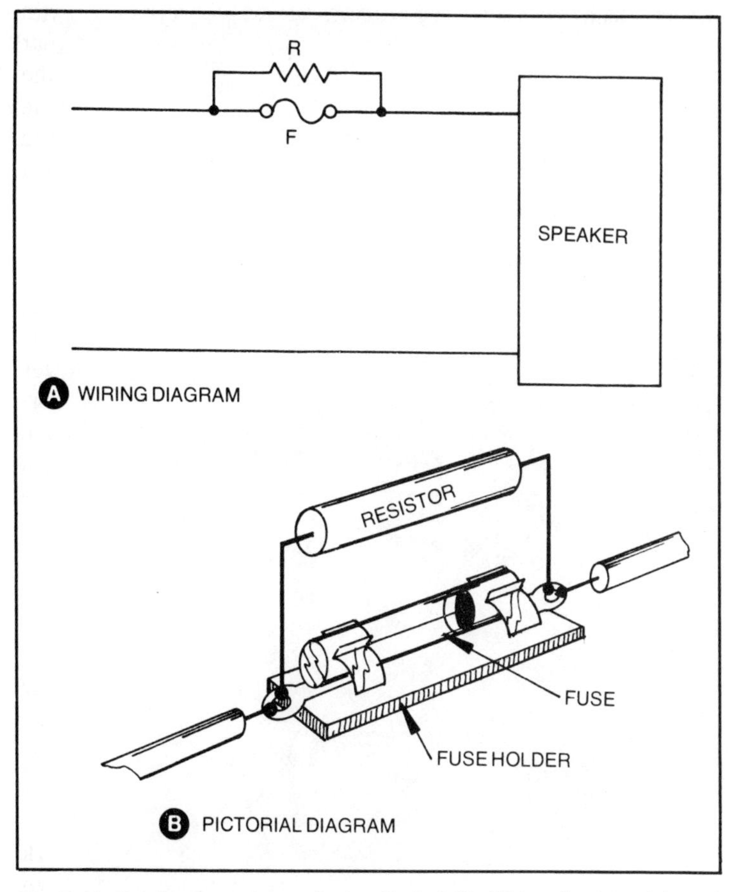

Fig. 7-10. Details of speaker protector, Project 23. Wiring diagram at A, and pictorial diagram at B.

When you have mounted the parts and wired in the resistor, run a wire from one lug of the fuse holder to one of the box terminals. Connect one lead from your receiver or amplifier to the other fuse holder lug; the other lead goes to the unconnected box terminal (Fig. 7-11). This completes the wiring.

Although Fig. 7-11 shows a slow-blow fuse in the fuse holder, this type isn't recommended. If such a fuse is the only one available and you want to listen to your stereo system, use one that has a rating of no more than one-half that shown in Table 7-3, preferably no more than one-quarter of that. Standard 3AG fuses are preferred.

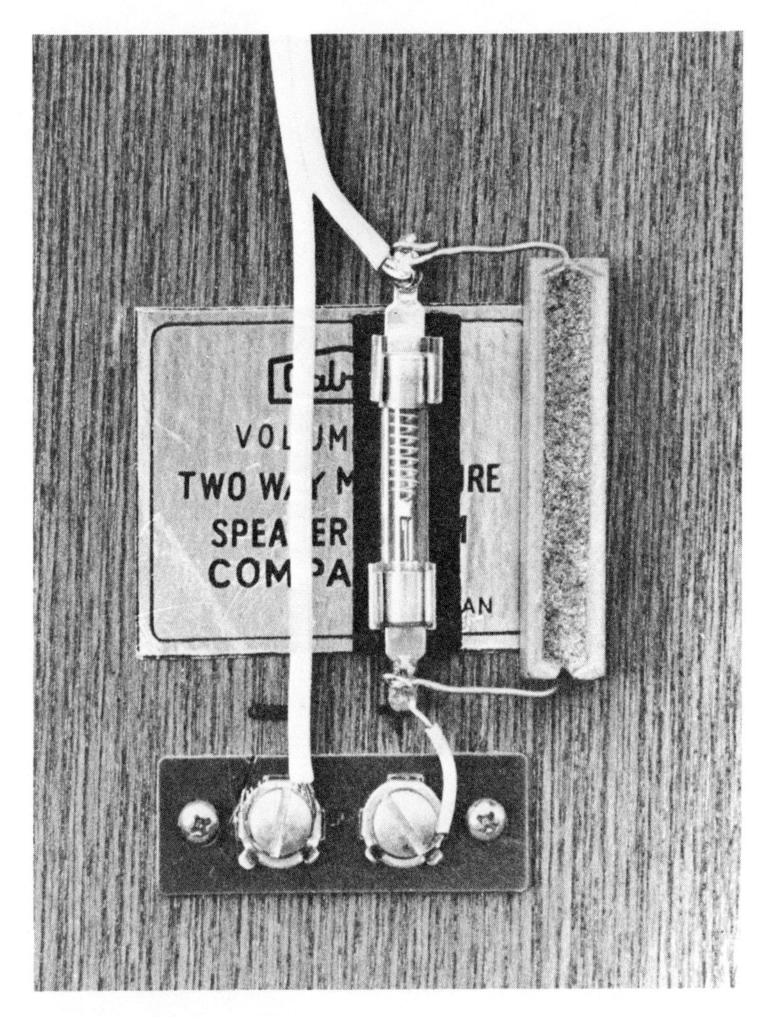

Fig. 7-11. Speaker protector in use. Slow-blow fuse shown here is less desirable than standard fuse.

To use Table 7-3, check the power rating of your speakers and their impedance and then find the fuse size from the table. For example, say you have a 20-watt speaker system that is rated at 8 ohms impedance. Table 7-3 shows a 1-½ amp fuse. But note that this is a maximum value that should never be exceeded and that you can get better protection by inserting a 3/4-amp fuse in the holder. There is no precise fuse rating that will protect your speakers at all times so if you value safety more than maximum sound output, use a fuse with a smaller current rating.

The speaker protector has one possible disadvantage: the resistance in the fuse can make a slight difference in speaker damping. To reduce this effect, make sure the fuse contacts are kept clean; use emery cloth to brighten them. Then install the fuse and listen. If you can hear a difference when the fuse is in place, as compared to the sound with a jumper wire across the fuse holder, you will have to decide which is more important—optimum speaker performance or the peace of mind you get from the protector.

PROJECT 24: SPEAKER PHASER SWITCH

Theoretically, speaker phasing should be a cinch—just hook up the same terminal on each speaker box to the positive terminal for each channel on the receiver. This doesn't always work, however. Sometimes the speaker box terminals are not marked as to phase, and in other cases the markings are wrong. Also, to make phasing tedious, some speaker wire isn't coded. The instructions often given with new receivers is to connect your speakers and listen; then reverse the connections to one speaker. You are supposed to choose the connection that sounds best.

Unfortunately, the human memory for sound quality is quite short. By the time you have disconnected the leads to one speaker and reconnected them again, you are probably wondering which connection really did sound better. Even employees at audio shops often get the connection wrong. Perhaps one reason is that some music can be deceptive; on other pieces, it is apparent to a careful listener that something is wrong, although that listener may not be able to identify the problem. For whatever the reason, I have noticed many stereo systems working with improper phasing.

There are several ways to ensure that your speakers are in phase. Most of these methods require you to disconnect your

Table 7-3. Suggested Sizes of Fuses for Project 23.

Speaker power rating	Maximum fuse rating. For better protection, use a fuse with ½ the current rating shown.		
	4 ohm speakers	8 ohm speakers	16 ohm speakers
10	2	1	½
20	3	1-½	¾
30	4	2	1
50	5	2-½	1-¼
75	6	3	1-½

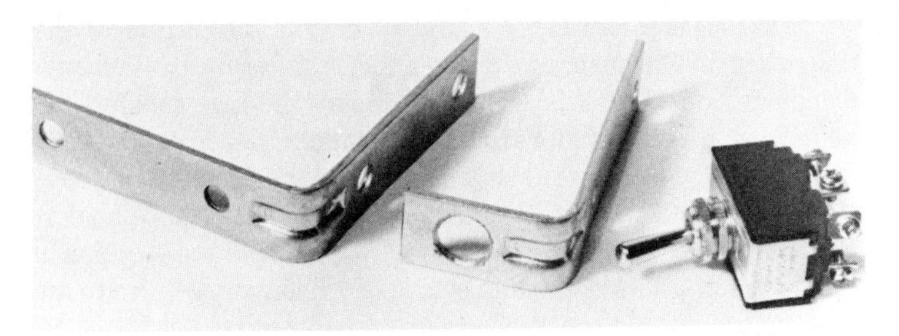

Fig. 7-12. Ordinary 3″ corner brace is cut off and bored out to hold switch.

speakers at least once during the phasing process. Some require external test equipment. By adding a switch at the rear of one of your speakers, however, you can immediately reverse the phase of one speaker without disconnecting it. This quick change phasing makes it much easier for you to identify the right hookup.

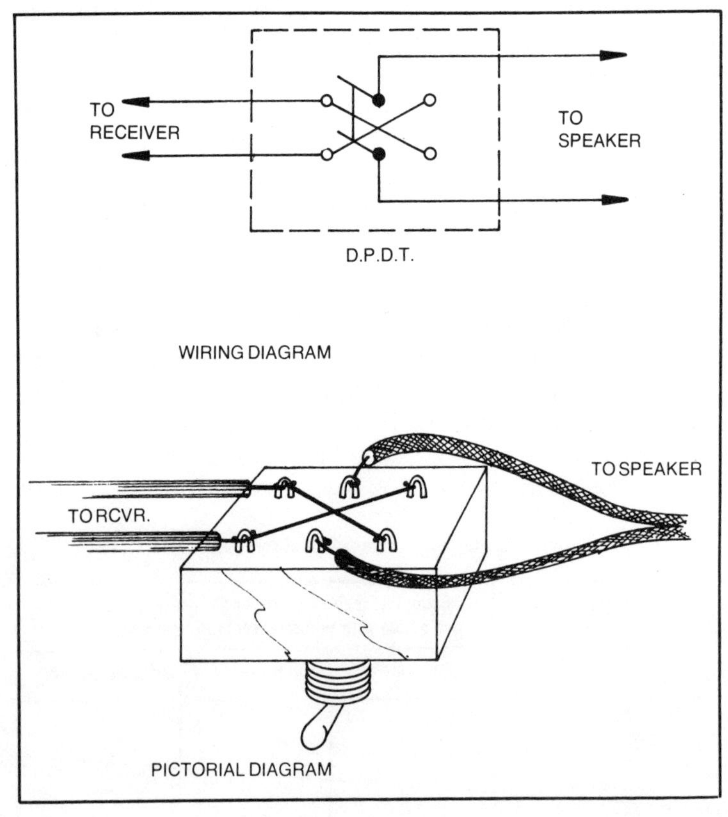

Fig. 7-13. How to wire a DPDT switch for phase reversal role in Project 24.

Fig. 7-14. Phaser switch installed on the back panel of a cabinet.

Construction

All that you need for this project is a double-pole—double-throw (DPDT) switch and an L-shaped bracket to hold it. You can make the bracket from a 3″ corner brace (Fig. 7-12). Cut off one leg of the brace about 1-1/2″ from the bend with a hack saw. Drill out the remaining screw hole in that leg to 1/2″ in diameter so it will take the switch (Fig. 7-12). Install the switch on the bracket. Then install the bracket on the back of your speaker cabinet. Wire the switch according to the schematic diagram in Fig. 7-13 and the circuit shown in Fig. 7-14.

Remember that you need only one switch for a pair of stereo speakers. If you were to reverse the connections at both speakers, they would still be in the same phase relationship to each other as before the switch.

Connect your speakers to your receiver. Place the speakers so they face each other with a small space between them. Set the receiver in the *mono* mode. Choose program material that has a full bass complement and flick the phaser switch from one position to the other as you listen. Leave it in the position that permits better bass range and a higher bass output. Mark the position of correct phasing for future reference in case children move the switch lever.

As a matter of self-education on the importance of phasing, try reversing the phase of your speakers during various stereo programs. You will begin to appreciate the need for proper phasing and will probably become more adept at identifying out-of-phase systems operated by your friends.

If your speakers are ever repaired or any kind of change made in them, make sure you do the phasing test before using them again. With the instant phaser on one speaker, it's a cinch.

8
More Speaker Improvement Projects

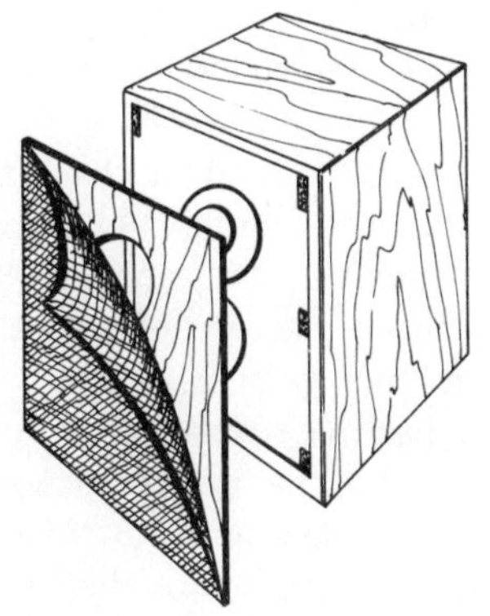

Here are some projects which directly affect the frequency response of your speakers. Starting with a *contour network* that can smooth the response of a rough speaker or modify the enthusiastic midrange of many full-range speakers, the following projects including *upgrading* and *dual woofer add-on* that can literally transform many bass-shy speakers into top-notch systems at modest cost. As with the other chapters in this book, you will have to read the descriptions of each project and then check your own speakers to see which ones that fall within your budget are most likely to make a significant improvement in your stereo set.

PROJECT 25: CONTOUR NETWORK

Although most of the emphasis in conversations on how to improve high-fidelity music systems is on frequency range and how to increase it, the midrange is extremely important for realistic sound. A system that is weak in the mid-frequencies seems to lack body. Even if you can hear tinkling highs and impressive lows, you must have good solid midrange for the sound to be credible. But most systems with a midrange flaw have the opposite problem: too much midrange. It is typical of all kinds of speakers to be quite efficient in the very band where the human ear is most sensitive. This makes many speakers that are rather good in most respects sound a bit harsh.

There are several solutions to this problem. If the speaker is composed of simply a woofer and a tweeter, you can install a midrange speaker and, with controls on it and the tweeter, adjust the level of each band to get proper balance. But in some cases you may not want to go to the expense of an extra driver plus the bother of changing the enclosure to make space for it. And what do you do with a single full-range speaker? To change it to a three-way system might be too costly and generally impractical. The answer is a *contour network*.

If you read the chapter on crossover networks, you know that a choke in series with a speaker limits high-frequency response, while a series capacitor limits the lows. When a choke and a capacitor are wired in parallel, they form a circuit that is resonant at the frequency where the inductive reactance equals the capacitive reactance. At this frequency, the current in the choke is equal to that in the capacitor but out of phase with it. With a pure capacitance and a pure inductance in parallel, no current circulates in other parts of the circuit at the resonant frequency. The impedance of the parallel resonant circuit to current flow through the two parts of the resonant circuit appears to be infinite, although there is a series current from the coil to the capacitor and back again. This series current appears only in the tank circuit formed by the paralleled members, not in the line outside them.

So much for the current at the frequency of resonance. For other frequencies, some current flow is in the line, and the farther we go away from the frequency of resonance, the greater the current is. At frequencies where the reactance of the capacitor is greater than the reactance of the choke, more current flows through the choke than through the capacitor. The line current is the difference in the current flowing in the choke and the out-of-phase current flowing in the capacitor. At frequencies where the reactance of the choke is greater, the opposite effect occurs, and again the line current is equal to the difference in current through the individual members of the tank circuit.

If you wanted to eliminate sound at a certain frequency from your speakers, you could use a tank circuit tuned to that frequency in series with each speaker. With real chokes and capacitors, no frequency would be entirely eliminated because all practical chokes have some resistance and capacitance. And capacitors are not pure either; nonpolarized electrolytic capacitors of the kind used in most speaker networks have some inductance. But a network consisting of a capacitor and a choke in a tank circuit would be selective enough and would cut the response over a narrow band deeply enough that it probably wouldn't be practical, unless you wanted to eliminate a spike in the response curve of a speaker. The kind of filter we want here is one that will reduce the output of a speaker over a broader band, a *smoothing filter*.

When a parallel resistance is wired into the tank circuit, the resistor carries some of the current without respect to frequency (Fig. 8-1). If the value of the resistance is several times greater than the reactance of the other components at resonance, the resistor will have little effect on the current flow in the circuit. As the value of the resistance is reduced, however, an increasing amount of the current flow will go through the resistor. This current increase broadens the response of the filter. When the resistance is reduced to zero, it shorts the tank circuit completely out of the picture, and the frequency response of the speaker will be the same as it was before any reactors were put in the circuit.

Knowing this much about the way a simple parallel resonant circuit can affect the frequency response of a speaker, we can alter the sound character of a speaker by controlling the shape of the response curve fed to the speaker. If we knew exactly what the response curve of the speaker looked like, we could feed an inverted version of that to the speaker and produce a system with perfectly flat response. But such precision isn't necessary. Also, it would not necessarily work out right for every room. A more practical approach is to play around with filters in the room where the speaker will be used, automatically compensating for speaker and room deficiencies at the same time.

Construction

The first step is to make an estimate of the frequency range you want to control. Most single-cone speakers have above

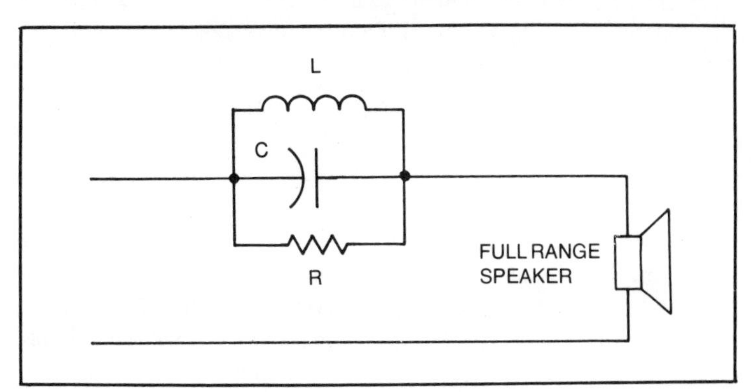

Fig. 8-1. Circuit diagram of full-range contour network.

Table 8-1. Filter Component Values for Resonance at Various Frequencies.

L (mH)	C (μF)	f (Hz)
0.5	1	7000
	2	5000
	4	3500
	6	3000
1	4	2500
	6	2050
	8	1800
	10	1600
	12	1450
2	4	1800
	6	1450
	8	1250
	12	1000

average efficiency in the 1000-Hz to 2000-Hz range, so that is a good place to start.

The most inconvenient part to change is the choke, but the frequency range of the filter can be altered with a single choke just by changing the capacitor (Table 7-1). To make chokes of 0.5 mH, 1.0 mH or 2.0 mH, you can look up the data in the coil winding data in Appendix A, but here is a summary of coil specifications. For a 0.5-mH choke, use 46 feet of No. 22 gauge wire to wind 144 turns on a ¾″ core. For the 1.0-mH choke, use 68 feet of No. 20 gauge wire to wind 175 turns on a 1″ core. And for the 2.0-mH choke, use 120 feet of No. 18 gauge wire to wind 203 turns on a 1-½″ core.

If you want some formulas to use to find a suitable combination of inductance and capacitance for any frequency, here are two:

$$C = \frac{1}{100\,f}$$

and:

$$L = \frac{1}{(40)\,(f^2)\,(C)}$$

Here is an example. Suppose you want to make a filter that will peak at 4000 Hz, a frequency not listed in Table 7-1. You could probably get close enough by guessing at the right values; it looks like a 0.5-mH choke and a 3-μF capacitor would come close enough. But to work out the problem:

$$C = \frac{1}{(100)(4000)}$$

$$= \frac{1}{400000}$$

$$= 0.0000025\,\text{F} = 2.5\,\mu\text{F}$$

$$L = \frac{1}{(40)(16000000)(0.0000025)}$$

$$= \frac{1}{1600} = 0.000625\,\text{H} = 0.625\,\text{mH}$$

So to follow the formula, you would go to the coil winding charts in Appendix A and make a coil that would have an inductance of 0.625 mH. It would be used with a 2.5-μF capacitor and a resistor. These formulas give values of filter elements that have a reactance of about 16 ohms at the frequency of resonance. A typical value for a resistor would be about 20 ohms, but you can experiment with various values to get the degree of cut you want.

Don't feel that these formulas give the only correct values, or even the optimum values, for a filter. Fortunately, the trick of shaping frequency response can be much more casual than that. If you are even close to the right value for the choke, you can probably find a combination of capacitor and resistor that will smooth the response of your speaker. The values suggested by Table 7-1 and the formulas are rough guides to get where you want to go without endless experimenting. But you should do some experimenting to choose the best path.

If you want a variable control on the filter, use a rheostat or potentiometer for the resistance. A variable control for rear auto speakers is available at most electronic stores. These usually have a resistance of 20 ohms and are heavy enough to be used with low-power to medium-power speakers. For heavy-duty speakers, however, you should get a *rheostat*. The problem with rheostats is that these cost enough to justify considerable experimentation with *fixed* resistors to choose the best value.

When testing the filter, try to adjust the speaker volume so that you hear the speaker at the same loudness with the filter in the circuit as with it out. If you don't do that, you will almost certainly prefer the sound without the filter. That's because of a human tendency to choose the louder of two speaker systems as the better, and when the filter goes into the circuit, it cuts speaker volume because it depresses part of the frequency range of the speaker. A system of switching in a resistive network to equalize the sound can be worked out, but because your filter should be tested with various cut levels by varying the value of the resistor, this approach isn't very practical. Just have someone adjust the volume control on your receiver to compensate for the change in loudness level when the filter is switched into the speaker circuit. It is important that you adjust the filter at your normal listening level.

If you have no test equipment and have no idea of what frequencies you might want to depress, you can build a universal network such as that shown in Fig. 8-2. This network has the advantage that you can easily adjust it to center at various points in the frequency range without winding a new choke for each frequency. You can make minor frequency adjustments with any choke just by changing the value of the capacitor in the network, but when you work with only one variable element, the extent of the range of adjustment is limited. Any capacitor and any choke will resonate at some frequency, but the point may be off the audio band band if the value of one element isn't chosen with respect to that of the other. Another limiting factor in adjusting filter frequencies with just one choke is that the impedance of the filter at resonance may not be appropriate for use with a typical loudspeaker.

For the universal network, only a single choke is used, but it is tapped at two intermediate points in the coil so that it can have any one of three possible values. In choosing the most practical choke to make, we considered that it would be convenient to use a pound of magnet wire to make up two filter networks, enough for a set of stereo speakers. It also seemed desirable to keep the resistance of the choke low so that it would have little or no effect on speaker damping. This mandated a heavier wire gauge than might be used in some marginal crossover networks. Optimum wire gauge should be chosen by considering the total wire length. For the choke used here, a length of 100 feet is necessary. Using No. 18 gauge magnet wire, the total resistance in 100 feet of wire of about 0.6 ohm, or under one-tenth the impedance of an ordinary 8-ohm speaker. Keeping the resistance to no more than one-tenth the impedance of the speaker is considered a conservative design figure, so No. 18 gauge wire is adequate.

When winding a coil for the universal network, there are two ways to choose where to put the taps without testing: you can measure the wire lengths between taps, or count the number of

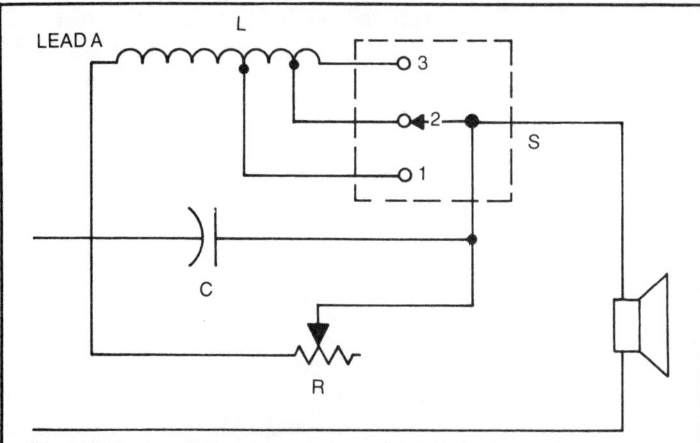

C—4 μF (To double filter frequency, use 1 f. To halve it, use 16 f.)
R—50 ohm rheostat, such as Ohmite 0149.
L—Coil made with ½ lb. of 18 gauge magnet wire on 1-½" core, as follows:
S—Optional switch

Tap	Number of Turns	Approximate Filter Frequency
1	100	3500
2	150	2500
3	Full winding	2000

Fig. 8-2. Universal contour network.

turns on the coil. The specifications in Fig. 8-2 give turns, but you can get just as good results by placing the first tap at about 55 feet and the second one at 80 feet. The exact points are not at all critical; the final performance of your speaker will depend more on how much time you spend in testing the filter at various settings than on how accurately you wind the coil.

To make the coil, you can follow the same instructions on making coil forms and using them as for crossover coils, with one exception: extra wire-exit holes for the taps. These holes should be drilled in the coil side in the position that will permit the coil to be wound neatly over the connection. To place the holes at the proper points, drill them *after* you have wound the coil up to the point where it should be tapped. For example, when you have wound 100 turns on the coil, stop, temporarily tape the coil so it won't unwind, and drill an exit hole adjacent to the circumference of wire already on the coil. Scrape the insulation from the magnet wire at the tapping point and solder a short lead to the wire. Feed the lead out through the prepared exit hole and label that wire with a piece of tape marked No. 1. Go on winding the coil until you get to the next tapping point at 150 turns. Then follow the same tapping procedure, except that this tap should be labeled No. 2 (Fig. 8-3). Don't depend on your memory to distinguish between coil leads; you could end up with a filter turned to a far higher frequency than desired if you happen to interchange either tap No. 1 or No. 2 for tap A at the center of the coil.

Figure 8-2 shows a couple of refinements that cost extra money, components that add convenience but can be dispensed with if your budget is tight. First, the switch to select the various coil taps makes possible instant comparison at different filter frequencies, but it does add some expense. Because the switch wouldn't be used once you find the proper frequency, you might want to connect the various coil taps by hand to check speaker performance with each value of inductance in the circuit. The other luxury is the 50-ohm rheostat. It permits you to make a fine adjustment on the degree of cut that the network can apply, but you can save several dollars by buying various values of fixed resistors, such as 10-ohm, 20-ohm and 30-ohm, and trying each one in turn. Get resistors that will have adequate power rating—at least 10 watts or higher.

Contour networks were once considered appropriate only for inexpensive speakers. When you consider that even the best speakers are far from perfect, it's a safe bet that most speaker systems can be improved by adding one of these networks—not just any network, of course, but one that has been carefully tailored to match the requirements of the speaker. By careful listening and much patience you can supply the tailoring.

PROJECT 26: WOOFER IMPEDANCE EQUALIZER

When you see the impedance of a speaker listed as 8 ohms, or some other single number, that means the impedance is equal to that number of ohms at 400 Hz or some similar frequency. Impedance varies with frequency, so it isn't unusual for the impedance of a woofer to rise to 50 or 60 ohms or more at high frequencies (Fig. 8-4). The rise in impedance is one reason speakers with long voice coils have poor high-frequency response. For crossover networks to perform properly, the impedance of the various drivers should theoretically be constant over their frequency range and even beyond their normal operating range. An audible difference in sound can be made in most speaker systems by using an impedance equalizer, especially for the woofers.

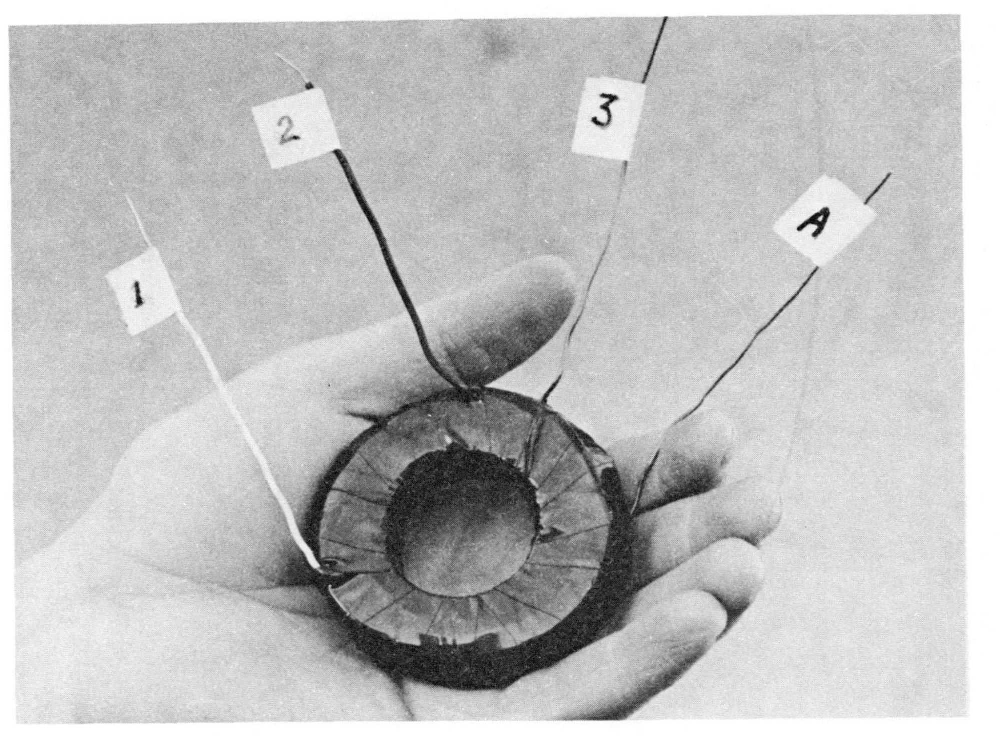

Fig. 8-3. Tapped choke made for universal contour network.

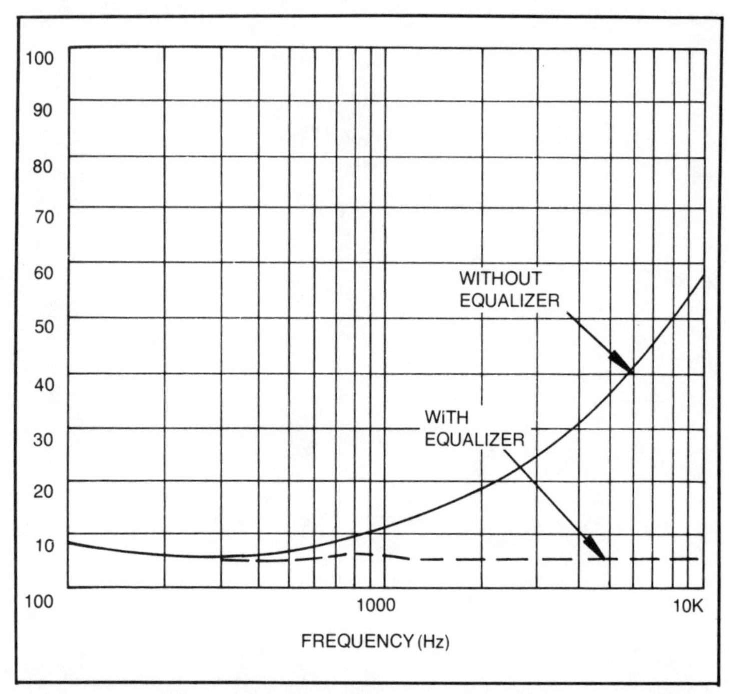

Fig. 8-4. Impedance curve of 8″ woofer before and after connecting impedance equalizer.

The equalizer consists of nothing more than a resistor in series with a capacitor across the woofer (Fig. 8-5). There are several ways to choose the correct values of resistance and capacitance: The choice can be made by calculation, by test or by an educated guess. We will give instructions on all three methods. For practical filters, the value of the resistor is the same in almost all cases, about 7.5 ohms is right for an 8-ohm speaker. To get the value of capacitance by formula, you must know the inductance in the voice coil of the speaker. If you know that, you can solve the problem by this formula:

$$C = \frac{\text{voice coil inductance}}{(\text{voice coil DC resistance})^2}$$

Unless you have a speaker from a rare manufacturer who gives out such information, you probably don't know the inductance of its voice coil. The second way is to run an impedance curve on the speaker, install the filter and adjust the value of *C* until the impedance curve is flat. You can find detailed instructions on several ways to run an impedance curve in TAB book No. 1064, *How To Design, Build, & Test Complete Speaker Systems.*

So we turn now to the third method, the *guess*. This is not as hard as it may seem because there is a normal range of inductance values beyond which few speakers will go. The typical 8″ woofer will require from 10 μF to 20 μF in capacitance. If you have an 8″ speaker that was designed to be a full-range speaker, choose the smaller value. If the woofer has heavy cone and a roll suspension, it probably also has a longer-than-average voice coil and needs the 20-μF value. The typical 10″ woofer can require capacitance from 10-μF for a very few old full-range speakers up to 30 μF or 40 μF for acoustic suspension woofers. For 12″ and 15″ woofers with accordian suspensions, the capacitance required is about 20 μF to 30 μF; with roll suspensions, it is from 20 μF to 50 μF. Note that it is the voice coil inductance, not the suspension type that counts, but longer voice coils usually go with roll suspensions.

Construction

Simply wire the filter across the voice coil terminals (Fig. 8-5). If you are choosing the capacitor values by test or by guess, just hook up the filter temporarily and listen. If the filter makes the sound smoother, either leave it in the circuit or check a different value of capacitance. When you think you have it right, solder the connections with rosin-core solder.

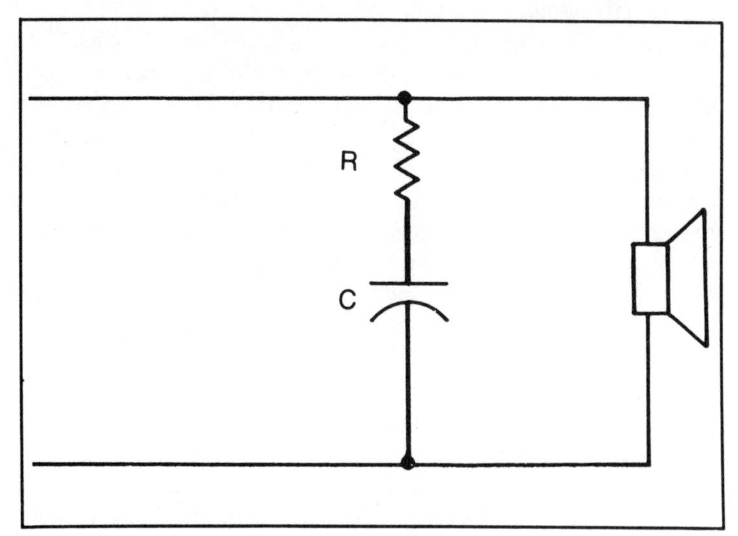

Fig. 8-5. Wiring diagram of impedance equalizer.

Woofers are the worst offenders in having rambling impedance curves, but midrange speakers and even tweeters can sometimes benefit from impedance equalization. For 8-ohm speakers, the resistance will always be 7.5 or 8 ohms. For midrange drivers, choose a capacitance in the 2 to 8 μF range, and for tweeters, 1 to 2 μF.

In choosing capacitors for these filters, the same rules apply as for crossover networks. Use nonpolarized capacitors for the larger values and mylar capacitors for the smaller ones.

PROJECT 27: DRIVER UPGRADING

If you have tried various cures on a set of ailing speakers, there comes a time when upgrading is the most sensible step. It is a good move if you have a set of good cabinets, because the cost of cabinetry can be a major part of the price for a speaker system.

Upgrading can take several paths, depending on whether you have full-range speakers, woofers and tweeters, or a three-way system. Upgrading is easy with systems that use a single speaker for each channel, a speaker that must reproduce the *whole* frequency spectrum. Such speakers are usually designed with a compromise between optimum bass and extended treble response. This kind of system obviously cannot offer the wide range of multiple drivers, each designed to provide top performance for a restricted frequency band. It has other advantages, however, such as perfectly integrated sound.

Most full-range speakers come in 5″, 6″ or 8″ sizes. The typical model has a whizzer in the center of the cone for good high-frequency range. The magnet can vary from the size of a thimble to a more than adequate piece of iron. In replacing a full-range speaker, you best clues to quality are a large magnet and a high compliance suspension. Bass response falls off rapidly below the fundamental bass resonance, so a high-compliance speaker, which will have a lower frequency of resonance than a speaker of the same size with a stiff cone, will have a much more extended bass response (Fig. 8-7). Such a speaker is also likely to be better damped.

Another aspect to consider in upgrading full-range speakers is power rating. Many speakers of moderate price will have no specified rating, but you can estimate the ability of a speaker to handle power by checking the diameter of the voice coil. The larger the voice coil, the better it can dissipate heat. Speakers with the smallest voice coils, from ½″ to 9/16″ in diameter, can be

Fig. 8-6. Use a tie point such as the lug shown here when wiring impedance equalizer across woofer terminals.

Fig. 8-7. In upgrading your speakers, look for high compliance, pointed out here, and a large magnet.

considered 5-watt speakers. These are useful for service in small radios, intercoms and TV sets, but have no place in a stereo system. For a much more useful full-range speaker, try to get one with a voice coil diameter of at least 1″ or more. If you like your music loud, look for a speaker with a voice coil that is *at least* 1-½″ in diameter.

For easy installation you will probably want to find a full-range model that will fit the cutouts in the speaker board of your old cabinets (Fig. 8-8). If you can alter the speaker holes to take a speaker of any shape, consider an oval driver. Oval speakers have some advantages over round speakers of the same price range, including better horizontal dispersion when the long dimension is placed vertically on the speaker board. Because the diameter of a speaker, for horizontally displaced listeners, is the narrow dimension, a 6″ x 9″ oval speaker disperses as well as a 6″ round model. In cone area, such a speaker is equivalent to an 8″ round driver. If you are replacing a 6″ round speaker, get a 5″ x 7″ oval one, or for a 5″ model, a 4″ x 6″. Oval speakers often have a flatter response curve than round drivers because the variation in radius around the cone helps prevent accentuation of certain frequencies and cancellation of others by internal cone reflections. Such faults can be avoided in round cone speakers by careful design, but if you are watching you budget, you probably won't be buying such a speaker. For many uses, such as extension speakers, you can get satisfactory performance at a reasonable price in an oval speaker.

There is at least one situation where the reverse switch—a round speaker for an oval—is likely to make sense. Although there

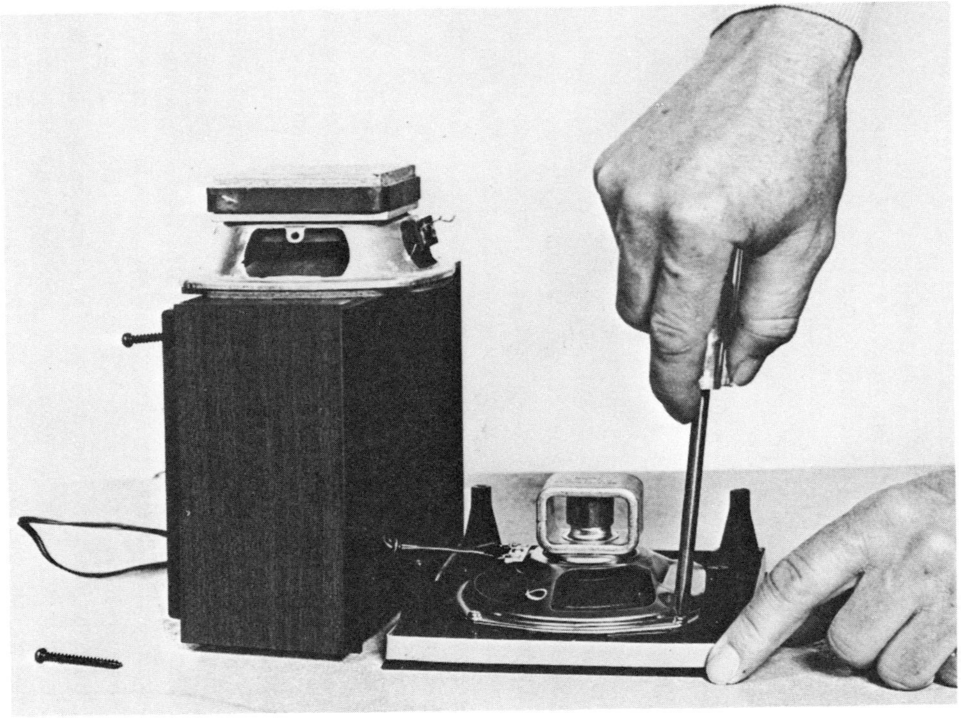

Fig. 8-8. For easy replacement, the new speaker must have the same mounting holes and cone diameter.

are plenty of good quality 6″ x 9″ oval speakers available, very few small ovals have high-compliance or large magnets. This suggests that you use round, high-compliance speakers for small boxes.

Here is an example of one switch of that kind. A friend wanted some cheap extension speakers. We found a pair of vinyl-covered boxes with working speakers for $3 at a Salvation Army store. The speaker cabinets needed reworking (Project No. 16) and the 6″ x 9″ speakers, while in good condition, had stiff cones and small magnets. The interior volume of the boxes was about 0.2 cubic foot, much too small for use with 6″ x 9″ speaker. It was obvious that a more extended bass range could be achieved with a high-compliance 5″ speaker, so instead of simply switching speakers, we installed a new speaker board cut for the smaller speaker. There were many good round 5″ speakers to choose from, all high-compliance models with large magnets, so we obtained a pair of these as upgrade replacements for the original larger speakers (Fig. 8-9). The new speakers, in combination with the reworked enclosures that were sealed and filled with damping material, gave smoother sound and a much more solid bass response than the original speakers.

Woofer Upgrading

If your speakers are installed in a sealed cabinet, you can upgrade the woofers by getting a higher quality woofer that will acoustically match the available cubic volume of the box. Note that the match should be made on acoustical terms, rather than just another speaker of the size originally installed in the box. Precise matching requires either careful testing or attention to the recommendation of the manufacturer. For closed box speakers with highly complaint suspensions, however, you can use the data in Table 6-1 to make an approximate match that should work out well. Don't try to substitute woofers in a ported box unless you are aware of the price requirements in box volume and tuning for the speaker you buy.

In choosing a new woofer, look for a roll suspension and an adequate magnet size for the speaker. An 8″ woofer, for example, should have a magnet weight of at least 10 oz., and a 16-oz. magnet would probably be better. In a 10″ woofer, look for a magnet of about 16 oz. or more, and a 12″ woofer should have a 20-oz. to 30-oz. magnet. It is difficult to specify just how heavy the magnet should be for a speaker of a given size; the weight depends on moving mass and magnetic gap construction. As a rule of thumb,

Fig. 8-9. This speaker system was greatly improved by installing a 5″ high compliance speaker with a heavy magnet in a reworked cabinet that had held an inexpensive, stiff-coned 6″ x 9″ speaker.

the woofers with the heaviest cones will work best in a sealed box, having lower resonance than those with similarly suspended lighter cones. The heavy-coned models are relatively inefficient but have a smoother response. Magnet weight should be scaled to cone weight, so it is typical to find the woofers that have a heavy magnet (for their size) to have a heavier cone. The magnet is the most expensive part of most speakers, so magnet weight gives a good indication of woofer quality.

Midrange Upgrading

If you shop for a midrange driver, you will find a wide variety of speakers offered. The rule here is: Look for a driver that is the same size or smaller than the one you are replacing. Many people associate size with quality, so this advice may raise the question, "Why smaller?" There are several reasons why small midrange drivers usually give better performance than large ones. They have better dispersion in the upper part of their range. Another reason is midrange drivers usually have a sealed back to make installation easier; that is, you don't have to build a subenclosure for them. The larger the cone of the driver is, the greater the cubic volume required behind the cone for good performance. In fact, the cubic volume behind the cone is proportional to the *square* of the cone area, and the area is proportional to the *square* of the radius. Because of the rapidly escalating volume requirements for larger cones, a typical cone on a 6″ speaker requires over twice the cubic volume of a typical cone on a 5″ speaker. Some of the least expensive midrange speakers have an 8″ cone, a sealed speaker frame which gives little volume behind the cone and a small magnet. This combination has nothing for which to recommend it.

One advantage of buying a smaller midrange driver than the original one is that you can make an adaptor and fit it into the same mounting hole. Install the new speaker on the front of the adaptor, if possible (Fig. 8-10).

One of the latest developments in midrange drivers is the availability of *soft dome* midrange speakers. The use of the dome construction permits a large voice coil which can dissipate heat easier than a small coil. Another recent innovation is the use of a *magnetic fluid in the magnetic gap*. This also permits greater power handling by acting as a heat sink.

Another point to consider when upgrading midrange speakers is the recommended frequency range of the new model. If you are going to use the crossover network that came with your speakers, you must consider the crossover points provided by it against the requirements of the replacement speakers. Some of the small midrange drivers have a recommended lower frequency limit that is higher than you might expect.

It is sometimes less expensive to upgrade midrange speakers by buying small full-range speakers and using them for midrange duty. Get 8-ohm speakers and build a subenclosure so that pressure from the woofer won't cause distortion or damage in the midrange drivers. For a 4″ or 5″ speaker used only for midrange duty, a 6″ x 6″ x 4″ deep box (inside dimensions) will usually be adequate. Fill the subenclosure with damping material before sealing it.

Tweeter Upgrading

Most low-cost speaker systems have cone tweeters that look like a typical small speaker, except that the back of the frame is sealed. Such tweeters come in various sizes, but the most popular is probably the 3″ frame size. In the early days of high-fidelity, many speaker systems had 5″ tweeters which were supposed to handle more power than the 3″ variety. All of these relatively large cone tweeters are likely to have poor high-frequency dispersion

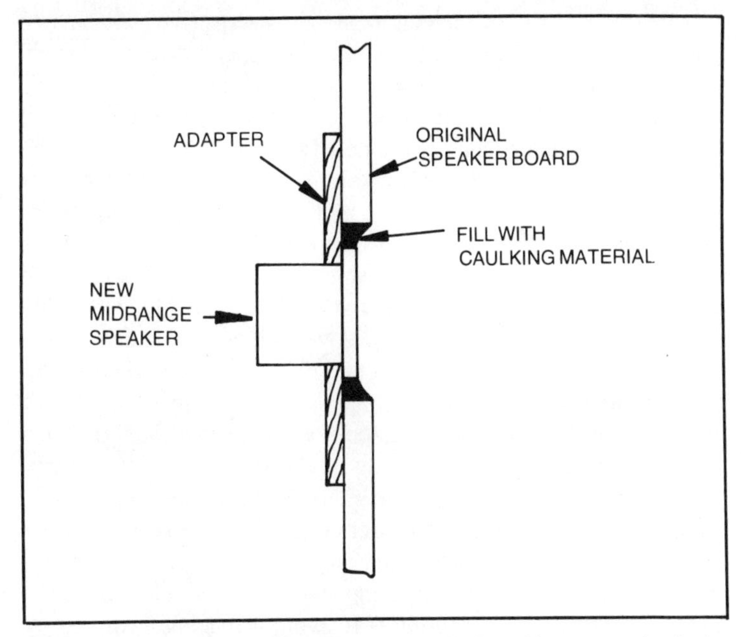

Fig. 8-10. How to use an adaptor to mount a smaller midrange speaker in the position of the original midrange speaker.

characteristics as compared to more modern tweeters. You can recognize an up-to-date tweeter in two ways: by the small diameter of its radiating surface and by its magnet, which will be larger than that of a bargain tweeter (Fig. 8-11).

As an alternative to upgrading your tweeter, you might want to retain the tweeter you have, relegate it to upper midrange duty and add a super tweeter. Inspect your speaker system with that in mind and, if it makes sense, consider Projects No. 30 and 31. In these projects, you will find instructions on how to add a

Fig. 8-11. You can get better treble dispersion and more extended highs by replacing old cone tweeters (left) with dome tweeters.

piezoelectric tweeter or a conventional *super tweeter*. You will have to juggle crossover components to limit the upper range on your old tweeter, but that step is explained in Chapters 4 and 5.

In upgrading tweeters, and to some extent other drivers, make sure the impedance of the new tweeter will be right for the crossover network. This will rarely be a problem because most tweeters you find will have an impedance of 8 ohms, sometimes less. Many British-made drivers have traditionally had a higher impedance—usually 15 ohms—which could create a problem if used in an 8-ohm crossover network. For example, an 8-ohm tweeter in a 6 dB-per-octave crossover network would require a 5-μF high-pass filter to cross over at 4000 Hz. If you substituted a tweeter with about twice the impedance that the network was designed for, the crossover frequency would drop to 2000 Hz. This kind of change could feed energy to the tweeter at a frequency below its safe operating range. The best course to follow in upgrading a set of drivers is to work out a new crossover network that will match the requirements of your new speakers. You can probably use some of the components in the old crossover network. The main precaution is to avoid damaging a tweeter by using a crossover point that is too low for it.

PROJECT 28: DUAL-WOOFER ADD-ON

If you have a pair of speakers that give good midrange and treble performance but are a bit weak in the bass, this may be just the project to add the punch of an extra octave or more of bass to your stereo system. Many good receivers are capable of much better bass performance than the speakers connected to them. While the typical solution for a pair of speakers that are too small to give full bass is to replace them, there is a less expensive path to better sound. In fact, this solution to the weak bass problem can give you a great bass capability at very moderate cost.

It has always seemed a terrible waste to dispose of a pair of perfectly good speakers just because they lack some range at one end of the audio spectrum. With that in mind, we offer the dual-woofer add-on, a mixed bass speaker system. By mixing the bass from two stereo channels in a single enclosure, you can use your original speakers as stereo satellites. The dual-woofer system in the bass enclosure handles the low end of the audio spectrum from both stereo channels. If you're worried about losing the stereo effect, don't be. Stereo imaging depends on your ear locating the direction of certain sounds, but the lowest frequencies

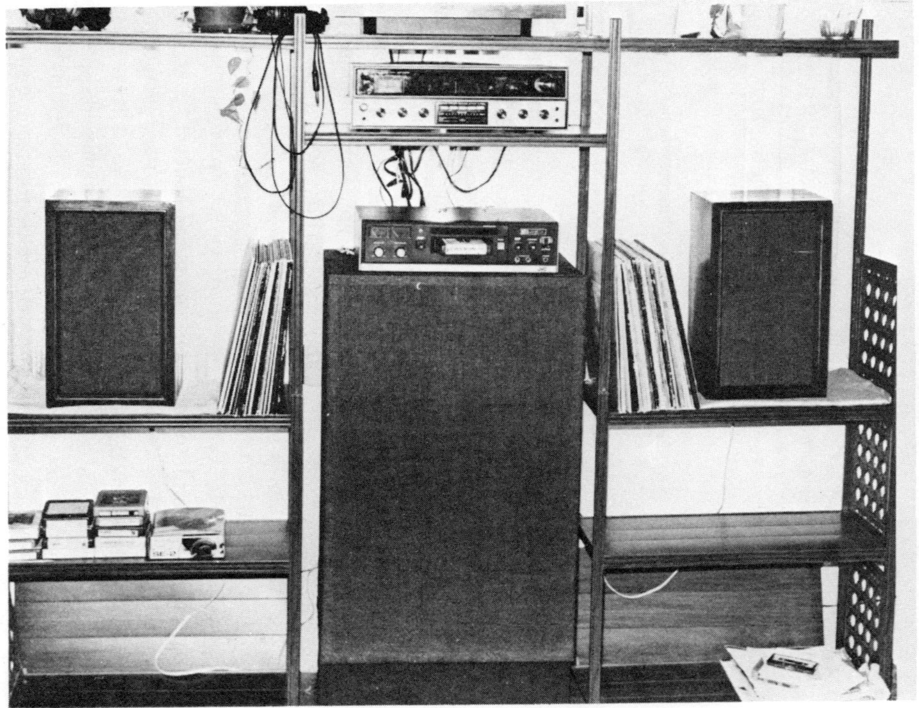

Fig. 8-12. This dual woofer system greatly extended the bass range when the original speakers were reduced to satellite duty (photo by Mark Rundell).

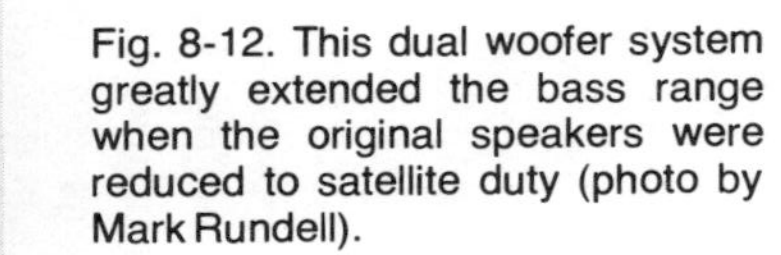

are so omnidirectional that they contribute little to the stereo effect. This means that you can locate the dual woofer cabinet almost anywhere in the room if the crossover frequency is low enough. From a practical standpoint it is easier to use a somewhat higher crossover frequency and put the woofer box midway between the small speakers. At that location, it will help fill the hole in the middle effect that plagues some stereo systems.

Mark Rundel, a college student now in law school, built a dual-woofer system based on the plans provided for this project (Fig. 8-14). His original system consisted of a used 1971 95-watt Kenwood KR-4140 receiver and two Kenwood compact speakers. He was satisfied with the performance of those components except for the limited bass response of the small speakers. After completing the system he made the following evaluation of it in a recent letter.

> "I've lived with the subwoofer for a month now and I can report that it is more than I expected and more than the neighbors like. I've always had to watch how loud I played my system but I can happily report that it has a marked improvement over the entire volume range. Even at lower volume levels the bass makes its presence known without booming out. My friends have noticed the improvement in the quality of sound. Also, the uniqueness of having three speakers for stereo has been something I've had to explain on several occasions.
>
> "I've enjoyed building the speaker so much that I may never purchase a speaker again. There's a certain pride you get when you turn on your stereo and good quality sound comes out. Also, there's that throw out your chest pride you get when your friends comment on that 'big speaker' and then you get to tell them that you built it yourself."

Rundel's experience is typical of many first-time project builders who discover that building or improving speaker systems is not only a practical possibility, but also great fun. This project is more ambitious than most of those in this book; that is, it costs more. But for the person who wants superior bass, it's worth it. As a rule of thumb, if your speakers have woofers smaller than 8″ in diameter, this project will probably add a new dimension to the low-frequency performance of the system. For speakers with 8″ woofers, it will undoubtedly give improvement, but how much depends on the quality of your 8″ woofers.

This add-on bass system consists of two 10″ woofers installed in a cabinet with an interior cubic volume of about 3 cu. ft. The woofers used in this project were obtained from Speakerlab; specifically, each was a Speakerlab model W1008B, a 10″ woofer with a heavy cone, a high-density foam roll suspension, and an 8-ohm overhung voice coil. Any good 10″ woofer designed for closed box operation can be used, but the performance may vary

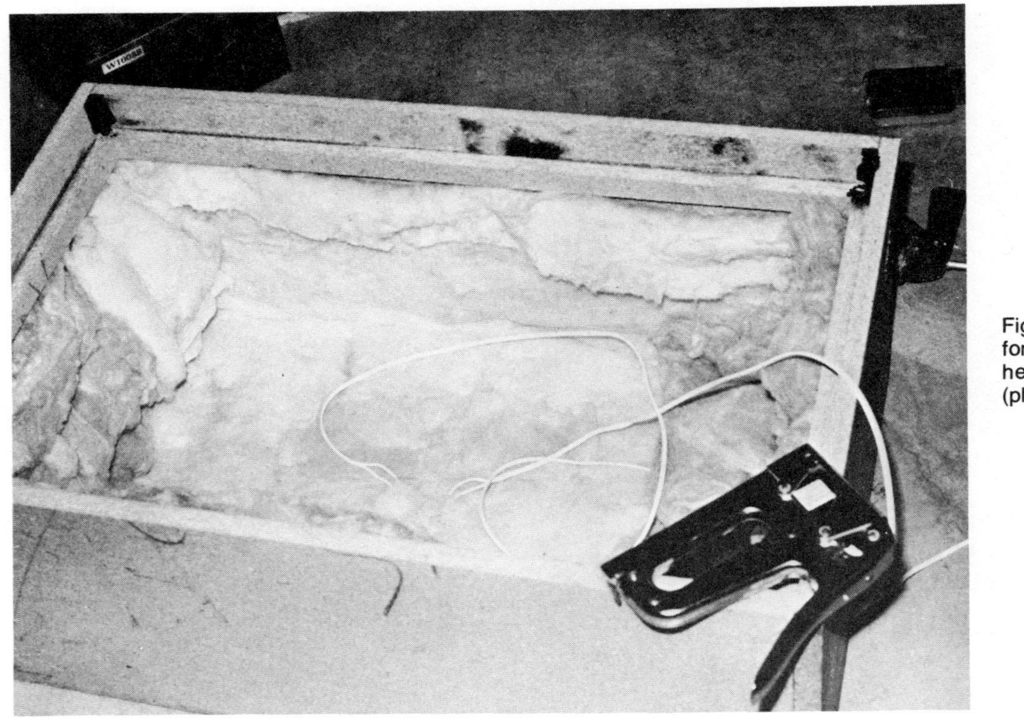

Fig. 8-13. Woofer enclosure ready for speaker board installation. Note heavy use of damping material (photo by Mark Rundell).

somewhat. Another Speakerlab woofer, the W1004B, would be a good choice too. With its 4-ohm impedance, a smaller inductance could be used to cut off the high-frequency energy to the woofer, but in that case, some care must be used to see that the capacitor chosen to feed the small speakers is right. Carelessness here could put the woofers effectively in parallel with the small speakers at low frequencies, dropping the net impedance of the system below 4 ohms. If you have no test equipment to check the impedance of the system, it's probably a good idea to restrict yourself to 8-ohm woofers.

Although Rundel installed his woofer enclosure between his small speakers (Fig. 8-12), you may be able to find a location where you can hide the large cabinet. Some people have even moved the bass speaker into an adjoining room or closet. One stereo experimenter who did this gained some fun by having friends compliment him on the full bass response of his small speakers, the only ones visible. Regardless of where you put the big speaker, if you have a set of decent small speakers but you miss the deep bass of a big system, this project is for you.

Construction

Because this is a large floor cabinet, ¾" material is mandatory for the walls (Fig. 8-14). For economy you can use particle board, but choose a board made from small particles. That kind has greater density than the flaky particles used in low-grade board. Simple butt joints, as shown in the plans, are adequate. Use plenty of wood glue in the joints.

Early high-fidelity speaker cabinets always had backs that were attached by screws so the back could be removed. In those days, all speakers were installed behind the speaker board, so if a speaker had to be removed, the only way to reach it was through the back. Current practice is to mount the speakers from the front of the board. If the speaker is installed with silicone rubber cement, the cement automatically forms an air-tight gasket under the speaker frame and yet permits later removal. To remove the speaker, all you have to do is place the point of a standard screwdriver blade under the frame and pry it loose. This change in speaker mounting means that you can install the back with glue, just like the other cabinet panels. This kind of construction makes a more rigid box with less chance of air leaks or rattles. Permanently glued panels are a secret weapon of the newer speakers in comparison testing against old speakers.

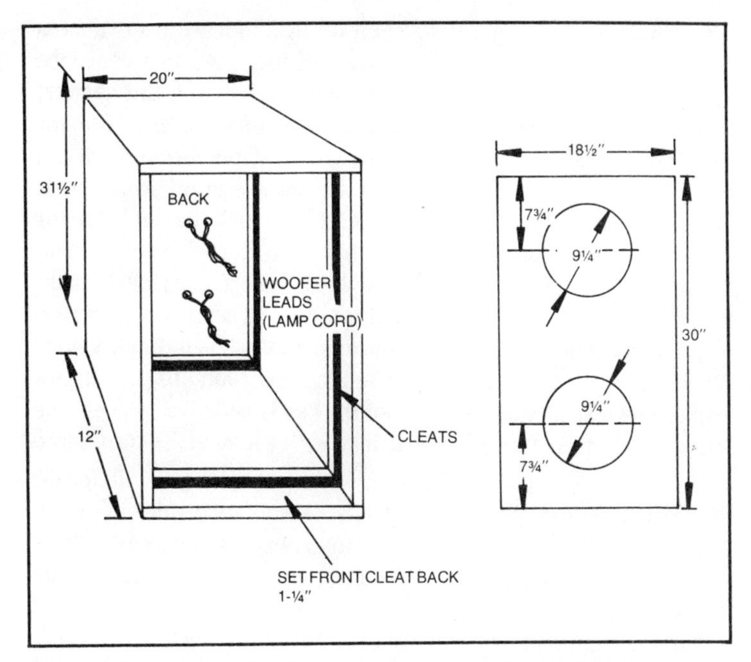

Fig. 8-14. Construction plans for dual-woofer add-on (Project 28).

Cut out the parts according to the plans. You can use extra pieces of particle board to make the cleats that hold the back and the speaker board in place. Mark a line 1-¼" from the inside front edge of each cabinet wall. Glue and nail a cleat to each of the longer panels (sides) at this point. To get an air-tight fit, it is better to leave the top and bottom cleats until after the cabinet walls are assembled. Mark a line ¾" from the inside rear edge of each panel. Glue and nail a cleat to each of the longer panels here too.

Set the top and the bottom panel on a firm work surface and start a line of 6-penny finishing nails along the side edges, indented about ⅜" from the edge of each board. These nails should be driven into the boards until they barely penetrate the bottom surface. Set the bottom panel over an inverted side wall and drive in two nails lightly to position the pieces as they should fit together. Then remove the bottom panel and smear the matching surfaces with glue. Set the bottom back in place and drive home all the nails. Follow this same procedure with the other side and then with the top panel and the two sides. Cut pieces of material for top and bottom cleats, front and back, so they fit the space between the side cleats. Then install them with nails and glue.

Now you have an empty shell of a cabinet with no speaker board or rear panel. While the glue is setting, you can prepare the rear panel and the speaker board. Cut two holes in the speaker board for the woofers, located as shown in the plans. Mark a location on the back panel for two screw-type terminal strips, placed directly behind each woofer. At each location drill two ¼″ holes, spaced to match the distance between the lugs on the terminals you will use. If you have no screw-type speaker terminals, you can simply drill a single ¼″ hole at each location and feed a piece of lamp cord through the hole to carry the electrical signal to the woofers. If you use the hole method, make sure you fill it with latex or silicone rubber caulking compound to prevent air leaks. It is also a good idea to knot the cord inside the back so that the woofer terminals cannot be damaged by a jerk on the cord.

At this point, you will have to decide whether you will locate the choke coils that block the highs to the woofers inside or outside the box. Putting them inside makes for a neater assembly, but this makes any later changes in these components a major operation. Rundel installed his chokes under the cabinet in the space made by a 5″ base he built under the speaker. You might want to tinker with the crossover frequency, so it's a good idea to put the coils under the enclosure or in some other accessible spot. However, if you want them in the enclosure, they can be wound and glued to the interior of the back panel near the lamp cord access holes. Each coil will be wired in series with its woofer.

After drilling the holes for the terminal strips, bring the ends of an 18″ length of lamp cord through the holes, from inside surface to outside. Solder these leads to the lugs on the terminal strips and glue down the terminal strips to the back panel. You can use either latex or silicone rubber compound for this job, but drive some carpet tacks through the mounting holes in the terminal strips to hold them while the glue is setting. Fill the interior of the holes with caulking compound to make them airtight.

Start a line of nails around the perimeter of the back panel, spaced about ⅜″ from the edges. You should place a nail about every 5″ around the panel. Glue the matching surfaces on the rear edge of the cleats and on the interior of the back panel, install the back panel and then nail it down. Caulk the joints between the back and the cleats with latex caulk. Line the interior of the box with 2″ to 3″ damping material, such as fiberglass or polyester batting (Fig. 8-13). Then install the speaker board in the same way you put on the back. If necessary, caulk the speaker board joints from the front edge of the board.

You can cover the outside of the box with plastic veneer material, the kind that comes in a roll with adhesive backing, or with any other material that pleases you. For an easy-to-apply finish, paint it flat black.

The simplest way to install grille cloth is to make a grille frame from ⅜″ plywood. This kind of grille frame fits inside the walls of the box in the ½″ space left at the front of the speaker board. Rundel wanted a grille that would cover the front edges of his box, so he used ¾″ particle board and cut a 45-degree angle along each edge. To attach this larger grille frame, he cut some ¾″ x ¾″ x ½″ plugs from excess material to fit into the inside front

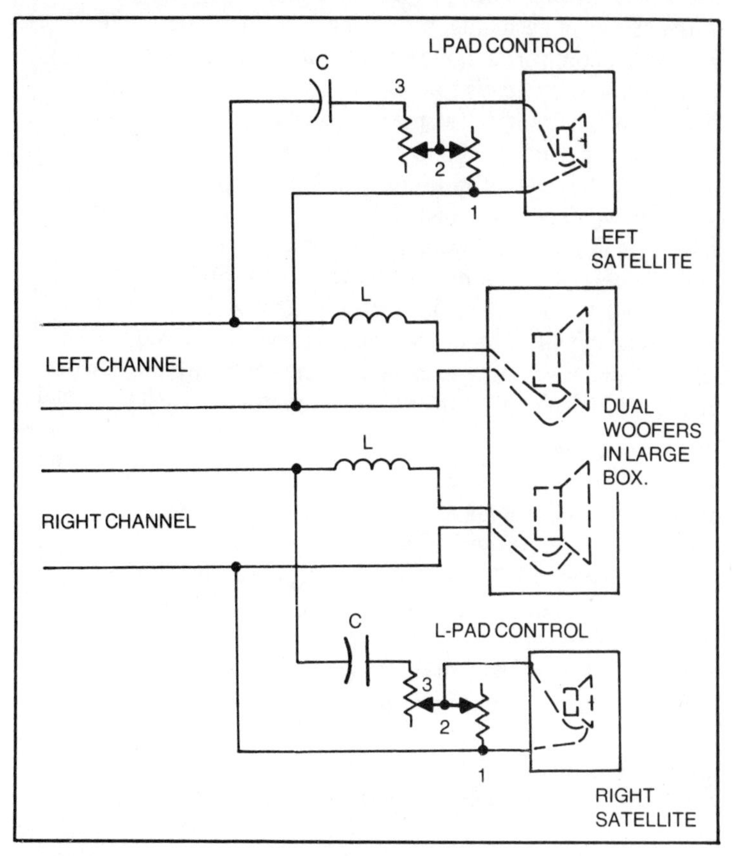

Fig. 8-15. Wiring diagram for satellites (original speakers) and dual-woofer add-on.

corners of the box. Then he glued and nailed these plugs to the rear of the grille frame at the proper locations to hold it onto the box. In fact, to make sure the plugs would fit into the proper location for each, he left a small space and then enlarged the plugs with electrical tape around each one until the frame fit nicely. If you plan to use this kind of grille frame, the speaker board need not be recessed at all. You can install the grille frame with Velcro, or small nails, and avoid any cavities between the speaker board and the grille frame.

For an inexpensive grille cloth that gives a good appearance, get some decorator burlap and cut a piece that will overlap the four sides of the grille frame by about 2 inches. Fold it over the edge and staple it at the center top. Stretch it to the bottom and staple it at the center bottom. Then pull it around the edge of the grille frame at the center of each side, stretching it again as you staple. Work your way to the corners, stretching it in each direction as you go. Finally, you can fold each corner as you pull the cloth taut around the back of the frame and staple it. You will have a double thickness at the corners, unless you cut away the excess. If you cut off any material at the corners, glue down the loose edge.

The woofers can be installed at any time after the paint on the cabinet is dry. Set the woofers face down—one at a time— near their respective holes in the speaker board. Strip the insulation from about ¼″ of each lamp cord conductor and connect it to the woofer terminals. Solder the connections. Then run a bead of silicone rubber sealer around the hole and set the woofer down into place, twisting it slightly to ensure a proper seal.

Crossover Network

The crossover network that divides the frequency range between the add-on woofers and the small speakers consists of a single choke coil in series with each woofer and a capacitor in series with each small speaker (Fig. 8-15). To wind the coil, make a form to the dimensions shown in Fig. 8-16. A single form can serve for two coils if you drill a hole through the wood core and use a bolt to hold the hardboard sides together while you are winding it. After winding the coil, dismantle the form and tape the coil for stability; or you can glue the pieces together permanently if you have two forms. Don't use any magnetic material, such as iron screws or nails, to hold the parts together on a permanent coil form.

To wind the coil, thread about 6″ of magnet wire through an exit hole in the hardboard side near the core. Wind the first loop by

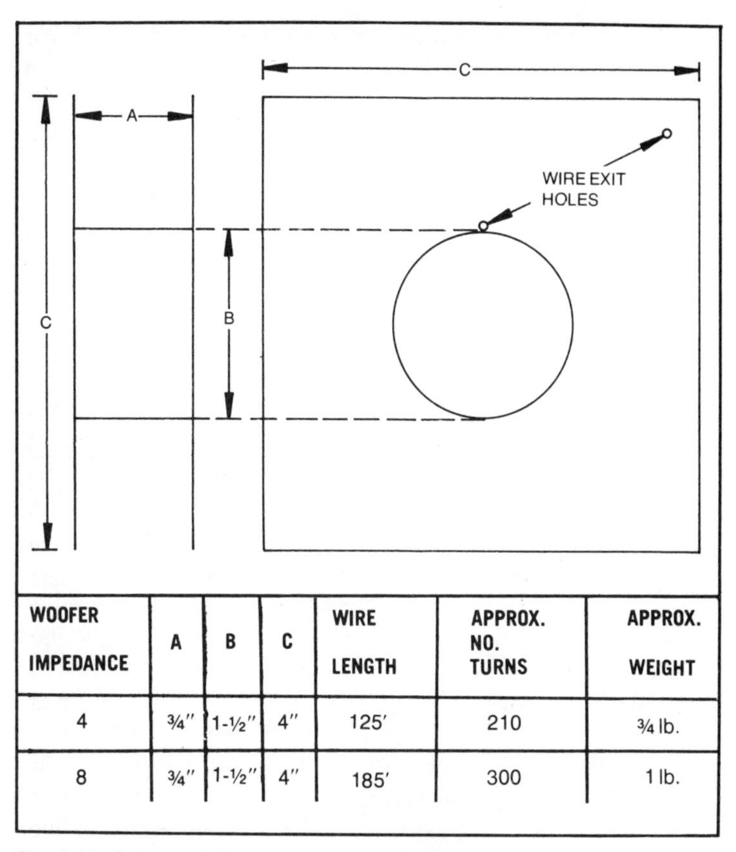

WOOFER IMPEDANCE	A	B	C	WIRE LENGTH	APPROX. NO. TURNS	APPROX. WEIGHT
4	¾″	1-½″	4″	125′	210	¾ lb.
8	¾″	1-½″	4″	185′	300	1 lb.

Fig. 8-16. Coil specifications for 300-Hz crossover frequency for dual-woofer system (Project 28). Use No. 18 gauge magnet wire.

pressing the wire tightly against the hardboard side with the exit hole. Wind the next turn as close to the first one as possible, and so on. Keep your layers as neat as possible. Even scramble-wound coils work well, but if the winding is too messy, the layers of wire may occupy too much space that will affect the value of the coil. For this coil you will use a pound of No. 18 wire. The inductance of such a coil will be about 4.25 mH, or slightly greater, which will produce a crossover frequency of 300 Hz with an 8-ohm woofer. To cut the low-frequency signal to the small speakers, wire a 66-μF non-polarized capacitor in series with each satellite speaker (Fig. 8-14). You probably won't find 66-μF capacitors, but you can use a 50-μF and a 16-μF in parallel or any other combination in parallel that adds up to 66 μF.

To provide better balance between the level of bass output and that of the upper frequencies, you will probably want to install an L-pad control on each small speaker (Fig. 8-14). If you don't want to disturb the cabinets to install the controls in the back, you can mount them on L-shaped brackets, similar to that used for Project No. 24. Mount the controls and then wire them, connecting the lead from the filter capacitors to pin No. 3 of each control and carrying on from pin No. 2 to the positive speaker box terminal. Split the lead to the other box terminal and connect the ends to pin No. 1 to put the L-pads across the line as well as in series with the speakers.

Connect the speaker system to your receiver. Turn on the set but keep the volume low until you are sure everything is working right. Check with your ear at the big speaker to see that lows and only lows come from it. Then listen to the small speakers to see if they are reproducing midrange and treble. Rotate the controls to see if a clockwise rotation brings up the middle and high frequencies. Rotate the controls fully counterclockwise. Then bring them up just enough to balance the sound from the bass speakers without becoming too prominent. It helps to have someone else do this while you listen at a distance of about 10 feet or so. When you are satisfied with the balance, you are ready to test the system with a good recording or your favorite FM program.

You will probably hear bass you didn't know was in the music. Once you get used to solid bass—the foundation of the whole sound spectrum— you won't want to go back to limited-range speakers again.

9
High-Frequency Improvement Projects

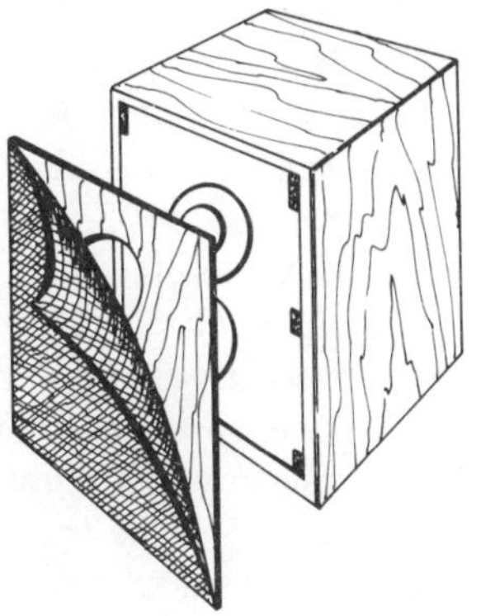

If you have ever listened to a pre-high-fidelity radio or phonograph, you know how dead one sounds to ears that are conditioned to modern sound systems. Those old console radios had virtually no response above 5000 Hz, and the rolloff usually commenced at 3000 Hz or lower. In addition to having narrow-band components, those sets had a single-knob tone control that did nothing except choke off the weak highs that had escaped the obstacle course presented by the electronic components and the large single-cone speakers. What one had with those sets was aptly described, depending on how you set the knob, as "low-fidelity" or "no-fidelity."

Audio fans sometime wonder about those old radios. Why did they need a tone control that served merely to eliminate the already depressed highs? Were listeners different then? If you look at the history of high-fidelity, you will find in the literature a number of technical papers, published about the time of World War II, that reported the results of experiments designed to determine the optimum frequency range of home music systems. These experiments seemed to verify that listeners preferred equipment with limited frequency response to that with more extended range. The conclusion was not lost on radio manufacturers, who continued to make radios and phonographs with no attention to high-frequency response, except to make sure there was none.

The most telling blow to experimenters who were tinkering with tweeters and wide-range amplifiers came when one study showed that professional musicians were *even less tolerant* of reproduced highs than the average listener. A report by Chinn and Eisenberg in the Proceedings of the IRE in 1945 gave the results of a study where listeners could choose between three bandwidths: narrow (150 to 3500 Hz), medium (100 to 5000 Hz) and wide (50 to 10000 Hz). The average listeners picked the narrow band even after they were told it was "low-fidelity." But what was even more surprising was that the professional musicians also liked the narrow band, with 73 percent of them choosing it. Only 5 percent of those musicians liked the wide-band sound.

Before you condemn those listeners for their tonal ignorance, remember that there were good professional musicians in 1945 too. This experiment has relevance to us today because when the truth was learned, it showed certain problems that must be avoided at all costs. For the truth, today's stereo listeners can thank Dr. Harry F. Olson, who was in charge of acoustical research at RCA Laboratories.

Dr. Olson analyzed the published report and concluded that there were only three possible reasons for the lopsided results: people were conditioned to a narrow range, musical instruments are improperly designed or—and this one he suspected as the real reason—*distortion is less objectionable in a system of limited range*. Olson decided to perform another experiment, one that would eliminate distortion that might skew the results. Since the best amplifiers of that day had a certain residual level of distortion, how could he do it? Olson's solution, like many classic approaches to complicated problems, was simple. He would do away with reproduced sound and instead use a live orchestra with an optional acoustical filter between the orchestra and the audience. The filter imposed a sharp cutoff at 5000 Hz, and the whole thing was hidden behind a curtain so the listeners didn't know when the filter was in operation. After a thousand listeners had been tested, the concept of high-fidelity was exonerated. About 69 percent of the listeners favored the sound when there was no filter.

Those experiments suggest that you should never try to extend the frequency range of your equipment unless you can do it with low distortion. We know now what low-distortion, high-frequency performance can do for a stereo system; it adds an airy quality that makes the sound more open and natural. We also know the importance of good high-frequency dispersion. A speaker that beams the highs in a narrow path will be too "hot" if you're in the beam and dead if you are out of it. Poor dispersion makes hole-in-the-wall music, the kind that puts the musicians behind a barrier instead of out in the open and well placed in relation to each other. And if the dispersion is poor, the stereo image will suffer so that small changes of listener location can give an altered balance as well as changes in apparent source position. One of the ironic facts about speaker systems is that it is the ability of the small tweeters to disperse the highs that most account for the larger-than-life, or expansive, effect of a large system.

The projects in this chapter are designed to improve the highs in your speaker system, either by improving the drivers you have or by adding an extra tweeter. Even if you are a bass freak, the treble end of the musical scale is extremely important to realistic sound. You can have a somewhat pleasant background kind of music without extended highs, but you can't have high-fidelity!

PROJECT 29: TWEETER RESPONSE SHAPER

Some low-priced tweeters have too much response at the low end of their range and too little at the top. The typical listener response to this kind of treble is to turn down the tweeter, a move that muffles the upper highs and reduces naturalness. One way to solve the problem without buying new tweeters is to feed a signal to the tweeter that has an inverted frequency characteristic to that of the tweeter itself. This is somewhat similar to using a contour network, as described in an earlier project, but here no choke is necessary and the affected frequency range is different.

You will need only a resistor and some capacitors to make such a network. The final network will consist of a single capacitor in parallel with the resistor. The resistor will pass all frequencies with equal attenuation, but the capacitor favors higher frequencies in its leg of the circuit. The net result is that the upper highs are favored a bit. If your tweeter appears to have a peak within its operating range, you should build a contour filter as suggested in Project No. 25, to remove the peak. You can use both networks to modify the frequency behavior of your cheap tweeters. This project will not improve the dispersion characteristics of your tweeters, but they may appear to have better dispersion because the upper highs will be stronger.

Construction

To choose the correct capacitor value, it would be nice to have an exact frequency response curve for your tweeters. Since you are unlikely to have that, particularly for cheap tweeters, you must estimate the frequency at which you want the filter to take effect. A frequency test record can help in this estimate. Play the record at low volume and listen for the point at which highs drop in volume. A low volume level is best for this kind of testing because the ear is more sensitive to minor changes at a low volume. If the level is near the threshold of hearing, you will easily notice when it drops below that level and you can no longer hear any response.

When you have an idea of what frequencies you want to boost, check the information in Table 9-1 for the right value of capacitor. If you can't decide what frequency would be best for the crossover point in your curve, you can just try several values of capacitance and use the one that most improves the sound. For example, you might start with an 8-ohm resistor and a 2-μF capacitor in parallel. If this combination works well, use it. But even if you notice an improvement with the first value you choose, it's a good idea to try others. By starting with a low value, such as a 2-μF capacitor, you can wire a second 2-μF capacitor in parallel with the first and test the system. By doubling the capacitance, you will have halved the frequency at which the boost begins. If that appears to be too low, replace the extra 2-μF capacitor with one of 1-μF capacitance. Choose the combination that makes for the sweetest, most extended high-frequency response with no apparent hole in the lower end of the range of the tweeter.

You can install the extra components in the speaker wiring at any point after the crossover network. Make sure that for each connection you use a solid tie point, such as a lug on a terminal strip. One of the tweeter terminals can serve as one tie point. Connect the leads mechanically. Then solder the connections with rosin-core solder.

Table 9-1. Values of Capacitance to use for Tweeter Response Shaper Network for Various Turnover Frequencies with an 8-Ohm Tweeter.

Boost Frequency (Hz) Starts At:	C (μF)
5,000	4.0
6,000	3.3
7,000	2.8
8,000	2.4
9,000	2.2
10,000	2.0
12,000	1.6

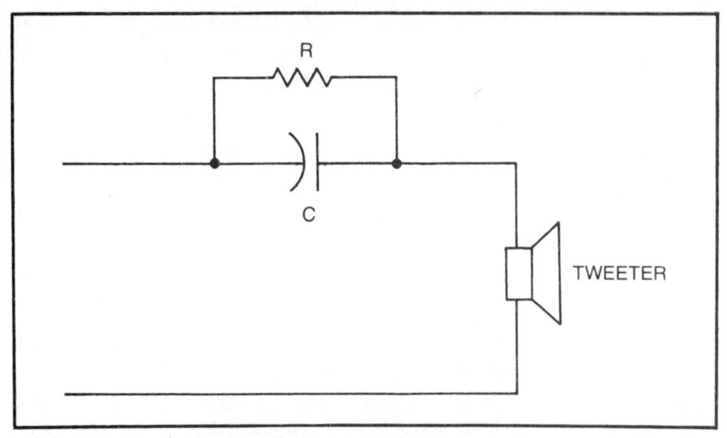

Fig. 9-1. Response shaping filter for tweeter with too much output at lower end of its range. R should equal tweeter impedance.

PROJECT 30: PIEZOELECTRIC TWEETER ADD-ON

Here is a low-cost project that can extend the high-frequency range of your speaker system to the limits of audibility and beyond. One of the reasons for its low cost is that, because a special kind of tweeter is used, no crossover network is required. The tweeter is a *piezoelectric tweeter (P.E.T.)*.

These tweeters work on a principle previously applied to crystal phonograph cartridges, ultrasonic sensors and underwater sonar transducers. They have no voice coil and no magnet, saving considerable weight and cost. And a P.E.T. has such a high impedance at the bass end of the audio spectrum that it is essentially out of the circuit at those frequencies. Even at 1000 Hz, the impedance is about 1000 ohms. The impedance characteristic of these tweeters is just the opposite to that of speakers with voice coils; it drops, instead of rises, as the frequency increases. Because of the high impedance at low frequencies, you can wire a P.E.T. directly across the speaker line, putting it in parallel to the woofer in your speaker system. With no crossover elements in the circuit, and because it has extremely low dynamic mass, the tweeter gives excellent transient response.

As with other kinds of tweeters, some kind of control is usually needed to balance the output of a P.E.T. to that of the typical woofer. One way to control the tweeter is to use a capacitor and a resistor to roll off the response of the tweeter at the low end of its band; the capacitor is placed in series with the tweeter, the resistor is across it. Typical values are 5 μF for the capacitor and 8 ohms for the resistor, but these can be varied to fit the situation. For a more versatile balance, install the capacitor and follow it with an L-pad and the shunt resistor too (Fig. 9-2). If you want to be theoretically correct and if you know precisely where to place the crossover frequency for optimum performance, you can select a capacitor whose reactance at the crossover frequency is equal to the value of the shunt resistor. To do that, you would use the crossover design chart in Chapter 4. Don't forget to use the shunt resistor; the impedance of the tweeter at the usual crossover frequencies will be several hundred ohms, and that value will be reflected to the crossover components if you leave the parallel resistor out of the circuit. After adding the shunt resistor, you can figure crossover values by considering the impedance to be that of the resistor.

A series resistance is not usually recommended with piezoelectric tweeters because it produces a rolloff in the high-frequency response, but there is one condition where it must be used. If you connect one of these tweeters to your stereo system and the amplifier becomes unstable, the cause is probably the low impedance of the tweeter at ultrahigh audio frequencies, perhaps at 50,000 to 100,000 Hz. When this problem appears, you can insert a 20-ohm to 30-ohm resistor in series with the tweeter and control the oscillation without too much effect on the upper highs.

Although the methods mentioned above give good balance, there is a less expensive way to balance a P.E.T. to the typical speaker system that also improves dispersion, at least at some frequencies. This is to install the P.E.T. on the back of the speaker enclosure, facing slightly upward so the firing angle will bounce the highs off the wall behind the speaker. If your speaker has a glued-on grille cloth, this offers the added convenience of easy installation in the back where you can modify the box without changing its appearance. Because of these obvious advantages, this installation method is the one that will be described.

Construction

The construction notes here will be keyed to the use of the *Radio Shack* radiating P.E.T., stock number 40-1382. If you use another model, you can adjust the instructions to fit the dimensions of your tweeter. It is assumed that you can remove the back panel from your speaker.

Start by locating the tweeter position high on the back of the box so that the upper rim of the tweeter frame will lie about a couple of inches below the top rear edge. Draw two concentric

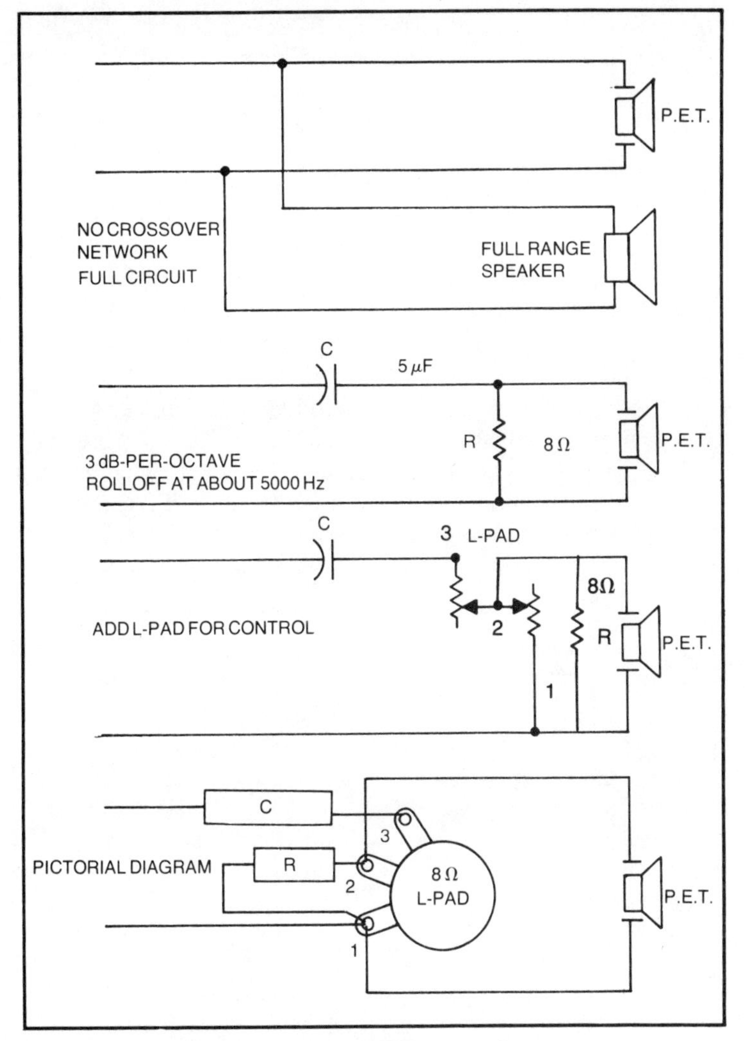

Fig. 9-2. Various ways of connecting a P.E.T. in a speaker system.

circles around the center point of the tweeter position, an outside circle 4″ in diameter and an inner circle of 2-½″ in diameter. Use a circle-cutting saw or a sabre saw to cut out the inner circle.

Next, remove some material from the upper half of the remaining circle with a wood chisel. Remove this material in increasing depth as you approach the upper part of the circle. The greatest depth should be about ⅜″ at the upper point, tapering so that material is removed midway down the circle. Now, cut a piece of plywood or other material into a 3-¾" square and cut a 2-½" hole from the center of this square. Depending on the kind of tools you use to cut the hole, you may find it easier to cut the hole first and then the square section. If you have a large table saw, the next step is easy. Cut two tapered sections from the piece of plywood (two if you are working on a *pair* of stereo speakers). Each section should be ⅜" thick at the thick end, tapering to zero thickness at the midpoint of the circle (Fig. 9-3A). If you have no table saw, you can use a hand saw, but it helps to clamp the plywood in a vise or clamp it to a bench or sawhorse with a C-clamp. Don't worry if you do a sloppy job on this; you can fill any gaps in the wedge with caulking compound.

Use some small brads and wood glue to glue and nail the wedge to the lower half of the circle marked on the back panel (Fig. 9-3B). This piece and recessed part of the panel should make a fairly smooth mounting platform for the tweeter. If there are any holes, fill them with caulking compound, such as silicone rubber sealant or latex caulk. Set the back panel with the interior side

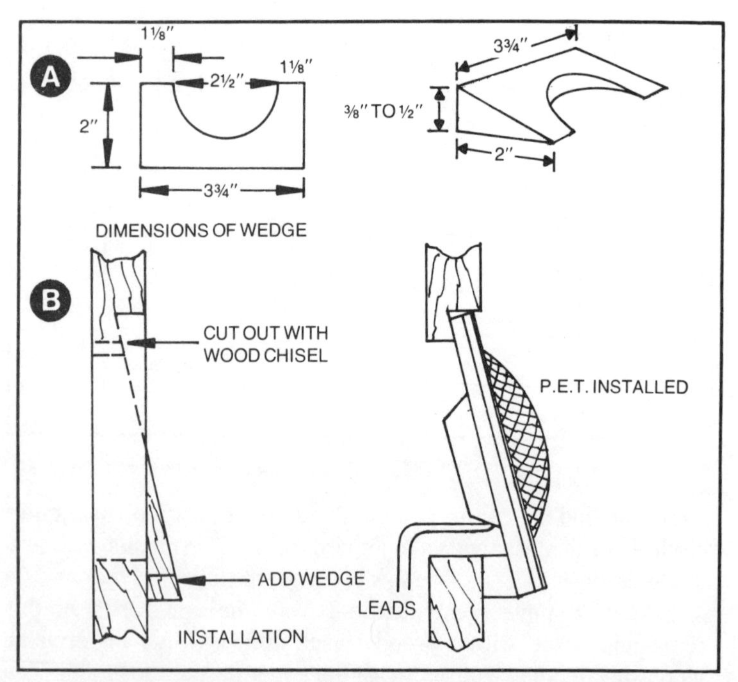

Fig. 9-3. Details of how to mount a P.E.T. in the back panel of the enclosure so it fires at an upward angle onto the rear wall. Dimensions at A, and installation at B.

down on a bench or other support to mount the tweeter. Spread an even bead of caulking material around the circle and set the tweeter in place. Leave it undisturbed until the sealant has set, usually overnight.

When the caulking material has set, invert the back panel and wire the tweeter. All you have to do is run a piece of lamp cord from the two terminals on the panel, or, if you have a speaker with no box terminals, to the full-range speaker. Note that you must wire the tweeter to a speaker that carries the full range; don't hook it to a woofer that has a choke in series with it. If the woofer is fed through a crossover network, you must find the point where the speaker line connects to the crossover network and connect the tweeter across the line at that location. There should be no crossover elements between the amplifier terminals and the tweeter.

If you cannot get the back panel out of the box without major surgery, but you can remove the grille cloth and the woofer, you can change the procedure slightly. Instead of working with the back panel out of the box, you can cut the hole and do the other preparatory work with it in the box. To wire the tweeter, you can remove the front-mounted woofer and work through its hole in the speaker board. Still another method is to connect the tweeter by means of an exterior piece of lamp cord run down the back panel to connect to the box input terminals. You will have to cut grooves under the tweeter mounting hole for leads and wire the tweeter before installation. Fill any gaps with caulking material.

Regardless of the method you use to install the tweeter, make sure you replace the damping material behind the tweeter if it was removed. It should also cover the rest of the interior of the back panel. If there was no damping material there, see Project No. 16.

When you have replaced the back panel, set the speaker in its permanent position and give it a try. You will find that minor changes in the distance to the wall or other surface behind the speaker will alter the reflection pattern. The rear mounting position diffuses the highs around the room so that they are much easier to take than a direct radiating tweeter with no control (Fig. 9-4). Reflected highs often sound more natural than direct highs, depending on the reflecting surface. If you think the tweeter isn't working, try covering it with a thick piece of felt or other damping material, and you will find that the sound has lost something. Even though it wasn't obvious before you added the P.E.T. that it was needed, you won't want to be without those upper highs.

Fig. 9-4. Radio Shack's piezo dome installed on an enclosure in the methods described in Project 30.

BONUS PROJECT 31: SUPERTWEETER ADD-ON

In this project, we will approach the problem of getting better treble by installing a small *supertweeter*, one that can be installed in the front panel of your speaker system to add an extra dimension of overtones with superb dispersion. Some audio fans question the necessity of a tweeter that can respond to frequencies to the limit of human hearing and beyond, mentioning that there are no music notes that high. The answer to that argument is that the overtones produced by each musical instrument, and by the human voice, are necessary in order for us to recognize the kind of instrument in use and for naturalness in any kind of sound. The high-frequency range is also important because of transients. The better the high-frequency response is, the better its transient response is. And a system with smooth response can reproduce transients better than one with ragged response. A system with poor transient response will exhibit a lackluster attack at the start of a sound and an echo-like decay when the signal cuts off. One reason that small supertweeters sound better than older cone tweeters is that small tweeters have *low moving mass*, and this, plus an adequate magnet, insures good transient ability. Another advantage of the small dimensions of supertweeters is their improved dispersion at high frequencies.

In the early days of high-fidelity, there was a rule of thumb that the product of the low-frequency response and the high-frequency response of the system should equal a certain number, usually about 500,000. This theory was developed to ensure perfect balance. If followed, this means that any speaker with a bass range that cuts off at 100 Hz, and many do this, should have no highs beyond 5000 Hz. Or if the bass goes down to 50 Hz, the highs can be extended to 10,000 Hz and so on. There may have been reasons to follow this theory when equipment had considerable distortion, but it no longer should apply to any stereo system that is good enough to classify even remotely as high-fidelity. After adding a supertweeter to a system that cuts off at 5000 Hz, there will be a tremendous improvement in realism even if the bass goes no lower than 100 Hz. So if you have heard those figures on frequency range balance, forget them.

If you have a speaker system with no tweeter, or with a cone tweeter that is 3″ or more in diameter, you can usually work wonders with its high-frequency response by adding a tweeter with superior range and dispersion. To choose a tweeter, look for a small radiator with a large magnet. Those tweeters with larger cones will give only fairly good dispersion at the upper end, and their transient response may be impaired by the mass of that cone. Their only advantage is in power handling at the lower reaches of the treble range, but this depends in part on the diameter of the voice coil, as well as the size of the cone. A dome tweeter can have a voice coil that is the same diameter as the radiator—1″ is the usual size—so it can often handle more power than a cone tweeter that is larger in cone diameter but smaller in voice coil dimensions.

About as important as the choice of tweeter is the way you install it. Early hi-fi systems had the speakers, including the tweeters, installed behind the speaker board. This is especially bad if the panel is thick because it makes a resonant chamber in front of the speaker that can be excited by the driver. This kind of resonance can add a peak to what is otherwise an exemplary tweeter. The ideal mounting position is flush with the front of the speaker board, but front mounting is acceptable.

Almost any speaker system will have space on the speaker board for one of these small drivers, but if your box has no room, you can install the tweeter in a separate box and set it on top of your speaker cabinet. One slight advantage to this kind of installation is that you can experiment with phasing the tweeter—moving it forward and backward on the top of the enclosure until you get the best position. Audio experts often debate whether this is worth the effort. Radio Shack supplies one supertweeter that comes with its own case for top-of-the-cabinet mounting.

After trying both methods, we came to the conclusion that most listeners would prefer to install a tweeter on the speaker board instead of placing it in a separate box. This kind of installation requires only an extra hole in the speaker board plus some changes in the crossover network, if there is one. For best performance from you supertweeter, it is important that you cut off the highs to any other speaker that can mask the superior qualities of the highs of a supertweeter. This means adding a crossover network to a full-range speaker or converting the tweeter in a two-way system to a midrange, or upper midrange and low treble range, driver by putting a choke in series with it. You will probably want to use a control on the tweeter. For full instructions on how to make these crossover changes, Chapters 4 and 5 deal with crossover networks.

Construction

Inspect your speaker cabinet to see if you can easily remove the rear panel. If it is permanently installed, take off the grille cloth and see if the woofer is front mounted. If it is, you can probably make the wiring changes by removing the woofer and working through the large hole in the speaker board. If the woofer is installed behind the speaker board—that is, on the interior of the board—you must remove the rear panel. For procedure on this, check Project No. 16 in Chapter 6.

After removing the grille cloth, you can get an idea of the best location for your supertweeter. Ideally, the drivers should be installed in a vertical line, but this is rarely possible because of lack of space. The likely location will be off to one side of the vertical center line between the drivers already on the board. To help you locate the best position, find a can which has the same diameter as the required hole for the tweeter and move it about on the board (Fig. 9-5). With the can as a pattern, you can see immediately how much board will be left when the tweeter hole is made. As a rule of

Fig. 9-5. Use a can with the same diameter as the required hole to locate position of supertweeter on speaker board.

thumb, there should be at least 1-½″ of board on each side of the hole to avoid weakening the panel too much.

When you have decided on a good location, draw a circle by using the can as a guide. Cut out the circle with a circle-cutting saw or a sabre saw (Fig. 9-6). This hole will usually be quite small—about 2-½″ in diameter or less—but for some tweeters it should be elliptical to provide space for the terminals on the tweeter frame. To insure a good fit for elliptical holes, make a cardboard pattern first and use it to mark the required cutout on the board. One tweeter that needs this kind of mounting hole is the Radio Shack soft dome model, which requires an oval hole that measures 3-¼″ in the long dimension and only 2-½″ across the short direction.

If you are planning to wire the tweeter without removing the rear panel, connect the leads before mounting the tweeter on the board. In fact, it's a good idea to complete the wiring before installing either the tweeter or replacing the woofer. If the back is removable, it makes no difference whether you wire the tweeter first or last.

Many tweeters are sold with an attached capacitor that serves as a high-pass filter designed to roll off the lows at 6 dB per octave below some appropriate frequency. A typical value for this capacitor is 4 μF, which provides a crossover frequency of 5000 Hz for an 8-ohm tweeter. You may want to substitute another crossover network, such as one of those outlined in Project No. 15, one which provides better protection for the tweeter with a sharper cutoff filter. Following the plan in Project No. 15, you would use the same crossover point as suggested by the value of the capacitor installed on the tweeter; however, to make the network into a 12 dB-per-octave filter, you would have to change the value of that capacitor, as well as add a choke. Here is another possibility: Leave the capacitor in place and use a choke which would give a lower crossover frequency. This would save buying another capacitor and installing it, and the reduced crossover frequency would be acceptable because of the sharper cutoff rate with the added choke across the tweeter. This procedure might not be advisable if you drive your speakers to the limit, but it will work well in most cases.

For example, the Radio Shack soft dome tweeter comes with a 4-μF capacitor attached. Simply adding a 0.5-mH choke in parallel to the tweeter would reduce the crossover frequency of that tweeter down to about 3500 Hz, but the cutoff would be sharper. To

Fig. 9-6. Cut out a hole in the speaker board with a sabre saw or circle-cutting saw.

see how this works, check Chapters 4 and 5. For maximum tweeter protection, you would use Table 5-3 or Table 5-4 to design a sharp cutoff filter that is effective at 5000 Hz, the original crossover frequency. For this particular tweeter, that would mean a 2.8-μF capacitor and a 0.36-mH choke in a 12 dB-per-octave filter.

Regardless of the crossover frequency and the sharpness of the filter you choose, you will probably want to put a control on the tweeter. This means an L-pad, typically an 8-ohm L-pad. While purists talk about a permanently set balance for tweeters and other drivers, in the practical world it's a good idea to have a variable control that permits you to match the output of a tweeter to that of the rest of the system in the acoustical environment where it is used. The most useful place to put the L-pad in most speakers is on the back panel because it will be out of sight there. In some cases, it can be installed on the speaker board. If the panels aren't too thick, the control can be installed by drilling a mounting hole in the panel. If the panels are ¾″ thick, cut a small window out of the panel and glue a piece of ¼″ tempered hardboard over the window to hold the L-pad. Typically, the L-pad appears in the circuit as the last element before you run wires to the tweeter, but it is possible to place the L-pad before the crossover capacitor, if necessary. Just make sure one side of the tweeter line goes to pin No. 1 and the other side goes to pin No. 2. Note that the input goes to pins No. 1 and No. 3. If you get the wiring right, the tweeter output will be reduced by twisting the control knob counterclockwise.

When you have finished the tweeter installation, set your speaker in its usual performing position and give it a trial. Turn the control knob to the extreme counterclockwise position. Then bring up the tweeter level until the sound blends with that of the other speakers. When the sound appears to come from a single speaker in each cabinet, you have it right.

This is a project that can put the final touch on an otherwise satisfactory sound system. A good supertweeter adds the airy quality of silky smooth and well dispersed highs, giving life to the music.

10 Stereo Equipment Cabinets

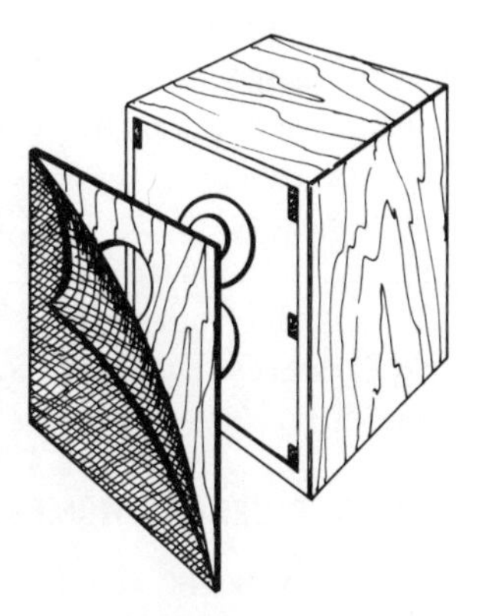

You may be surprised to find a discussion of equipment cabinets in a book of stereo improvement projects. An equipment cabinet is usually considered a completely passive member of a stereo system, contributing nothing or taking nothing from the ability of the active members of the stereo chain to produce a satisfactory performance. That this is not necessarily true was brought home to me recently by some friends' special problem with a stereo system.

The trouble began after they moved into a small apartment in a neighboring state. They called me by long distance to report that the stereo rig was misbehaving. I asked for symptoms. They said, "The speakers are vibrating." Further enquiry revealed that by "vibrating," my friends meant that the woofers were undergoing excessive excursion because of some kind of instability. Long distance diagnosis being what it is, we discussed several possible reasons for the trouble before I remembered that these people had no equipment cabinet. I asked if the problem occurred only while using their turntable. It did. And where was the turntable now located? It was on one of the speaker cabinets. I suggested that they move it to another piece of furniture and see if that cured the ailment. They did, and it did.

When I later talked to my friends about their trouble with acoustic feedback, they said it had never occurred to them to isolate the various components from the speakers because their parents had owned a department store stereo set that had all the components, *including the speakers,* in a single cabinet. Such all-in-one stereo sets usually have a limited low-frequency range in both the speakers and the phono cartridge, so acoustic feedback isn't a big problem. Although such sets give the effect of adequate bass to the casual listener, it is only because of false resonances and frequency doubling. If one must install other components on speaker cabinets or in any structure that makes contact with the speaker cabinets, there is a danger of acoustic feedback. One solution is to mount the turntable, or other component that is picking up the vibration, on a foam support. Another possibility is to suspend it from springs. But there is an easier and better way to install stereo components: put them in a separate equipment cabinet.

BONUS PROJECT 32: HOMEMADE EQUIPMENT CABINET

You can find ready-made equipment cabinets, but you can make a cabinet out of particle board that costs less than one you buy. The easy way to finish the particle board is to paint it flat black. That hides any minor mistakes you may make and is a good choice if you have nothing more than hand tools for the job. But even with hand tools and with patience, you can build a beautiful birch cabinet that will grace any living room.

In planning an equipment cabinet to match your needs, you may decide to custom tailor it for the equipment you now own. This would make an attractive system with no gaping extra spaces, but it might be shortsighted if you are a typical audio fan who trades equipment every few years. When you trade up or add something, our tightly planned cabinet might have to be junked.

Another consideration in planning a cabinet is to position the equipment for easy use. Don't place items that require careful adjustment at floor level. A receiver placed there, for example, would require you to stoop, squat or even get down on your knees every time you tuned it. The bottom space in your cabinet is a good place to store records. By allowing about 15″ or more vertical shelf space there, you will raise the other components high enough for accessibility without stooping. The top of the cabinet is a logical place for a record turntable because records should be handled with care, and it is easier to place a record on a turntable if there are no low overhanging shelves above the turntable.

Sometimes you must reach a compromise between logical component placement and attractiveness. If you think only of

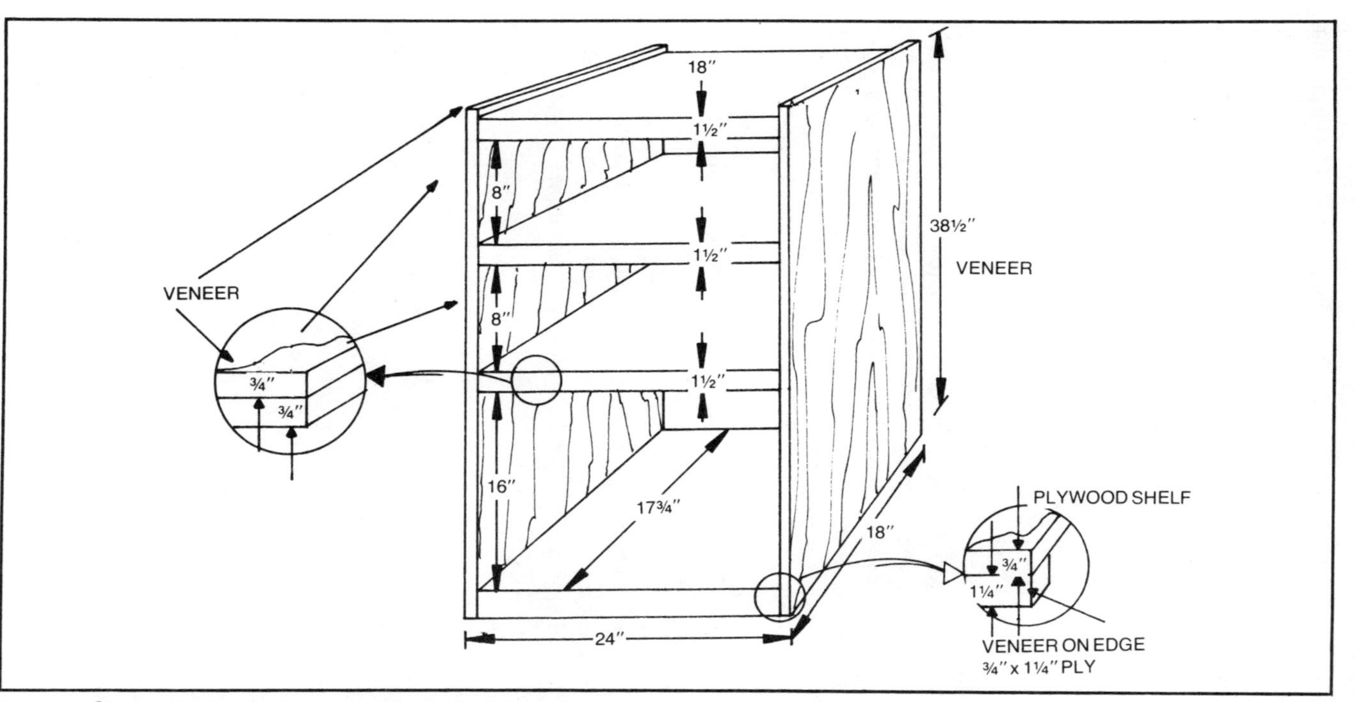

Fig. 10-1. Construction details for universal equipment cabinet.

accessibility, you might end up with an ugly piece of furniture that other members of the family will reject. Bring them into the planning. When planning proportions, consider the rule of the *golden mean*. According to the rule, the height-to-width ratio should equal about 1.62. This ratio was used in the plan shown in Fig. 10-1 for universal appeal.

The design shown here is quite universal in the sense that it will take almost any brand of turntable, receiver and tape deck. Unless you go for something very expensive or unusual, you can consider it completely universal. Some components that sell for about $2000 each, for example, will not fit in the shelves. This brings up a point about some brands of stereo equipment that make no sense. On a visit to a receiver factory a few years ago, I was told that the next year's models would be considerably larger in size because surveys had shown that people would pay more for a larger receiver even if it were electronically identical to a small one. If you have a special reason for using components that are too large to fit in this cabinet, you can either enlarge the cabinet to fit them, or if you have the money available, hire a cabinet maker to do it.

A more likely reason for needing additional cabinet space is that you might want to install an equalizer or other special equipment. In that case, you can add 9-½" to the height of the cabinet by lengthening the sides and adding an extra shelf. This will give an extra 8" x 22-½" x 18" compartment. Or if you have plenty of record storage in other locations, you can utilize some of the bottom shelf for the extra components.

Construction

Because a birch plywood cabinet requires more care in construction, we will describe the construction procedure with plywood. Cut out the parts to the dimensions in the parts list shown in Table 10-1. Label all the parts. The marking of each should be done in a place where it won't show later or by adhesive-backed paper labels that can later be pulled off. In labeling the parts, make sure you indicate which edge of each is forward, for the shelves and sides, or at the top, for the sides. Cover these raw edges by gluing a strip of ¾" wood veneer on them (Fig. 10-2). Some wood veneer comes with glue already installed on the back of the veneer. If you have that kind, check the directions with the veneer to see if it is pressure-sensitive or if you must use heat to make a satisfactory bond. For veneer with no adhesive, use contact cement. It makes a quick bond, and the excess can be rubbed off with no harm to the

Table 10-1. Parts List for Universal Equipment Cabinet.

Part	No.	Material	Dimensions (In.)
Sides	2	¾″ Plywood	18 x 38-½
Shelves	4	¾″ Plywood	17-¾ x 22-½
Back	1	¼″ Plywood	23-¼ x 37-¾
Rails	3	¾″ Plywood	¾ x 22-½
Bottom Rail	1	¾″ Plywood	1-¼ x 22-½

wood. Wood glue, if spilled on the visible surfaces, will seal the wood against further staining or varnishing. Because the glue is a liquid, it is hard to keep from spoiling at least a few spots with it, spots that will stand out because of their refusal to accept stain.

Next, dado a ¼″ wide by ⅜″ deep groove along the inside rear edge of the sides and the top to receive the back panel. If you are building a particle board cabinet, eliminate this step and use ⅛″ hardboard for the back panel. The panel won't extend back far enough to cause problems even though it is mounted on the back surface without grooves.

Begin assembly by gluing the ¾″ x ¾″ rails to the lower front edge of the upper 3 shelves and the ¾″ x 1-¼″ rail to the lower front

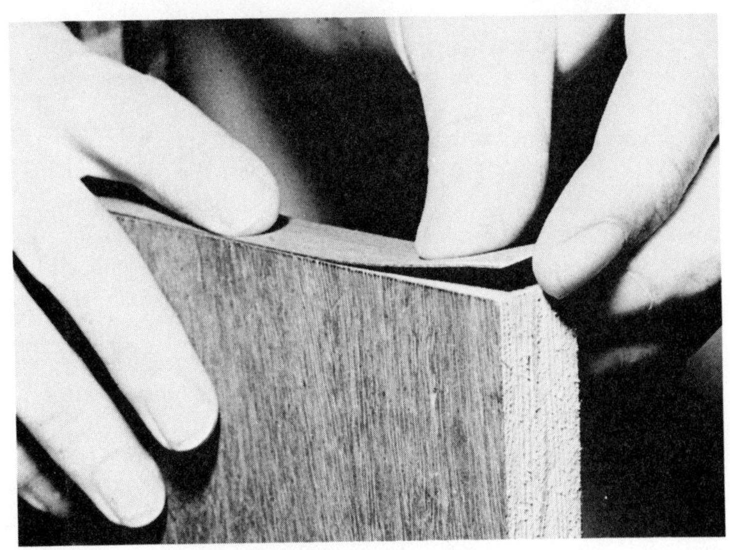

Fig. 10-2. Use ¾″ veneer to cover raw edges of hardwood plywood.

edge of the bottom shelf. Take care to glue a raw edge of plywood to the bottom of each shelf so the front surface of the rail will be finished plywood. It isn't necessary to use veneer on the bottom edge of the rails because the rails will not be noticeable to anyone standing or sitting in a chair, especially if they are stained with a dark stain.

Mark lines on the inside surfaces of the sides to help locate the shelves. Lay one side on your workbench with the exterior side up. Measure the center lines of each shelf position carefully and mark it lightly with a pencil. Start several 4-penny finishing nails along each of these lines, evenly spaced. If rather small nails are used in assembly, they can be hidden more easily. The strength of the cabinet comes from proper gluing, not the nails. Follow the same nailing procedure for the other side.

Now set one of the sides over a shelf, making sure you have the upper and bottom shelves identified so they will go in the proper location. Drive two nails in lightly to correctly position the shelf under the side. Do this for each shelf on both sides, laying the pieces down so they will not be mixed.

Smear a bead of wood glue along each matching surface at one side and one end of the shelves. Use the glue sparingly so it will not run across the shelves and stain them. Set the side over the end of one shelf and drive the nails home, starting with the two pre-set nails. Follow this procedure for each shelf. Then invert the assembled shelves and side. Some help is advisable for this step to protect the glued and nailed joints. Glue and nail ¾" x ¾" corner blocks under the end of each shelf. These glue blocks should extend from behind the shelf rail to the back end of the shelf. Quickly wipe off any excess glue with a slightly dampened cloth.

Set the back panel in place and lightly mark the position of the two middle shelves on it. Lay the panel on a solid surface, one that you do not mind drilling into, and drill some access holes for AC cords and audio cables. These holes should be large enough to accommodate AC plugs and should be located just above the position of the two middle shelves.

Use a fine sandpaper to remove any marks or roughness left on the birch plywood. Stain the wood to match your other furniture. When the stain has dried, apply a final finish of satin varnish. It's a good idea to use a single brand line of finishes, from stain to final coat, for guaranteed compatibility. Satin varnish has several advantages over a gloss finish. It hides imperfections, looks rich and needs no rubbing down after the final coat.

Your cabinet is now ready to use. Put your turntable on top of the cabinet and arrange your other components to give easy access to each.

AN EQUIPMENT CABINET FROM SCRAPS

Here we introduce a project of Robert Clifton, an audio fan who built his own equipment cabinet from scraps of hardwood (Fig. 10-3). The wood had been discarded by a furniture factory because the pieces had flaws in them or were too short or narrow for their usual purpose. Clifton obtained the wood from a pile at the city dump in a small Midwestern town. Many people of the area had hauled truckloads of the wood home to use in wood stoves and fireplaces, but Clifton saw more than kindling in the pieces of maple and other northern hardwoods.

He planned a stereo equipment cabinet, based on the space requirements of some high-quality British audio equipment he had bought while serving in Europe several years ago as a civilian employee of the U. S. Air Force. Clifton obviously believes in buying the best equipment one can afford, and in his case it was among the best available, and sticking with it without regard to the latest fad or innovation. Because of his adherence to this doctrine, he could plan an equipment cabinet that precisely fit his equipment with no fear that the cabinet would soon be obsolete. The only components that he can easily change are the loudspeakers, which fit into open wings at each end of the cabinet. But even there he has seen no reason to change his original policy of staying with his original models.

Clifton's next move was to turn the mixed bag of various sizes and shades of wood into the cabinet he had designed. With his table saw, he ripped each board to a width of about 1″. By ripping the boards, Clifton could glue them together to make boards of any width he wanted. This method of building board width has one great structural advantage: Boards made from narrow strips of wood are much more resistant to warpage or twisting than wide boards of one piece.

To glue the boards, Clifton used a 2′ x 8′ piece of plywood as a temporary base. He attached a rail along one edge and tipped the plywood at a slight angle so that gravity would hold the boards in place against the rail until he was ready to clamp them. Before gluing, he cut the ends of the boards at a 90-degree angle on his saw so that they would fit together closely, end to end. If you do any gluing of this kind, it's a good idea to check the squareness of the board ends with a try square.

Fig. 10-3. This equipment cabinet was built from free scraps of hardwood.

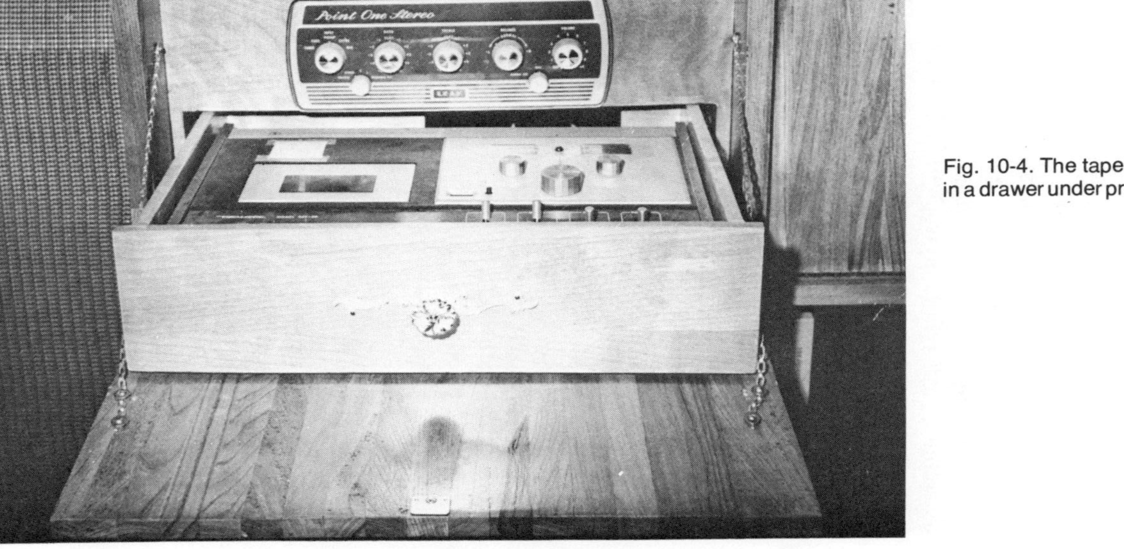

Fig. 10-4. The tape deck is installed in a drawer under pre-amp.

In placing the narrow boards on the plywood base, Clifton gave no attention to pattern. Instead, he picked up each board as he came to it, allowing random choice to make the pattern of various shades of wood. He glued the board and then set it in place against those he had already glued. After he had glued enough boards to make a piece about 1′ wide and 8′ long, he clamped the boards with pipe clamps, the kind of furniture clamps used by carpenters. The clamps were pressed tightly against the rail at the lower edge of the plywood base and against a long, straight clamping board which he placed against the upper edge of the glued-up pieces.

In making lumber from small boards, one could at this point leave the boards on the plywood until the glue had set or lift it off and install a new rail so that another board could be glued. Using either method, one would later continue the gluing process until the board was as wide as desirable.

After making several large boards, Clifton took them to a planning mill and had the boards planed. Now he had solid boards, about 2′ x 8′, which were planed like commercial lumber. He cut the boards into the sizes required by his cabinet plan and constructed the cabinet, using the glued lumber for every part except for some drawer fronts and an equipment mounting panel. These he made from ¾″ white birch plywood.

Two drop leaf doors in the center top section of the cabinet hide the electronics component area. These doors are mounted on sections of piano hinge at the bottom edges and drop down to make a horizontal panel where they are suspended by brass chains from each side. Clifton's FM-AM tuner and preamplifier are located in the upper left of this central area, and a tape deck is directly below them (Fig. 10-4). To gain access to the tape deck, he mounted it in a drawer. The turntable is also mounted in a drawer to the right of the tuner and preamp. Both drawers are installed with heavy-duty slides which work smoothly and with ease. They hold the drawers firmly in place during operation so that vibration is no problem. The power amplifier is installed below the drawer that holds the turntable. The space under the two central drop leaf doors is used for record storage (Fig. 10-5).

Clifton's cabinet is presented here as an example of what can be done at extremely low cost if you use a bit of imagination. While most of the people who watched the furniture factory dump scraps of wood saw nothing more than kindling, Clifton saw an attractive building material.

Fig. 10-5. Except for white birch in component panel and drawer fronts, Rober Clifton built this equipment cabinet from narrow boards that had been glued together.

EQUIPMENT CABINETS FROM OLD FURNITURE

If you don't want to build an equipment cabinet from scratch, you can easily adapt one from an old buffet, radio or TV cabinet. Some audio fans have used old cabinets made for acoustic

Fig. 10-6. Old phonograph cabinets can be used for stereo equipment but these are getting to be expensive.

phonographs in the pre-electronic age of the World War I era (Fig. 10-6). Those items are now hard to find and often very expensive.

Before you shop for old furniture that can be converted to equipment cabinet use, measure your stereo components and make a list of their dimensions. If you don't have the components yet, obtain the measurements of the ones you hope to buy from catalogs or brochures.

When you start to look for old furniture, take a tape measure with you. It will be more useful in measuring large pieces of furniture than the usual ruler or yardstick that you can borrow at a garage sale or secondhand store. Stop and think about how your components would fit into any cabinet you find before putting down your money. How much reworking of the piece will be required? Where will the cables run? Can the cables be hidden? Do your components need more space for easy use than their bare measurements show? And is there space in the piece for tape and record storage?

To get a cabinet at the lowest possible cost, try any church thrift shop in your area or the general stores of used merchandise operated by such nonprofit organizations as the Salvation Army. Watch for yearly auctions held in many communities by charitable organizations. Or visit garage sales and flea markets. If you go to used furniture stores, try the most run-down stores you can find.

When picking up used furniture buys, look for structural soundness but disregard scratches or other minor imperfections that can be eliminated by refinishing or hidden by paint. An example of the kind of adaptation that can be done, one audio fan found a buffet priced at only $20 because it was in bad condition. The top surface was especially bad, and the interior had been the home of several generations of mice. He bought the piece and completely removed the top finish, sanding the surface until the bare oak wood was exposed. He applied a coat of stain, followed by satin varnish to the top, but he limited his work on the rest of the buffet to the judicious use of stain and scratch remover. At that point he had a useful buffet (Fig. 10-7), but he then decided to adapt it to a stereo components and record storage cabinet.

His next step was to rework the interior of the cabinet. He removed the drawers and the slides and then covered the rough surface under the drawers with tempered hardboard to make a smooth shelf bottom. After dismantling the old drawers, he salvaged the fronts and installed them with cabinet door catches to cover the upper shelf spaces when the stereo equipment was not in

Fig. 10-7. An old buffet that the buyer restored with a refinished top.

use. Vertical slats were installed in the lower compartments to make space for record storage. Such slats, or dowel posts, are desirable to prevent the warp that can occur when records are left leaning against one side of the cabinet. This cabinet was put into use with only a turntable and a receiver, plus the record storage (Fig. 10-8), but there is an extra equipment compartment in the upper left of the cabinet for any future additions to the owner's stereo component stable.

In looking for old pieces of furniture that you can turn into stereo cabinets, try to find out what kind of wood was used in the piece and if it is solid or veneer. If veneer was used, check to see that it is tight and that there are no splintery corners that will show. Such faults reduce the value of a cabinet, but if the price is right, you can rework one of these by removing the loose pieces of veneer and filling the depressions with wood putty. You would then have to use paint or some other solid covering. Some audio fans have used such unlikely materials as wallpaper or glued on burlap to restore a cabinet in really bad condition.

One of the worst problems in cheap cabinets is numerous coats of thick paint over a hardwood base. It is a problem if you are tempted to restore the original hardwood finish because thick paint can be removed only by tedious work and many applications of paint remover or by a quick but risky sanding process. The sanding work can be done by a rough sandpaper installed on a disc in your electric drill. Unless you are careful, however, you will put grooves in the wood that cannot be easily removed by later fine sanding. The paint remover path is more sure. Have plenty of steel wool and good wire brush handy to clean the grooves and get into the corners. And do the work in a well ventilated place. You must remove all the paint remover with warm sudsy water before refinishing the wood.

ANTIQUING OLD CABINETS

Instead of removing thick paint or covering bad gouges with ordinary paint, antiquing makes sense for many old cabinets. When you antique a problem piece, you don't have to remove the old finish or even fill the holes unless they are large. Small holes or cracks only add to the antique effect.

The only materials you need are an antiquing kit, some paint thinner, a good paint brush and some clean cloths. You can even use a powder puff to apply the antique finish if you don't have a suitable brush. In choosing an antiquing kit, be aware of the latex materials that are easy to apply and dry more quickly than oil-based paints. A finish that dries quickly saves time and accumulates fewer dust particles.

Remove all the hardware from your cabinet and wash the surface with mineral spirits to remove the accumulated furniture polishes and other oil or grease. If the surface is too shiny, it should be lightly sanded for the antiquing paint to adhere to it.

Fig. 10-8. The same buffet converted to use as a stereo equipment and record storage cabinet.

Apply the base coat by moving your brush *with* the grain, then *across* the grain, and finally *with* the grain in long, light strokes. Follow the base coat with the glaze coat. Aim at making the large central areas of each panel lighter in tone than the edges. Allow plenty of glaze to fill in depressions such as grooves, carvings or crevices next to moldings. Wipe the glaze with a clean cloth, starting at the center of a panel and leaving it unwiped, or barely wiped, near the edges. Try to make the change from light to dark areas a gradual one by dabbing at the glaze in places where it is too thick. Wipe the high spots of carved sections clean, leaving the glaze in the deep grooves, where it will show off the pattern in bold relief. If the change in glaze thickness is too abrupt, use a clean brush to remove as much shading as you desire. The brush can also be used to lightly blend streaks. You can make minor corrections in the glaze even after it is dried by rubbing dark sections with steel wool.

Even if you have never done any antiquing, you will find it easy. One audio fan who antiqued several cabinets to house his stereo equipment and that of other members of his family told me that he had used various methods to antique the pieces he had reworked and all turned out well. Unlike a fine new furniture finish, it's almost impossible to do a bad job. That makes antiquing a perfect choice for old pieces of furniture in bad condition.

UNCONVENTIONAL EQUIPMENT CABINETS

The universal cabinet (Project No. 32) and the adaptation of old furniture follow more or less conventional patterns in equipment cabinetry. If you like something different, you can design your own equipment cabinet and make it do double duty in an unlikely way, such as serving as a coffee table or chairside table. The advantage of this method of housing your equipment is that you can use space that would otherwise be wasted.

One precaution to take in such designs is to allow enough ventilation. Modern solid-state equipment needs much less ventilation than old tube-type amplifiers, but the heat sinks on output transistors are obvious reminders that, while heat buildup is less, transistors don't like too much heat. The ventilation holes need not be conspicuous if you plan for them as you design the piece (Fig. 10-9).

If you have a closet near your listening room, you can probably find sufficient shelf space there for your components. To keep your speaker lead lengths within the reasonable bounds, you might have

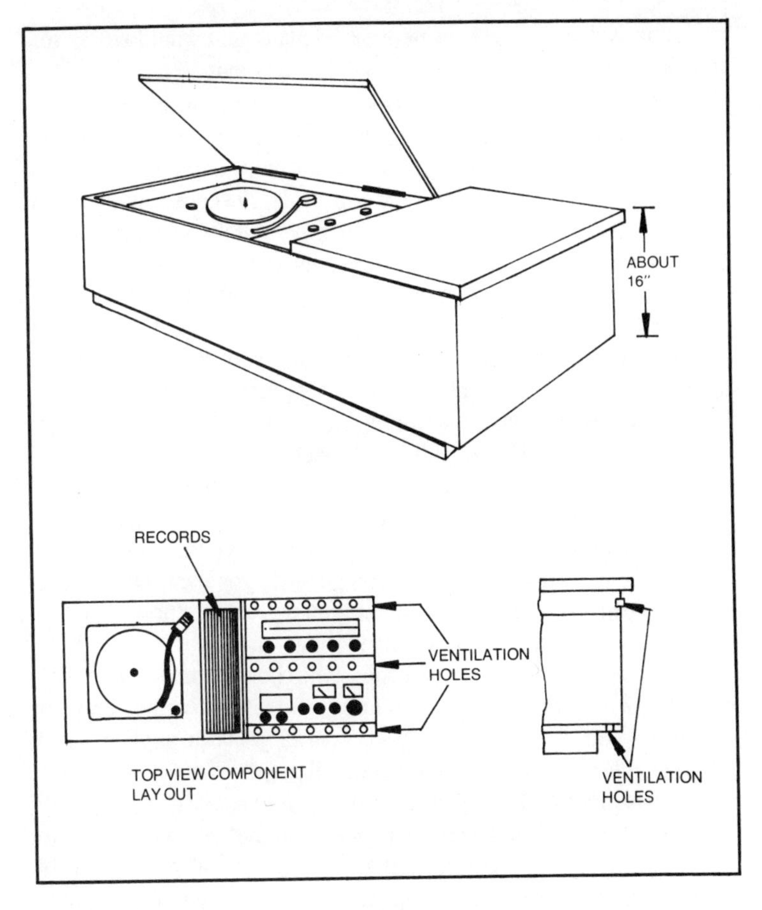

Fig. 10-9. Dual-purpose coffee table and equipment cabinet.

to place the speakers in the end of the room nearest the closet, regardless of the acoustical virtues of such placement. Another disadvantage is that you have to go into the closet each time you change a recording or tune your receiver.

THE ULTIMATE CABINET: A MUSIC WALL

If you own your home, you can add considerable value to it by installing a music wall. This is a project that must be custom designed to fit the available space and decor of your home. Because each installation requires special consideration of such factors as available wall space, presence of windows or other obstacles and even accessibility of house wiring outlets, we will make suggestions instead of giving a set of detailed plans that would work well only in some hypothetical house that doesn't exist.

The typical music wall consists of a lower counter, about 36" high with various storage spaces under it, and upper book shelves that cover the wall. The counter top normally extends about 24" from the wall to the front edge and will hold large components, such as turntables. Smaller components can be installed in the cabinet beneath the counter or, if deep shelves are installed above the counter, in the bookcase. For many installations, this above-the-counter bookcase will be relatively shallow. The choice depends on the number of components you own and whether it is desirable to mount any of them above the counter. In some cases, all components will be installed under the counter with such items as tape decks or turntables in pull-out drawers.

As in any installation, there should be some provision for ventilation of components. The amount of ventilation required by a system depends in part on the size of the air volume around the equipment as well as the kind of components you own. Obviously, if you have any tube-type equipment, you will have to provide more ventilation holes than if everything is solid state.

The kind of building materials you choose for a music wall depends on the kind of tools you have available as well as on your taste in wood. If you have a power saw, you can apply Robert Clifton's method of gluing together small pieces to make the panels for your cabinets. But if you have only a few hand tools, you will probably choose large sheets of plywood so that you won't have to carefully join pieces. Birch plywood is the typical preference of cabinet makers, who often stain it to match the color of various other woods, such as walnut. In solid lumber, pine is relatively inexpensive and easy to work. Its main disadvantage is its tendency to warp with changing room temperature and humidity.

In planning a music wall, try for verstility. One way to make your cabinets more timeless is to use adjustable shelving instead of permanently installed shelves. Another way is to plan the individual compartments in the wall so that each one will accommodate a larger piece of equipment than the one you now plan to install there.

In building the cabinets, don't forget to install lead-in wire for your tuner or receiver and lamp cord for speaker cables. If the cabinets cover a wall of your listening room, you might want to install the speakers in them. There is always the chance of acoustic feedback in this arrangement, but you can usually eliminate that by acoustically isolating either the speakers or the turntable on foam pads or on loose springs. Like many other aspects of stereo fine tuning, this can best be done by trial and error.

Above all, use your imagination in planning a music wall. Don't stop at the first ideas you get, but make several plans. You will usually find that the more plans you try, the more original are your ideas. Make rough sketches for initial plans, but make them to scale so you can judge the proportions of various parts of the wall. After you have developed several fairly complete plans, set them aside for a week or more to allow yourself time to cool to them and assess them more objectively. Then get out your ideas and inspect them; if you still like one of them, go ahead.

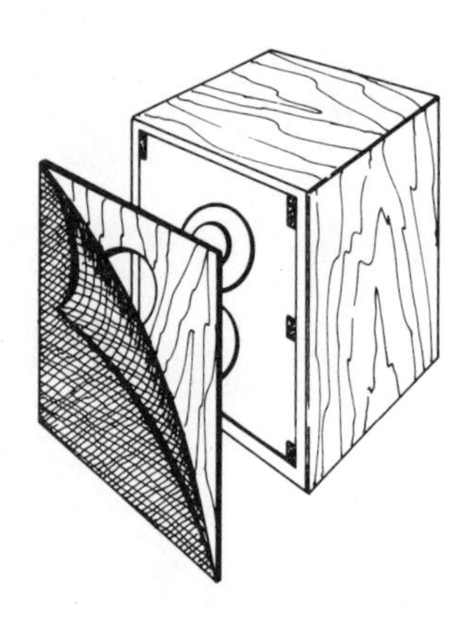

Appendix A
Wire Data

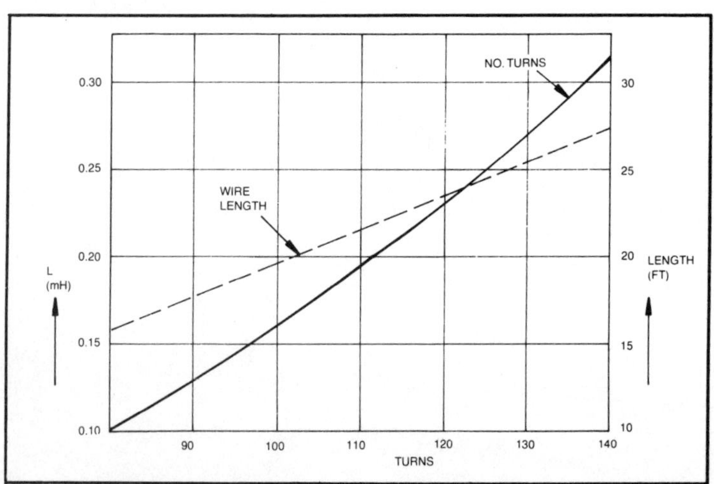

Fig. A-1. Wire length and turns for coils with an inductance from 0.1 mH to 0.32 mH. Use No. 24 gauge wire on a ½" core. Find the desired value of inductance along the left margin, then find the number of turns from the point where the horizontal line from the left margin crosses the solid line. To find wire length, move vertically from the point located previously on the solid line to the dashed line, then to the right margin.

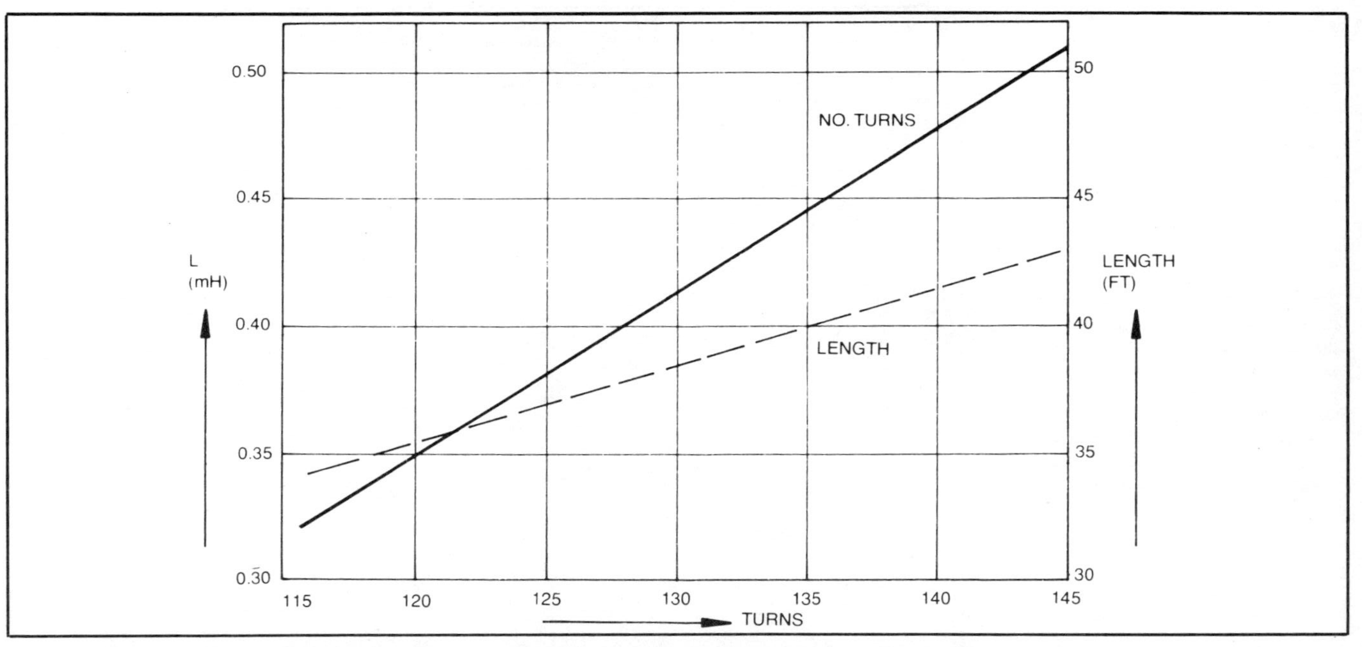

Fig. A-2. Wire date for coils with inductance from 0.33 mH to 0.51 mH. Use No. 22 gauge wire on 3/4". Find the desired value of inductance along the left margin, then find the number of turns from the point where the horizontal line from the left margin crosses the solid line. To find wire length, move vertically from the point located previously on the solid line to the dashed line, then to the right margin.

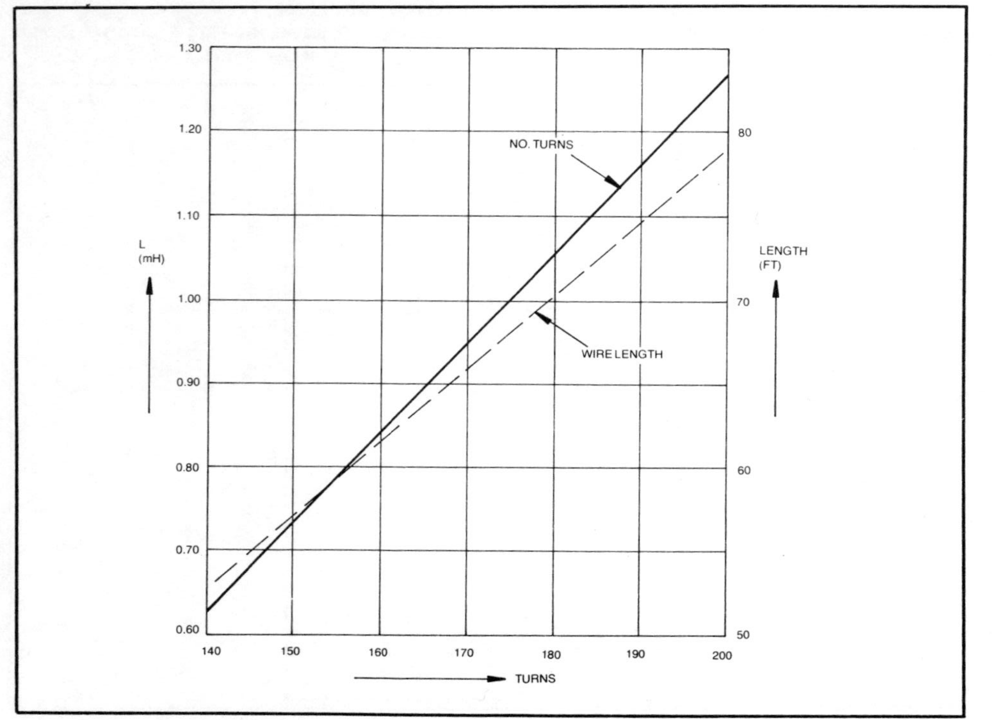

Fig. A-3. Wire data for coils with inductance from 0.60 mH to 1.28 mH. Find the desired value of inductance along the left margin, then find the number of turns from the point where the horizontal line from the left margin crosses the solid line. To find the wire length, move vertically from the point. Use No. 20 gauge wire on a 1″ core.

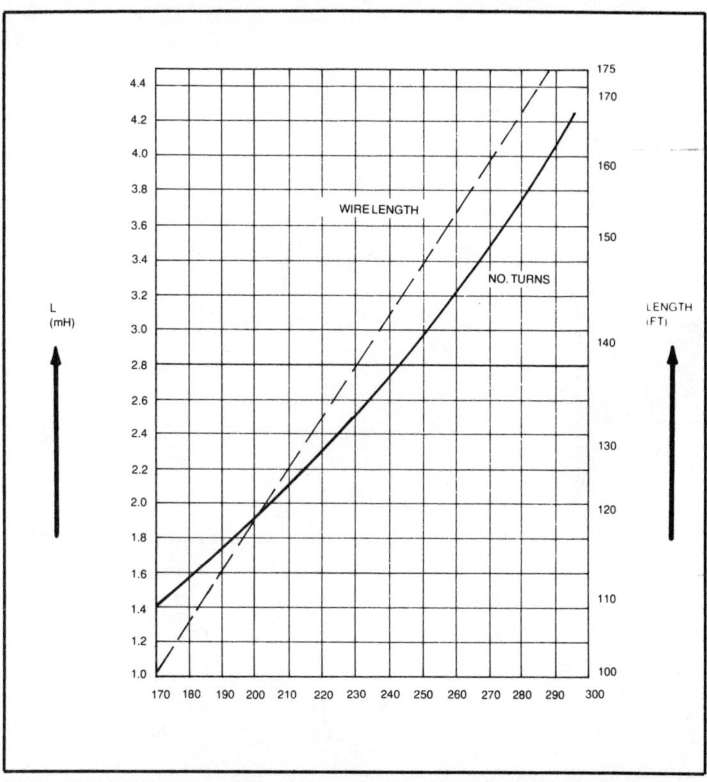

Fig. A-4. Wire data for coils with inductance from 1.40 to 4.3 mH. Find the desired value of inductance along the left margin, then find the number of turns from the point where the horizontal line from the left margin crosses the solid line. To find the wire length, move vertically from the point located previously on the solid line to the dashed line, then to the right margin. Use No. 18 gauge wire on a 1-½″ core.

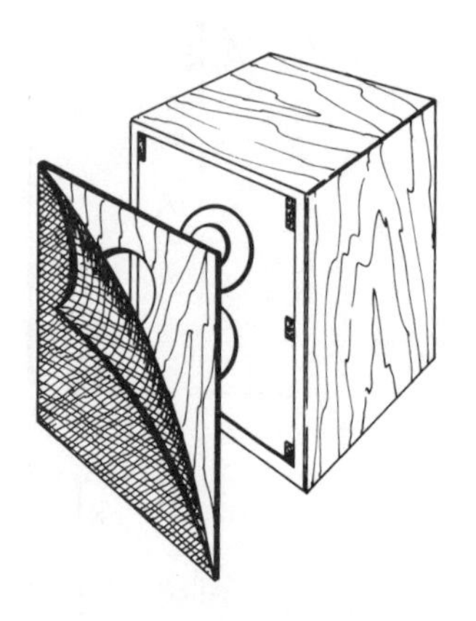

Appendix B
Parts Suppliers

Here are sources of electronic supplies, speakers and components. Some of these suppliers will send you a free catalog even though it carries a price tag of a dollar or so. If you write and ask for one, you will receive either a catalog or a notice of its price.

Gladstone Electronics
1736 Avenue Rd.
Toronto, ON M5M 3Y7

Wide range of high-fidelity speakers, including Altec, Celestion, KEF, Norelco, Peerless and others. Catalog.

McGee Radio Co.
1901 McGee St.
Kansas City, MO 64108

Extremely wide variety of speakers plus passive radiators. Some crossover networks and components. Catalog.

Olson Electronics
260 S. Forge St.
Akron, OH 44327

Line of house brand speakers. Catalog.

PREP'S
2635 Noble Rd.
Cleveland, OH 44121

Line of speakers, crossover components, enclosures, etc. Catalog.

Speakerlab
Department BRE
735 Northlake Way
Seattle, WA 98103

Wide variety of speaker system kits and separate drivers plus crossover components, such as capacitors and coils. Also various low-priced booklets. Catalog.

SRC Audio
3238 Towerwood Dr.
Dallas, TX 75324

A variety of name brand speakers such as Altec, Electro-Voice, CTS, Heppner, Panasonic, Peerless, Phillips, Pioneer and Polydax. Catalog.

Index

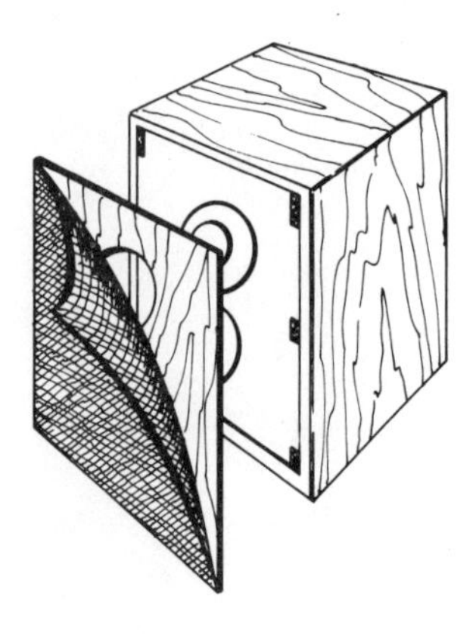

Index

T

U

V

W

Y

No. 1284
$16.95

DESIGNING AND BUILDING YOUR OWN STEREO FURNITURE

BY CARL W. SPENCER

FIRST EDITION

FIRST PRINTING

MARCH 1981

Printed in the United States of America

Library of Congress Cataloging in Publication Data

Spencer, Carl W
Designing and building your own stereo furniture.

Includes index.
1. Cabinet-work. 2.Loud-speaker cabinets.
3. Stereophonic sound systems—Amateurs' manuals.
I. Title.
TT197.S67 621.389'334 80-28365
ISBN 0-8306-9634-2
ISBN 0-8306-1284-X (pbk.)

Dedication

This book is dedicated to my wife,
Dorothea Ann Spencer.
Thank you for your patience and encouragement.
Thank you for the loving criticism.
Thank you for eating stereo furniture for
breakfast, lunch, and supper.
Thank you for being able to type 80 words per minute.
You are truly a Gift of God.

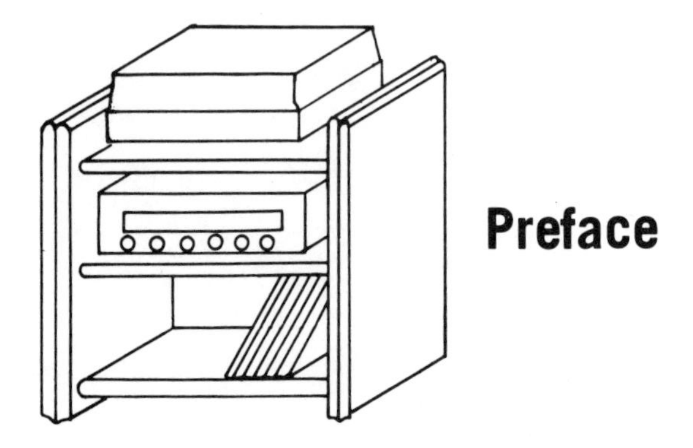

Preface

This book is for all component stereo owners who know that there has to be something better than pine boards and decorative blocks or a coffee table for mounting electronic equipment. This book is also for the thousands of consumers who would probably get more enjoyment from their sound system if they could only conveniently get to it.

When I was younger, my family had a rather nice hi-fi for the times. But in order to play the thing, one had to squeeze down beside the arm of an easy chair, angle his body a little sideways, open a cabinet door, bang his head against either the easy chair or the edge of the cabinet, and precariously balance on his toes just to put on a record. And once he got down, he became trapped and couldn't get back up easily. A thoroughly discouraging experience. Familiar?

This book is for interior designers and stereo systems installers whose professional reputation may someday depend upon how successfully they can hide wires, eliminate dust accumulations, provide good looking convenience, and allow for future changes—all at the same time.

This book is for do-it-yourselfers and wood shop students in search of a worthwhile and unique project to showcase their skills.

This book is for the stereo furniture industry, to demonstrate a potential for a truly specialized and evolving product superior to more shelving units. I have attempted to codify the standards and principles of the industry, both those that I developed myself and those I learned by studying the work of the industry pioneers.

There was a day when I wanted to find out all about building stereo furniture. The only printed material on the subject available at the time this project was undertaken was a vague chapter here and there in stereo installation books.

I was encouraged by Mr. Anthony R. Curtis of TAB BOOKS Inc. to try my hand at writing this book. It is perhaps appropriate here to mention that I am thankful to have had a very understanding publisher. Mr. Curtis patiently answered each one of my many and lengthy questions. Our correspondence alone would probably fill another whole volume.

In no way am I deluded that this book contains every aspect of designing and building stereo furniture. I see this work merely as a start—a foundation to build on, both by myself and others.

This book is hardly original information. Many, many people made contributions, both directly and indirectly. The principal benefactors were, and continue to be, my customers. Most of the really good ideas I have developed were the direct result of special problems and some clever solutions with which they presented me.

Jack Benveniste, as chief designer for Barzilay Company, is perhaps the patriarch of the stereo furniture industry. He kindly invited me to his plant for a visit and took the time to dig through old records and photos and to show me photos of new designs not yet released. I got a good rundown on the past and present of production stereo furniture, along with a view of the future.

Likewise, I was enlightened by James Sapp of Greenwood Forest Products (Mariani brand stereo furniture) and Maureen Tenenbaum of Nomadic Furniture in their particular niches of the industry. This is probably less than thrilling for most people, but for me or anyone interested in stereo furniture it was truly an exciting adventure.

Four key employees of my company (Hyacinth Ukwuagu, Kenneth Williams, John Wilson, Jr., and David Ream) stepped in and filled my shoes by virtually running Presidential Industries so I could finish this book. This was no easy task, but a necessary one since typing makes me too cranky to deal with the public, and in-shop questions invariably lose detail when yelled through a darkroom door. I am very lucky to employ such resourceful people.

There is one more major contributor who should not be neglected. While I did all of the camera work and most of the darkroom duties, Jerry William, a friend and neighbor, pinch hit for me several times in the darkroom with very professional results.

This book is not yet really finished. I am still learning. I am still getting customers with altogether new and different problems. The reader will undoubtably discover better solutions and techniques than I have presented as he gets more into designing and building stereo furniture. Please write me with your suggestions and criticisms to help make this book even better in the future.

Carl W. Spencer

Contents

Introduction to Stereo Furniture

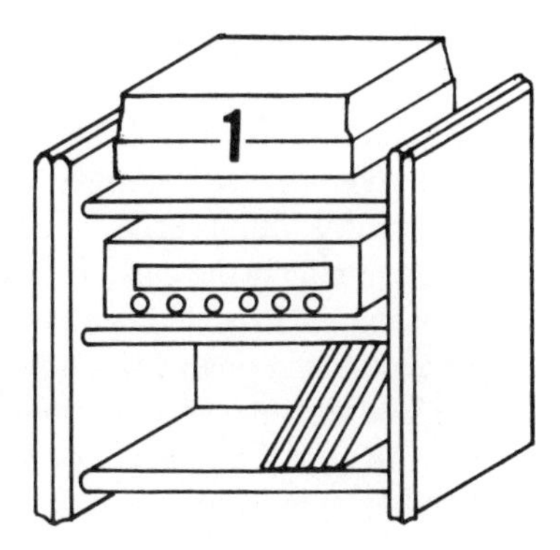

Man has built cases for his tools and possessions almost from the beginning of his existence. Casework provided a convenient organized place to store his equipment when it was not being used, and furnished a protective covering to retard or eliminate deterioration of his most valued possessions.

In some respects, the development of casing or cabinet storage is as important a basic human innovation as the wheel. The fact that casework has long been effective as a cultural tool is evidenced by the preservation to this day of many artifacts stored in a wide variety of cases found with the remains of Tutankhamen of Egypt (over 3300 years old) or in the ruins of the Roman city of Pompeii (catastrophically destroyed by volcano in 79 A.D.).

Whether or not any item is provided with a case can be a true test of its cultural value. Modern day families award specialized cabinets for their china and silver. Valuable documents are stored in filing cabinets or safety deposit boxes. Priceless art objects are displayed in protective glass cases. Musical instruments, firearms, clothing, food virtually everything man holds important to his life has its own case or cabinet.

THE DEVELOPMENT OF STEREO FURNITURE

Music reproduction equipment has been encased since its inception. Basic wooden boxes were utilized both with Edison's first production phonographs in 1888 and with the American Marconi radio receivers in 1916 (American Marconi was later bought by General Electric and eventually split off as RCA). In these instances, the casework was not only for protection of the equipment and esthetic appeal, but also as protection for the operator from the hazard of electrical shock.

The location of audio (and later video) equipment in the home became more prestigious in direct proportion to its importance in the modern lifestyle. Starting about 1925, the cabinetry became more and more ornate (Figs. 1-1 and 1-2). After World War II, whole rooms were arranged around a television with the *hi-fi* conveniently within reach. Thus, the family room was born.

Beginning in the early 1950s, the production of single cabinets combining various audio elements increased dramatically. The phonograph and tuner shared the same amplifier and speaker, and were sold as a unit in a specially designed piece of furniture (Fig. 1-3). Although separate components have always been available, at this time they were of chief interest mainly to hardcore audiophiles.

The design of the *consoles* evolved into a rather standard format with the commercial development of stereophonic sound around 1960. They became long, low sideboards with built-in speakers at either end, a turntable and tuner/amplifier in the center under a lift-lid, and either storage or a television centered below.

Towards 1970, the audio market became more sophisticated. People discovered that by buying the best turntable from one manufacturer, and the best tuner, amplifier, and speakers from another, that the combination was a far superior sound system, for the money, than any prepackaged console. Now, component systems, formerly the exclusive domain of the audiophile, have become commonplace at all levels of the market.

THE EMERGENCE OF CUSTOM INSTALLATION

The early hi-fi and hi-fi stereo components were rarely mounted into cabinets. Few craftsmen were around who could provide this new installation service. Probably the main reason for slow growth in this service industry was the predisposition of audiophiles to change equipment frequently. A permanent or semi-permanent mounting was a hindrance to this fast moving hobby. Another reason may be that to a true hi-fi buff, the tangle of the wires and the glow of the vacuum tubes was (and is) a beautiful sight.

Since about 1955, hi-fi shops and independent craftsmen offered the service of wiring sound systems into houses and offices. Part of this service included concealing the components in closets, end tables, and the like. One of the best books on custom installation remains TAB book No. 1186, *How To Install Your Own Stereo System-2nd Edition*.

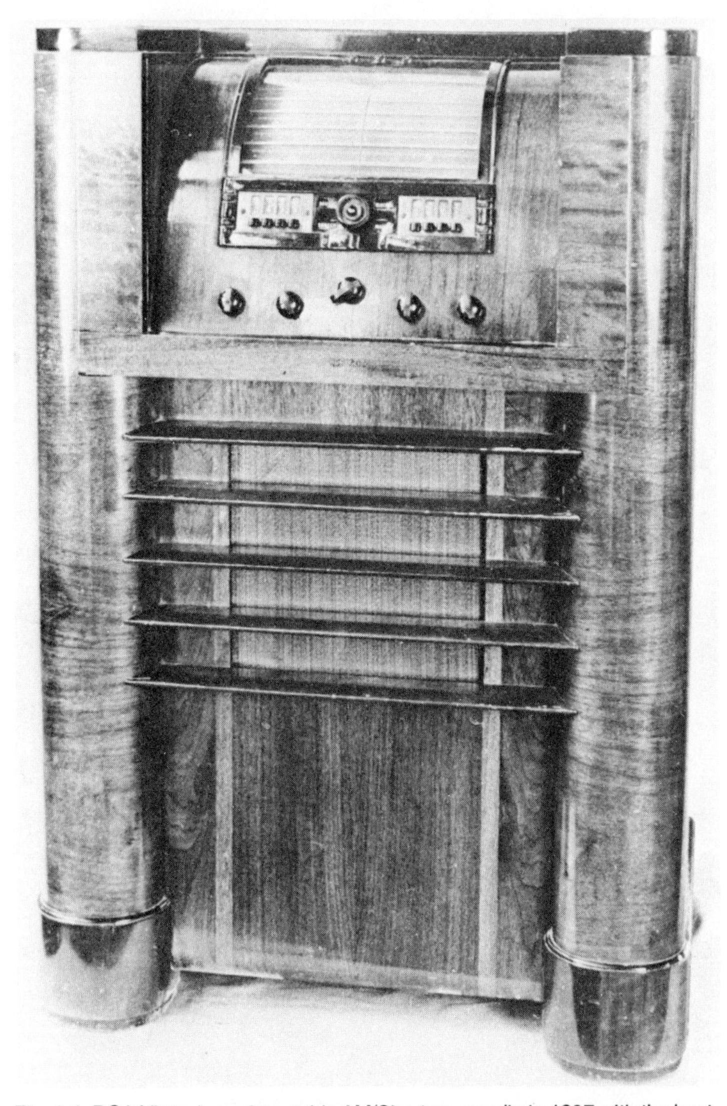

Fig. 1-1. RCA Victor brought out this AM/Shortwave radio in 1937 with the best electrical and mechanical features available at the time. It had 20 watts output through a 12″ "super-dynamic" speaker, optional armchair control, and a phonograph connection (courtesy RCA).

Fig. 1-2. The matched African Mahogany veneers on this chest-style traditional piece enclose post World War II state-of-the-art equipment. This very early console had an AM/FM/Shortwave radio and phonograph with a two-post record changer in an "all-in-one roll-out unit" (courtesy RCA).

Fig. 1-3. This twin console housed a three speed record player, tape recorder, and AM/FM radio in one unit; four speakers mounted in the other—the epitome of opulence in 1955 (courtesy RCA).

COMMERCIAL HI-FI CABINETS

After the custom installation business came into its own, several companies began to specialize in making casework specifically for hi-fi and stereo components. The production of stereo cabinets has waxed and waned several times since then. Until about 1975, most of the popular designs retained approximately the same format as the prepackaged consoles. More recently, other format cabinets have been eclipsing the demand for the old console designs.

Horizontal Format

The major advantage of the horizontal units is the ease with which they blend in with existing decor. Console cabinets are usually low enough to place a lamp on or hang a good-sized painting over, and appear not unlike various traditional sideboards. As a rule, pieces with long, low proportions tend neither to dominate nor disappear, but affect the overall impression neutrally. See Figs. 1-4 through 1-16.

This esthetic is at the expense of function. Although equipment mounted under a lift-lid is quite convenient to use, the lid itself precludes the use of a lamp on the console. The angle at which the equipment is mounted allows stagnant pockets of heat to build up. If the equipment is mounted down in the body of the cabinet,

Fig. 1-4. This horizontal cabinet is unusual in that the components are mounted in the body of the cabinet rather than under a lift-lid. Speaker housing is large enough to accept most bookshelf speakers.

Fig. 1-5. A seven foot cabinet partially resolves a placement dilemma inherent in horizontal designs.

even though the top may be used for a lamp, one has to kneel to operate the stereo. This disadvantage is partially offset by better ventilation.

Compromised speaker placement is another drawback of the horizontal format. In order to achieve the best stereophonic effect in most rooms, the speakers should be placed seven to nine feet apart. Most cabinet designs place the speakers only four or five feet apart.

Fig. 1-6. Only the turntable is placed under a lift-lid. This permits the balance of the cabinet top to be used for lamps or other decorations.

Fig. 1-7. Esthetics take priority in this Oriental-motif cabinet.

The apparent sound quality of stereophonic speakers is a direct function of the position of the high frequency transducer (the tweeter) relative to the position of the ear. This is why most stereo

Fig. 1-8. Speakers are concealed behind doors, a receiver and cassette deck are under the lift-lid. Space that would otherwise be wasted is utilized for cassette storage drawers.

Fig. 1-9. Speakers are left out of this six foot long console to allow room for a 19" portable television. Speakers are placed independently for optimum stereophonic sound.

Fig. 1-10. Another speakerless console features a receiver, turntable, top-load cassette deck under a lift-lid; a foam-damped open reel tape player mount; and combination cassette/read storage bins.

Fig. 1-11. Compartment at far left can hold over 300 record albums.

stores have their sound rooms laid out with the various speakers placed nearly at shoulder height. The tweeters are pointed straight at the ear and the speakers sound better.

Listening usually takes place sitting down in the home. Ideally, the tweeters should be situated somewhere between

Fig. 1-12. Country French lines show the diversity of style that can be used for stereo furniture.

Fig. 1-13. Receiver, turntable, and a dozen records are enclosed behind doors on the left side, a 19″ portable television and storage drawer on the right.

twenty-six and forty-two inches off the floor. The same speaker that sounds great in the store can sound quite mediocre when placed almost on the carpet as in most horizontal case designs. A good portion of higher frequency output soaks into the rug before it has a chance to be heard.

Fig. 1-14. Even a four foot wide cabinet can hold quite a bit of equipment.

Fig. 1-15. One way to avoid a long, bulky cabinet is to design it in two pieces. This is especially appreciated while moving it up stairs to a second or third floor room.

Of course, a cabinet could be built that satisfies ideal speaker placement both in height (for sound quality) and width (for separation). But can you imagine what a nine foot wide, four foot high monstrosity would look like? Now imagine what it would be like trying to get it down the hall and into the room after you built it.
separation). But can you imagine what a nine foot wide, four foot high monstrosity would look like? Now imagine what it would be like trying to get it down the hall and into the room after you built it.

Fig. 1-16. These modules may be placed side-by-side, across the room from each other, or even stacked.

Fig. 1-17. A commercially available stereo rack inexpensively holds many typical component stereo systems (courtesy Presidential Industries).

Fig. 1-18. Improved rack designs can incorporate dustproofing systems and concealed wiring (courtesy Presidential Industries).

Fig. 1-19. Turntable is conveniently placed on top.

As more inhabitants are born onto this planet, it is inevitable that individual space will become smaller and smaller. You can see this for yourself by looking at the average room in a house built fifty years ago and a house built today. Apartment and condominium living has been taking a larger and larger share of modern housing needs. Fewer people actually have the wallspace to spare for a long

Fig. 1-20. A seven foot tall vertical cabinet places the most frequently used components at nearly ideal heights (courtesy Presidential Industries).

Fig. 1-21. Modern electronic equipment and Early American style Maple furniture can be compatible in the six foot high design. Bottom drawer stores 150 record albums, top drawer holds accessories and cassettes.

Fig. 1-22. Vertical designs use up far less floor space compared to horizontal designs containing the same equipment (see Fig. 1-11).

Fig. 1-23. This armoire-type design is five foot wide, seven foot tall. The right side contains a 19″ portable television and video cassette recorder.

piece of furniture, and even if they have it now, they may move in the future.

Vertical Formats

For a very basic system, a low vertical format (Figs. 1-17 and 1-19) may be appropriate. These cabinets are usually inexpensive and can hold quite a few components, a turntable, and usually some records. Most commercial stereo cabinets or racks are of this type.

These can be tucked just about anywhere in a room. The convenience of eye-level mounted equipment, concealed wiring,

Fig. 1-24. A four-by-six design cannot hold many televisions, but can accommodate a very large stereo system or allow plenty of room for growth (courtesy Presidential Industries).

Fig. 1-25. The same basic proportions may be used with both contemporary and more traditional styles to fit in with other room furnishings.

dustproofing, and many other amenities of large cabinets are sacrificed for lower cost, but the cabinets do look considerably better than pine boards and decorative blocks.

Taller vertical format cabinets (Figs. 1-20 and 1-22) have three main advantages: they enhance equipment convenience, they require a minimum lateral wall space for the equipment they hold, and they tend to cost less to build than comparable horizontal units.

Fig. 1-26. Four-by-six cabinets allow all components in a large system to be placed at, or near, ideal heights.

Fig. 1-27. These cabinets offer lots of storage space.

A six or seven foot tall cabinet doubles the options for equipment height over any waist-high case. The probability is greatly increased that most or all of the equipment can be mounted at the optimum height for easy and comfortable operation.

Houses and apartments have average ceiling heights of 7½′ to 8′. A 3 foot wide, 6 foot tall cabinet will fit in virtually every situation, while a cabinet 6 feet wide and 3 feet tall may not.

Any cabinet taller than average eye level (60″ to 70″) need not have a completely finished top surface. This labor economy combines with a slightly better material economy to reduce the total cost per cubic foot of storage over a horizontal cabinet.

The trade-off is that a taller cabinet can look too tall and skinny for a given room arrangement and throw the esthetics off balance. You can expect cabinets of these proportions to attract more attention in the room, so plan to make them the focal point.

Generally, the speakers are left out of a vertical format cabinet. The listener is free to place them where they can extract the optimum quality sound and stereophonic effect.

Armoires

Some cabinet designs feature the advantages of both vertical and horizontal formats while minimizing the disadvantages. Be-

Fig. 1-28. Space under stairs may be salvaged with cabinets built at an angle. Equipment is placed for walk-up operation. Record storage is under the sections where kneeling is required.

Fig. 1-29. An unused closet space in an extra bedroom becomes an entertainment center.

Fig. 1-30. Simple lines and the aristocratic appearance of American walnut bring the room alive. Upper compartments store tape reels. Base section stores record albums behind doors. Note that speakers are raised to improve sound.

Fig. 1-31. A sixteen foot vaulted ceiling allowed these cabinets to extend ten feet high and twenty-two feet wide over the dining room entry way. Speakers are placed on top at opposite ends to reflect off ceiling angle.

Fig. 1-32. Stereo and television storage require much deeper casework than bookshelves. One solution to this is to extend out the stereo section. This adds interest to the design by breaking up the flat plane of the cabinet fronts.

cause of their size and proportion, these cabinets generally resemble traditional wardrobes or armoires. For our purposes, these pieces range from 4 foot wide, 6 foot high to 5 foot wide, 7 foot high. Generally, the 4×6 cabinet is the most economical. Going as large as 5×7 is only necessary where video equipment is anticipated. A 5×7 cabinet is large enough to enclose a 19″ portable television, but generally costs up to 30 percent more for materials.

Fig. 1-33. This stately built-in look is actually two detachable freestanding pieces. Made of Pecan lumber and veneers, it would weigh over three hundred pounds if moved in one piece.

Fig. 1-34. The bare essentials of television cabinetry are all represented in this efficient design. The television is located at eye level for comfortable viewing. The video cassette deck is mounted below to enable monitoring the screen while locating a section of interest on the tape. Storage holds about forty video cassettes (courtesy Presidential Industries).

These proportions avoid some of the esthetic difficulties of the vertical units because they are more similar to the so-called *Golden* or *Classical proportions* (A ratio of 1:1.618). The case amounts to two vertical cabinets side-by-side and has storage space accordingly. Equipment placement is simplified merely because more space is available at the ideal equipment heights than

Fig. 1-35. Here video placement principles are expanded. Tape deck is lighted and angled up at operator's face. Stereo is mounted behind doors on left (courtesy Presidential Industries).

either verticals or horizontals. Like the vertical units, speakers are usually not mounted in these cabinets. See Figs. 1-23 to 1-27.

The 4×6 cabinet is probably the best option for those with either a large sound system, or those who anticipate expanding their system in the future, beyond the fairly standard turntable, receiver, and cassette deck. The major drawback of this design format is its cost. If the space isn't needed, the cabinet costs an

Fig. 1-36. Freestanding modular-style dry bar encloses a 25″ console television with legs removed.

Fig. 1-37. Upper cases and base section may each be used separately at another time and place.

inordinate amount relative to its function. However, if the space can be used, this layout represents probably the most space per dollar.

Built-Ins

Sometimes it is desirable to conveniently store not only the stereo and the television, but also house books, large quantities of records, relics of past experiences, and serve other functions such as desks, bars, or reading centers.

The solution might be a more permanent household fixture—built-in cabinets. See Figs. 1-28 to 1-32. These may be designed to take advantage of the precise specifications of the room and squeeze out the last bit of useful space. Since built-in casework usually is only partially enclosed behind doors, the cost per square foot of space can be as little as 60 percent of a freestanding piece with doors. Since the casing is screwed to the framing of the house, the sheer bracing function of the back would be redundant, and may be left off for cost cutting or esthetic effect.

People who move must leave the units behind. This can be an advantage in disquise. The appeal of built-in cabinets can raise the value of a house more than the actual cost of the improvement.

For the people who want to take the cabinets with them, a compromise can be reached. A built-in type cabinet can be constructed so that it is freestanding. See Figs. 1-33 to 1-37. It costs slightly more to construct because of the additional bracing. The size should be kept reasonable since you may wish to move it in the future.

Equipment Placement Priorities

Each component of the sound system has a different function which is generally exhibited by its own set of controls, meters, or specialized mechanisms. These different functions each have different priorities within the system as a whole.

A typical system might have an open reel tape deck, a receiver, a cassette deck, and a turntable. The receiver controls are the most important since they are used every single time any other part of the system is used. It would be very rare for the open reel tape deck, turntable, and cassette deck to be used at the same time, or even equally. Most people use the turntable or the cassette deck much more than the open reel deck.

Occasionally components compete for the same ideal space. Under these conditions, frequency of use should dictate which component should be given preference.

Ideal equipment heights have been established to guide designers with equipment placement. See Table 2-1. These guidelines were determined by studying the function and controls of each component, and the ergonomics involved in operating them.

The table makes several assumptions. First, it is assumed that the equipment operator is in a standing position. This is justified by the observation that very few people actually sit immediately by their stereo when they listen. Normally they are doing something else such as reading, cooking or entertaining. Therefore, if they decide to change a record or tape, they walk over to the stereo, make their adjustment, and walk back to what they

Table 2-1. Standard Equipment Heights

COMPONENTS	PARAMETERS	OPTIMUM	OTHER PRIORITY
Receiver height (center line)	54" - 66"	60"	heat
Amplifier height (center line) (w/VU meters)	54" - 72"	60"	heat
Preamplifier height (center line)	48" - 66"	60"	
Tuner height (center line)	54" - 66"	60"	
Front-load cassette deck (center line)	48" - 66"	54"	horizontal for some decks
Top-load cassette deck (shelf)	30" - 60"	42"	
Turntable height (shelf)	24" - 40"	30"	
Electronic antennas	(36" plus --- above level of house wiring)		
Digital time delya, Dolby, DBX, switch boxes, etc.	30" - 66"	N/A	
Open reel tape deck (at heads)	20" - 72"	48"	
Television (shelf)	18" - 36"	30"	
Video cassette deck (shelf)	18" - 48"	48"	

were doing. If the stereo equipment were lower for seated operation, the same people would either have to stoop (which is uncomfortable) or sit down (which requires an additional expenditure of energy).

Secondly, the chart assumes an average height of 68" and an average eye level of 62". Although many adults, particularly males, are taller than this, it is a good idea to adhere to these standards. Any variation will more likely be shorter than taller. It is well established that people grow from the floor up.

Where seated operation is desirable (for instance, someone in a wheelchair), subtract twenty inches from all standard heights over forty inches.

AMPLIFIER, TUNER, RECEIVER

The amplifier is placed above all other components. The power transistors and transformer within this unit produce the only significant heat in the whole system. If the amplifier is placed below any other unit, the heat filters through the topmost equipment and can damage or at least shorten the life of it. An exception to this, the open reel deck, will be discussed later.

The tuner scale is probably the one adjustable setting that needs precise eye-to-hand coordination. Ideally, it is placed as close to eye level as possible. However, if the tuner placement competes with the amplifier, mount the tuner below the amp.

Another way to solve the tuner/amplifier space competition is to use a receiver instead. The receiver is a component that combines the tuner and amplifier into one unit.

Depending on the priority of esthetics over function, components may be placed outside their respective ideal parameters. This is quite common in horizontal format cabinets. Sometimes function is maintained by mounting the receiver in a horizontal panel under a lift-lid as in Fig. 2-1. By facing the front panel of the component up at the user's face, this set-up is almost as comfortable to use as a normal eye level set-up. A lift-lid does preclude setting lamps or other decorative objects on top of the cabinet. Ventilation is not as efficient with a horizontal equipment panel and should not be used with amplifiers delivering more than 50 watts per channel.

If you don't frequently use your stereo, or if you use it primarily for FM background music, there is no reason why you cannot go below the suggested heights. If you do so, you must be aware that you will end up kneeling to use the equipment, and all of us are getting older every year.

PREAMPLIFIER

The preamplifier is another competitor of the tuner and amplifier for the same elevation. Most of the controls on the preamp can be adjusted by touch, such as the click-stops on the knobs. The balance and tone controls can be set by ear and so can be placed below eye level.

The preamp should be placed as close to the amplifier as a short set of patch cords will allow. This guarantees purity of signal by reducing the net resistance of the wires and decreasing the chance of picking up RF interference. It should never be placed above the amplifier because of the heat.

Fig. 2-1. Components mounted under a lift-lid in horizontal designs are easy to operate from a standing position.

OPEN REEL TAPE DECK

The reel-to-reel tape deck is mounted so that the magnetic tape can easily be threaded across the heads. This will normally be with the heads at shoulder height.

An open reel deck is rarely used as much as the other components competing for the same placement. Because of this, its position is usually compromised first. In a vertical format cabinet, the open reel deck may be placed above the amplifier if it is out of the direct path of the heat. Most amplifiers extend twelve or thirteen inches back from the face, and have their heat producing elements in the rear. Most open reel decks extend only five or six inches back, and are not directly above the heat.

Most open reel decks are designed with the heads and VU meters at the bottom of the face. Mounting above the amplifier places the critical controls near eye level.

More than any other component, open reel decks come in diverse shapes and sizes. Remember to allow in the design for the possibility of changing to a deck of different proportions. If possible, allow clearance for a future 10″ reel unit should you decide to upgrade your equipment in the future.

CASSETTE DECKS

A front-load cassette deck must be low enough to load the tape and frequently adjust the level control (particularly if recording "live" material), yet must be high enough to visually monitor the VU meters. Mount it as close to eye level as possible without straying over shoulder height. This prevents fatigue by keeping the hands approximately the same level as the heart, and eliminates the eyestrain or sore back associated with stooping.

In a horizontal cabinet, the deck may sometimes be mounted face up under a lift-lid, as in Fig. 2-2. Do not make a commitment to this method without checking with the manufacturer of the tape deck. Gravity can affect the mechanism of some decks and cause unacceptable distortion.

Cassette decks are frequently used to record new records for convenient playback. Transferring each new disk to tape saves having to clean the records each time they are played. If this is your expected pattern, then place the cassette deck near the turntable so that both may be operated from the same standing position. This simplifies coordinating the start of the turntable with the tape machine.

Top-load cassette decks should be mounted lower than their front-load cousins. They may be mounted either on a horizontal shelf, as in Fig. 2-1 or angled to direct the VU meters towards the eyes. See Fig. 2-3. Experimentation will yield the best angle.

THE TURNTABLE

Of all the components, the height of the turntable is the most inflexible. Mount it too high and it is hard to line up the center hole

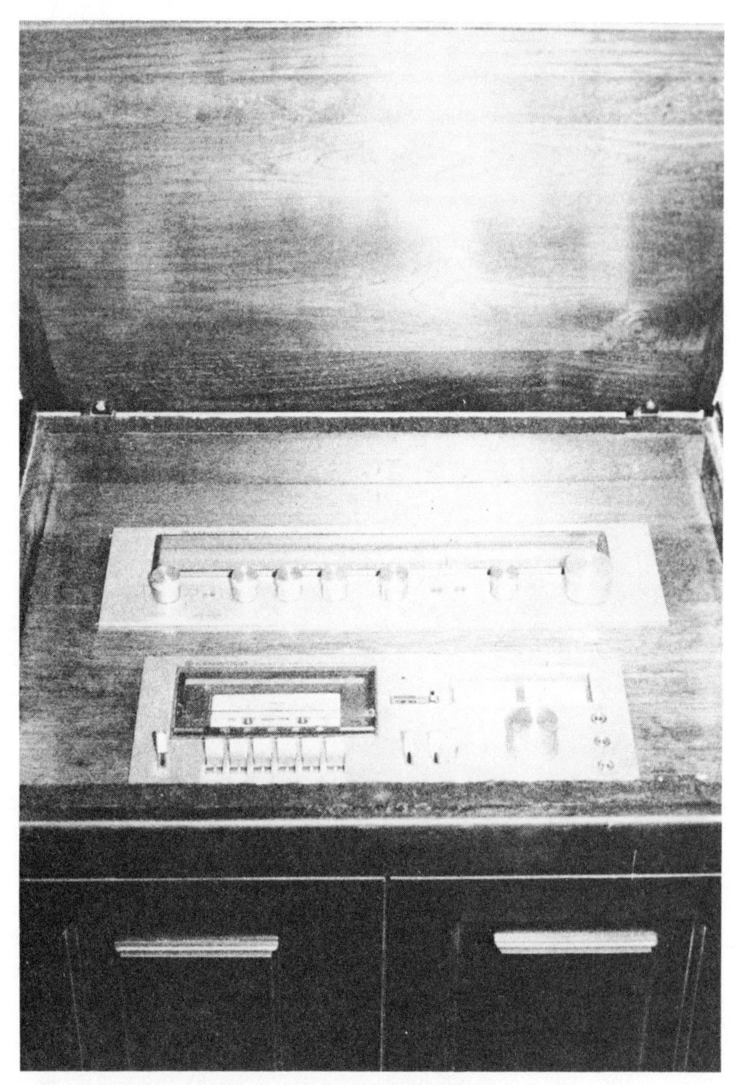

Fig. 2-2. Some cassette decks may be suspended from a horizontal equipment panel, but not all. Check owner's manual or write manufacturer to be sure.

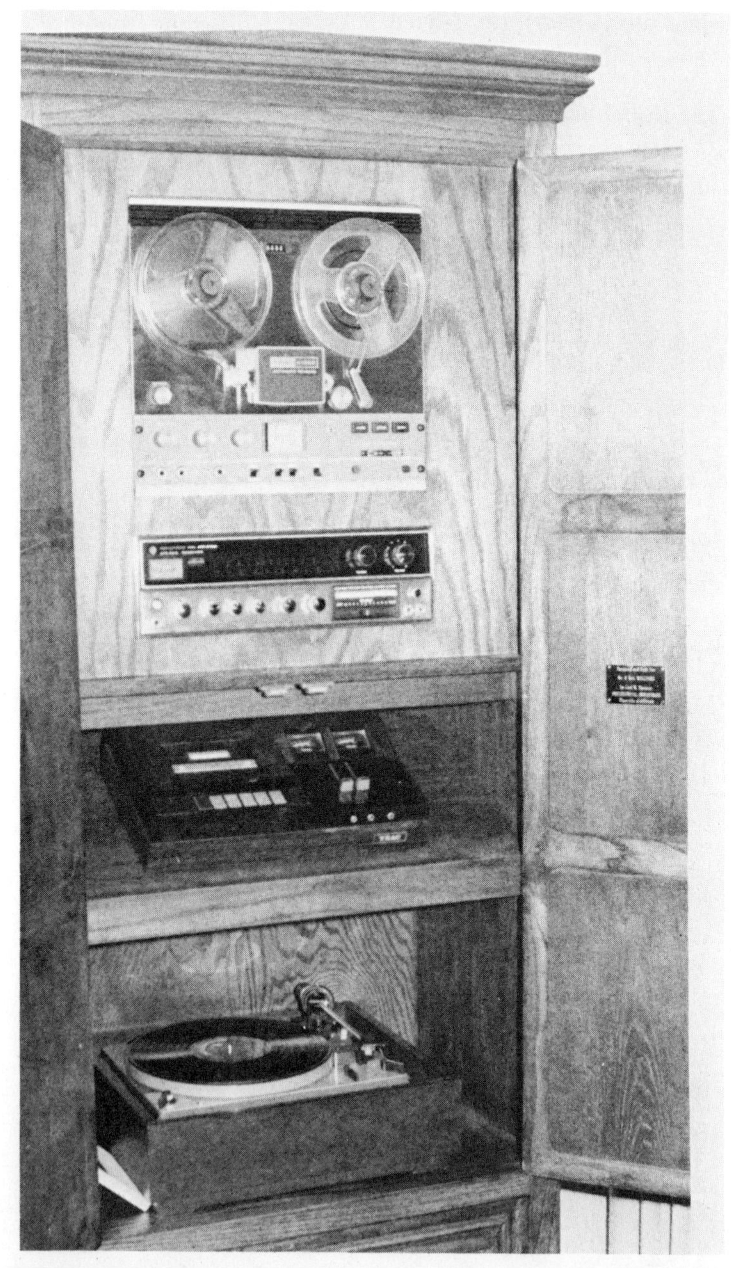

Fig. 2-3. Both appearance and function are heightened by mounting top-load cassette decks at an angle and lighting their compartments. Equipment layout here demonstrates principles of good design.

of the record with the spindle. Mount it too low and you have to stoop while your vision is blocked by the next shelf up. Waist height offers the best compromise.

The turntable should always be shelf mounted as in Figs. 2-4 and 2-7 and not panel mounted. Like the open reel deck, it is highly mechanical and will require maintenance. The state-of-the-art of turntables has been improving at a very rapid clip. You will face a greater temptation to replace your turntable in the future compared to the power equipment.

Turntables should never be mounted on pull-out drawers or shelves. You spend good money in order to buy the most carefully regulated speed and balance. Even the best engineered drawer slides have enough play in them to throw the mounting surface out of level each time it is moved. Sooner or later someone will push or pull the drawer too fast while the needle is on the record and skate it across the disk. If your goal is high fidelity, stick with a fixed platform.

AUXILIARY COMPONENTS

Every year new gadgets are thought up to improve the quality of your high fidelity stereo system. Digital time delays (actually

Fig. 2-4. A shelf mounted turntable under a lift-lid. The more mechanical components require more maintenance and should always be shelf mounted.

Fig. 2-5. Ideal television placement is at, or just above, eye level while seated in the viewing chair.

signal delays), bass enhancement, Dolby units, equalizers, switch boxes, antennas, and a host of other things will compete for space in your cabinet. Each of these has a different function and control requirements. In general, all are of secondary or tertiary impor-

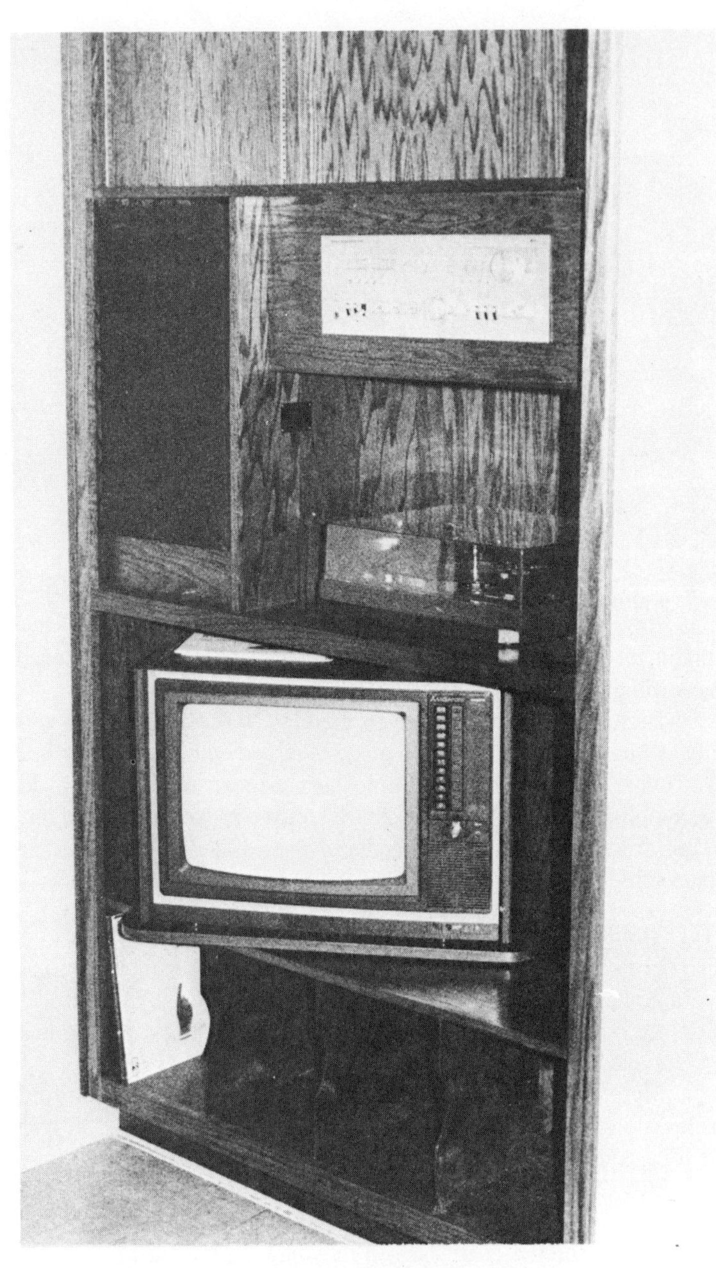

Fig. 2-6. Television is mounted on a lazy susan platform to best serve any one of several viewing chairs in the room. Note near-ideal placement of stereo components. Speaker is mounted slightly above ear level. The right speaker is mounted at the same level in another bookcase.

Fig. 2-7. In enclosed horizontal units without lift-lids, try to raise the components as high off the floor as possible.

tance to the aforementioned components and should be placed accordingly.

Perhaps it is appropriate to mention here that room should always be left for at least one piece of future equipment, whether you expect to ever buy one or not. The cheapest and easiest way to accomplish this is by allowing more storage space than called for. Then if future equipment is added a record shelf can be quickly converted.

TELEVISIONS

For some reason, optimum television height is more controversial than one would expect. Some people swear by traditional console television placement only inches off the floor, and others prefer eye level or higher.

My experience with my customers is that most people find television viewing comfortable if the screen is centered at eye level or slightly above when they are seated in their favorite chair. See Figs. 2-5 and 2-6. This can be explained by the fact that people are better able to relax while sitting in a normal position with their head resting on the chair back. If they have to look down towards the floor, their shoulders and neck eventually fatigue. I have found a high correlation between customers preferring a lower screen and people who customarily view while lying on the floor.

It is rare to place both television and stereo in the same cabinet. Frequently, people will set their stereos in their living rooms as background music for social entertainment, reading, or just working around the house, while they leave their television in the family room.

A television complicates the stereo cabinet design. It competes for nearly all the ideal locations of the components, and is both wider and deeper than almost any stereo component. The cabinet must be significantly larger than required for stereo alone, and because of additional materials, more expense is involved.

VIDEO RECORDER

The technology and hardware involved in auxiliary video equipment is changing so rapidly; I almost hesitate to comment. Consideration for this equipment is similar to considerations for top-load cassette decks and turntables.

It is a good idea to place these units either below the television screen, as in Fig. 1-35, or beside it so that the screen may be monitored while locating a segment in the middle of a tape or disk.

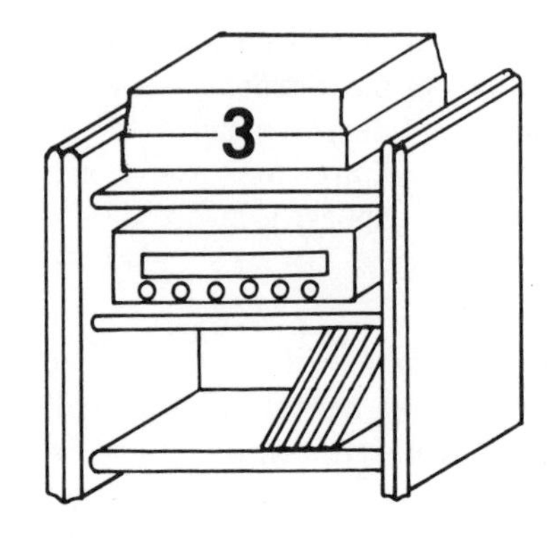

Dust Protection

Dust protection and ventilation are two sides of the same coin. The more thorough the ventilation, the greater the chance for a dust problem. The more dustproof the cabinet, the harder it is to achieve adequate ventilation.

WHY BOTHER

In another sense, the two problems are more intimate. The more dust that infiltrates the cabinet, the more dust that settles on the equipment heat sinks. See Fig. 3-1. As dust gradually builds up on the heat sinks, they radiate heat less efficiently. This gradually increases the internal operating temperature of the amplifier and can shorten its life.

A fair percentage of common household dust is really lint from drapes, carpets, bedsheets, and clothing. Lint is very flammable. If you have ever seen lint burn, you know the potential for a very vigorous fire.

Practically speaking, the insulating effect of dust and the related fire hazard are not anywhere near the risk they used to be, thanks largely to the efficiency of modern solid-state circuitry. But as long as you have control over the cabinet design, you may as well engineer these problems out altogether.

BENEFITS OF PANEL MOUNTING

No matter how good a housekeeper you are, dust enters with you every time you come in the door. It blows with the wind through cracks and joints around the windows, doors, electrical junction boxes, the plumbing, and many other paths.

The bulk of the dust settling on your equipment comes from the direction of the middle of the room. Enclosing the components

Fig. 3-1. Circulation around the heat fins of power equipment must be free. An excessive accumulation of dust can impair their efficiency and make the equipment run hotter.

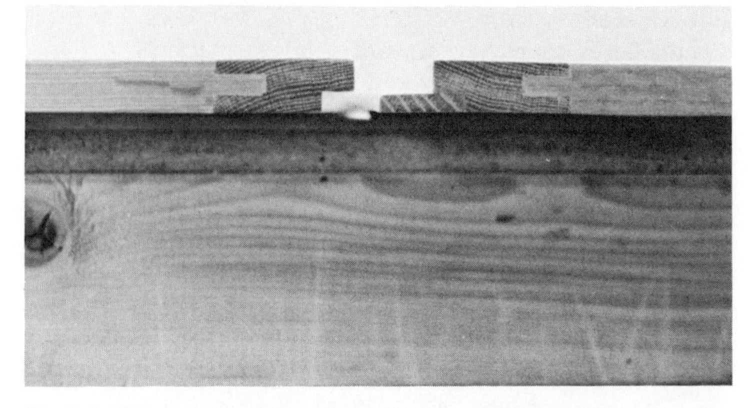

Fig. 3-2. A top view of unfinished doors shows construction of the dust lips. After the doors have been mounted on the cabinet, no center gap is left between the closed doors.

on all sides but the back will stop most of it. The best method of enclosing the equipment from the front side is by panel mounting it. This involves cutting a wood panel for an exact fit around the equipment (techniques of panel mounting are discussed in Chapter 8).

DUST LIPS

Some of your equipment probably won't be panel mounted. The turntable, top-loaded cassette deck, or other shelf mounted equipment need dust protection even more than the power equipment.

Dust is attracted by static electricity to the vinyl records, and manifests itself with pops and clicks. The stylus can grind these along the grooves and permanently affect the quality of the sound. Dust can build up on the lightly lubricated mechanisms in your tape deck, increase the abrasion on the tape heads, and cause a host of other problems.

The dust covers provided with these components or sold by the manufacturer as accessories are the best protection. These plastic covers have been designed to fit the housing with no gaps, yet raise to allow convenient operation. It is a good idea to allow room in your design to permit the continued use of these dust covers. Turntable compartments should be at least 18″ high and 17½″ to 18″ deep to permit the lifting of most dust covers.

Close fitting doors can provide the first barrier to dust, reducing the infiltration considerably. Doors should be used in pairs wherever possible. The larger the door, the greater the chance of slight warpage. This misalignment leaves a greater gap which lets in more dust. A very wide door, if leaned on, exerts significant leverage on the door hinge. This racking action can also destroy the perfect fit of the doors.

Double doors are narrower and more stable, but have a crack down the middle which must be protected. These should be constructed with a dust lip (Fig. 3-2) which can reduce by 50 or 60 percent the amount of dust sneaking past.

Fig. 3-3. A view of the false back shows intake and exhaust vent holes. The corresponding holes in the actual back have been offset from these, and therefore are not visible. Other holes are for wires (note the pull wire in bottom section).

Fig. 3-4. The holes in the dust panel set into the cornice molding are the reverse of the holes in the actual case top.

DUST BAFFLES

Amplifiers and receivers must be properly ventilated. Ventilation (see Chapter Four) inherently involves air flow, and moving air is the most common medium for dust.

But dust weighs more than air. Both gravity and centrifugal force can pull the lightest particles of dust out of the air. The secret of dustproofing the ventilation system is to baffle the airflow enough to let gravity separate the air from the dust.

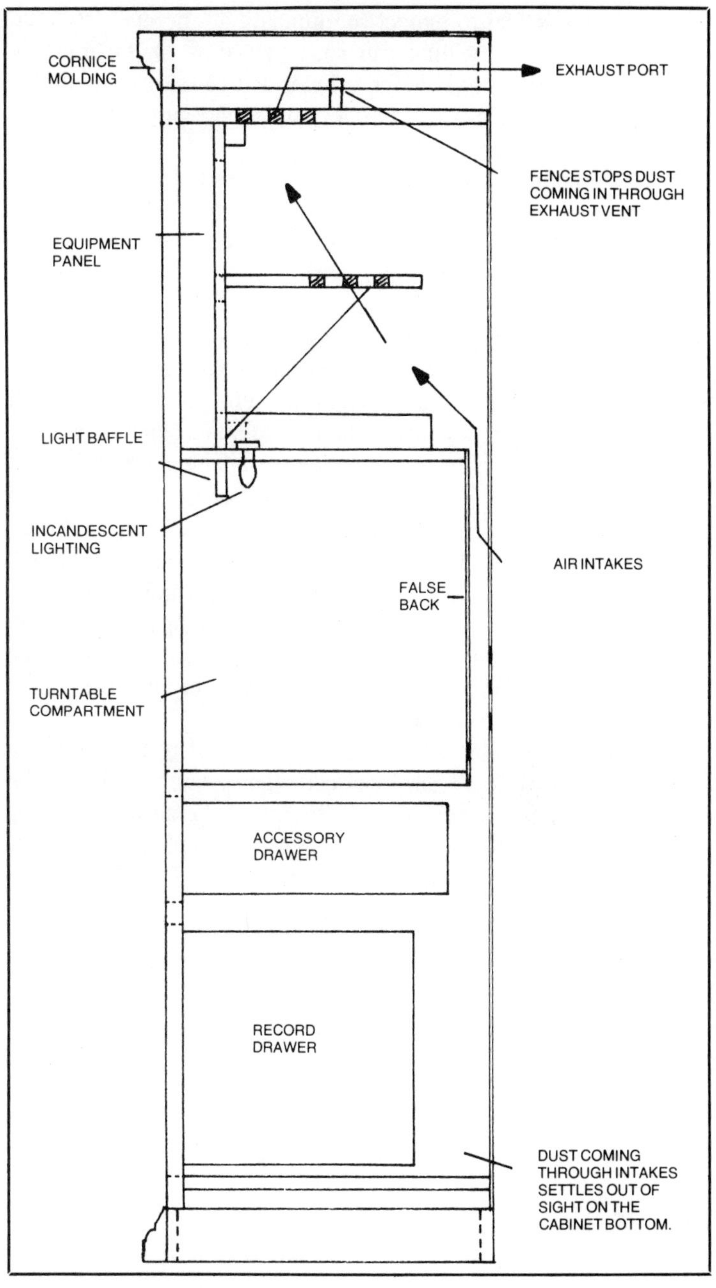

Fig. 3-5. An efficient folded double dust baffle system (courtesy Presidential Industries).

It is impractical to expect to eliminate all the dust from the system. Too much baffling can resist air flow to the point of insufficient ventilation. It is far simpler and cheaper to minimize the exposure and clean out the accumulation every four or five years with a vacuum cleaner.

The simplest dust baffle system involves nothing more than placing the intake and exhaust vent holes in a vertical surface such as the back. The top of the holes acts as an overhang and shields most of the hole from the dust settling out of the air. Another variation of a single baffle is a louvered back. These methods are a little like building a house with no glass in the windows. Unless the wind blows very hard, only a little rainwater comes in.

A considerably more efficient system uses two vertical baffles, one behind the other, with the vent holes offset (Fig. 3-3). Most higher quality stereo cabinets have a false back and a real back for concealing the wire runs, so these two backs can serve as the baffles. Just make sure that the vent holes in the different panels do not line up. What dust makes it past the first baffle will strike the second, drop down between the panels, and fall out of the way. Using the previous analogy, this is like adding a porch on the house with glassless windows. The water that comes in the porch windows ends up on the porch floor, and very, very little can come in the house.

This second system can be used in almost any cabinet with a false back/real back set-up. In a design going over eye level, the primary baffle can be folded over the top of the cornice molding (Figs. 3-4 and 3-5) for the same amount of dustproofing. This has an additional advantage of improving the convection draw efficiency of the ventilation system.

Ventilation

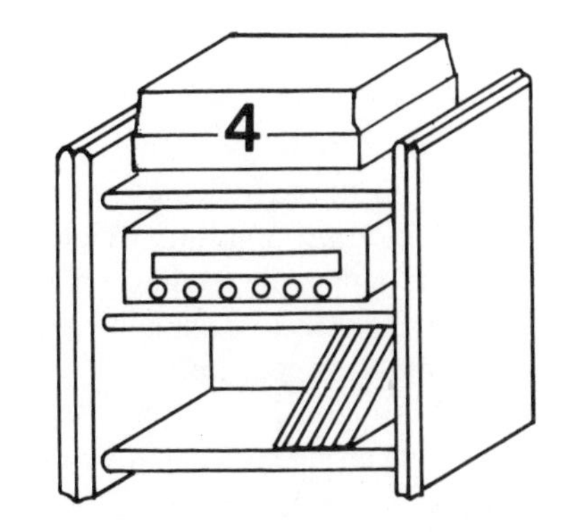

Amplifying components give off energy as heat. In general, the more powerful the amplifier, the more heat is given off. Old-style vacuum tube amplifiers give off very large amounts of heat. The newer transistorized units radiate much less. Improvements are being made all the time in the efficiency of power equipment, so the trend is towards less and less waste heat.

PROBLEMS WITH HEAT

If the heat is continuously drawn off the equipment and away from the other components, then there is little concern about it. Manufacturers have engineered the amplifiers to ventilate adequately when placed on an open shelf under normal room temperature conditions. A problem arises when the power equipment is enclosed in a cabinet for protective or esthetic reasons.

If the heat is left to build up in an enclosure, physical and chemical changes can occur to the detriment of the solid state circuitry. The temperature can build up to the point of melting insulation, shorting out high power circuits, and possibly causing a fire.

Concentrated heat is murder on wooden casework as well. Hot, dry air sucks the natural moisture out of the equipment side of the wood. The other side remains in contact with the more humid air out in the room. This moisture imbalance swells the outside surfaces or shrinks the interior. Warping, cupping, splitting, and veneer blisters are manifestations of this problem.

The movement of the wood also stretches and compresses the finish applied to it. Actually, this will happen anyway because of seasonal humidity changes, but improper ventilation speeds up the cycle from happening twice a year to every time the equipment is used. The lacquer or varnish can check or discolor years before its time.

CONVECTION SYSTEMS

Convection, or gravity, is the simplest way to circulate cooling air through the power equipment. Warmer air is less dense than cooler air, and thus weighs less. The warm air rises (or the cool air sinks, depending on how you look at it). This is the underlying principle explaining the way chimneys, hot air balloons, and politicians operate.

A simple convection system involves nothing more than providing an air intake in the cabinet back below the amplifier heat fins, and an exhaust vent above them. The hot air off the power transistors rises to the top of the case and out the upper vent holes, drawing cooler air in the intakes at the bottom. See Fig. 4-1. The circulating stream of air can be directed wherever needed by careful placement of the vent holes as shown in Fig. 3-5.

The actual heat output of any given power unit is a function of that specific component's efficiency, and is much simpler to determine experimentally than by a general formula. Start with fifteen to twenty square inches each of intake and exhaust vent holes (Fig. 4-2). In case you do not remember your geometry, the formula for figuring the area of a round hole is $A = \pi r^2$. If you use one inch holes, each hole amounts to three-quarters of a square inch.

If the air flow must negotiate more than one slight turn on its upward path, add five to six inches more. This area should be sufficient for amplifiers with a rated power output of up to sixty or seventy watts per channel RMS, assuming a typical double baffled dustproofing design. See Fig. 4-3.

Always doublecheck the system in actual service before trusting it. You can literally be taking your life in your hands with an inadequate system because of the potential for fire.

TESTING A VENTILATION SYSTEM

For the price of a couple inexpensive room thermometers, you can know for sure if the system works as designed. Temporarily tape one of the thermometers to the top of the equipment compartment with strapping tape. Do not use electrical or masking

Fig. 4-1. The rows of holes are air intakes. The slot under the back of the cornice molding is the exhaust. This is the system pictured in Fig. 3-5.

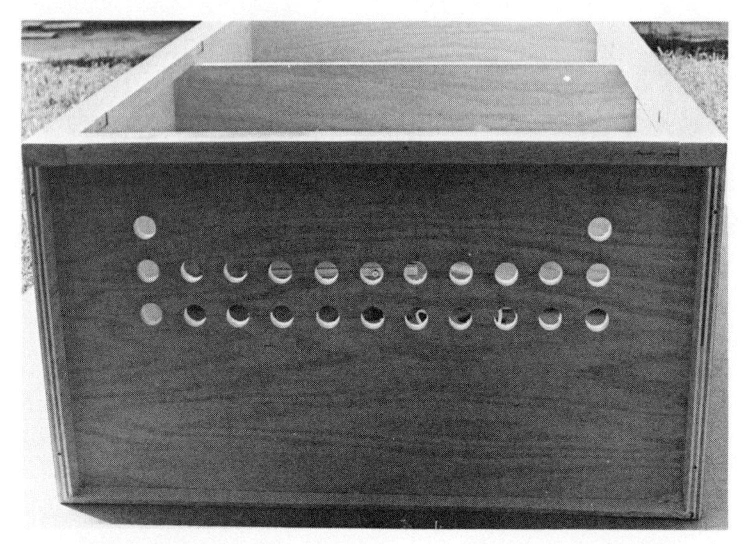

Fig. 4-2. Vent holes in the actual case top.

tape since the heat can soften the adhesive and drop the thermometer. It should be placed directly over the back of the amplifier in such a way that you can quickly open the wire access panel and read the temperature.

Fig. 4-3. In a case with a finished top, additional warm air exhaust may be concealed behind the top frame rail, above the equipment panel. Cabinet doors must remain open for this vent to work.

Fig. 4-4. Whisper fan should be mounted over exhaust hole of identical proportions. Other exhaust vents should not be used with a fan since they would dilute the suction at the intake ports.

Set the other thermometer on a nearby table out of drafts. Run the stereo at a moderately loud volume through all the speakers you might ever use simultaneously. After two or three hours (or when the police arrive, whichever comes first), check the

Fig. 4-5. Fan is insulated from case with rubber washers.

difference between the thermometers. If it is ten degrees Fahrenheit or less, the system will be okay. If not, increase your vent areas or check for obstructions. Make sure you increase the intakes and exhausts equally.

FORCED AIR

Amplifiers with power outputs in excess of about 75 watts per channel RMS should probably be used in conjunction with a forced air system. This may not be necessary for low volume usage with a single pair of efficient speakers, but it is a good idea to engineer the ventilation for the worst possible circumstances. You may add speakers later or turn up the volume for a party.

A forced air system is like a convection system, only a fan is substituted for the exhaust holes. Additional exhaust venting should not be used with the fan because air sucked in through these holes will dilute the draw and weaken the overall air flow.

The type of fan used is important. The best variety is the "whisper" fan designed for use with electronic equipment. These have been designed to maximize air flow with little or no noise or RF interference, and are available through most stereo stores.

Even a three hundred watt per channel behemoth can be effectively cooled with a 50 cubic foot per minute fan, so don't buy anything larger. Too much fan contributes a very annoying wind noise and wastes energy.

Fig. 4-6. Side view of how to mount the exhaust fan.

The fan should be mounted over a hole the exact size of the venturi, as shown in Fig. 4-4, and it should be insulated from direct contact with the case. The wood panels act as huge resonators to

Fig. 4-7. A forced air ventilation system may run horizontally.

Fig. 4-8. Entire component underside is ventilated with a perforated shelf. Do not drill the holes so close together that the shelf is weakened.

amplify the vibrations of the fan motor into a nasty hum. Don't forget to insulate any fasteners from the fan chassis.

Rubber or felt washers (Figs. 4-5 and 4-6) can reduce a good portion of this noise, but sponge rubber is much better. If the fan is

Fig. 4-9. A very heavy power amplifier needs strong support and excellent ventilation. Lattice-type shelf is reinforced with 1½" bar of solid oak in back to prevent sagging.

mounted in the top of the case, a foam rubber boot can be made to hold the fan over the exhaust port. Fasten the boot to the case, but let gravity hold the fan in the boot. This is the best solution to fan noise I have come across.

The fan should be plugged into a switched outlet in the back of either the preamp or receiver. This precaution guarantees that the fan goes on with the heat source. Make sure that the fan blows out of the cabinet, not in.

The circulation of a forced air system does not need to run vertically. It may be set up laterally since the fan pulls the draft, not gravity as shown in Fig. 4-7.

MINIMIZING OBSTRUCTIONS

The skids or shelves on which the power equipment rests can impede the air flow where it is needed most. These should be modified to suit the particular needs of the equipment they are supporting.

Examine the metal chassis bottom and see where the factory vented each component. Then either drill holes or cut slots in the shelves directly below these perforations.

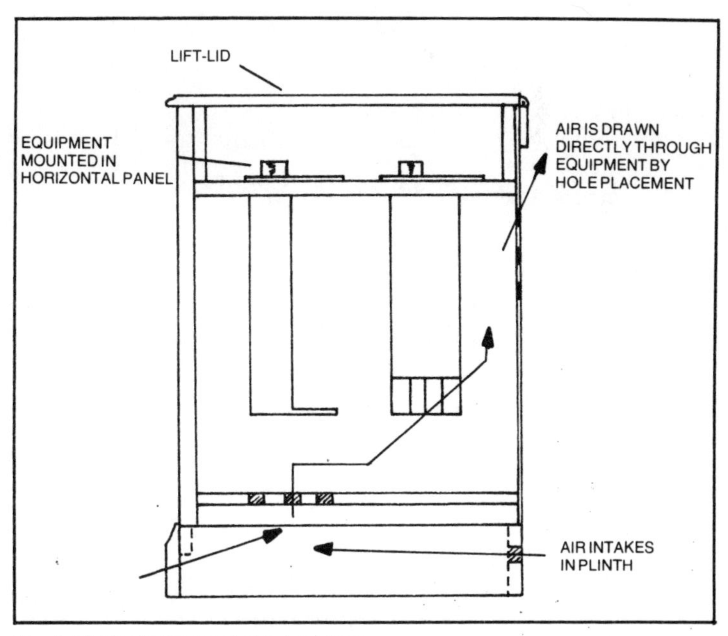

Fig. 4-10. Ventilating horizontal cabinets.

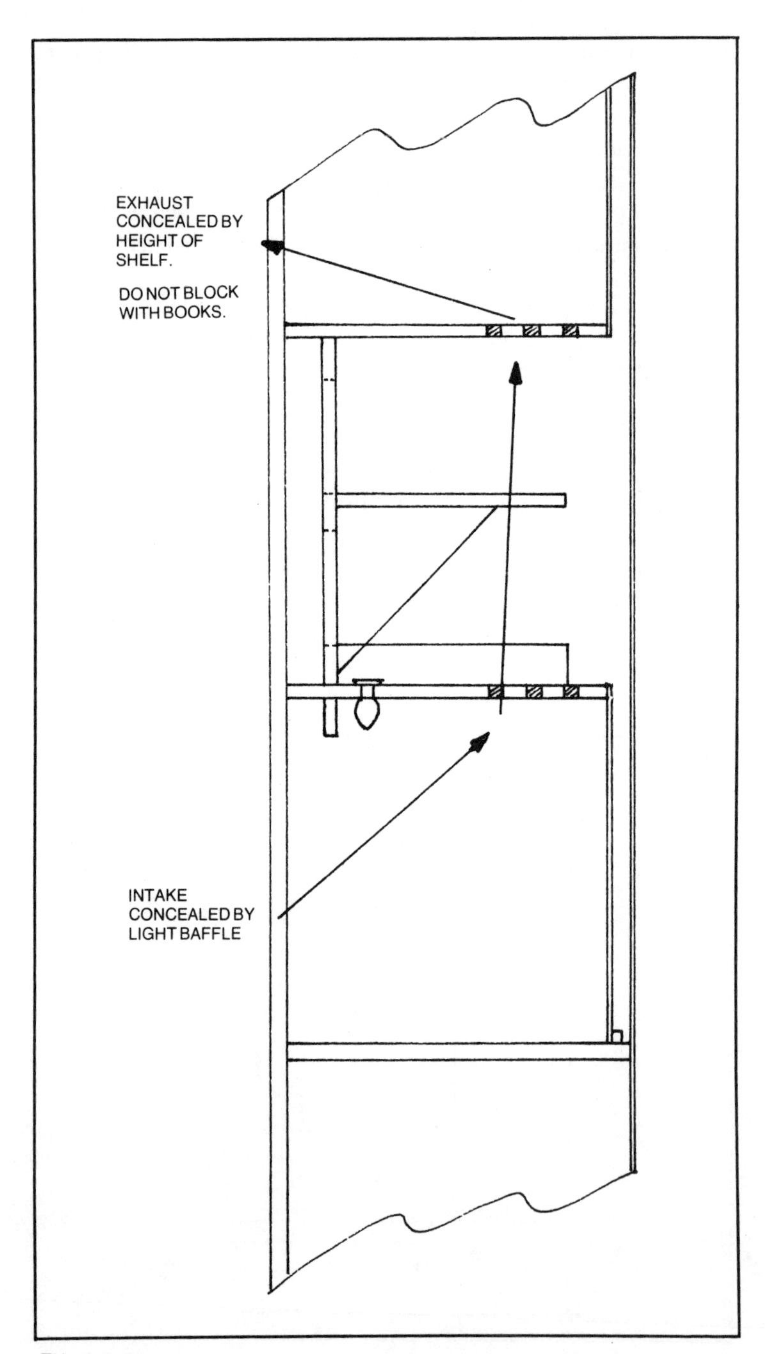

Fig. 4-11. Ventilating built-ins.

If the entire chassis bottom is perforated, you need to perforate the entire shelf under it (Fig. 4-8). Drill one inch holes approximately one inch apart. If you perforate the shelf much more than this, you will weaken it.

If power transistors are mounted on obvious heat fins, let the fins hang over the edge of the shelf altogether. A very heavy amplifier that requires ventilation throughout requires a strong, lattice-type shelf (Fig. 4-9). Also, see Figs. 4-10 and 4-11.

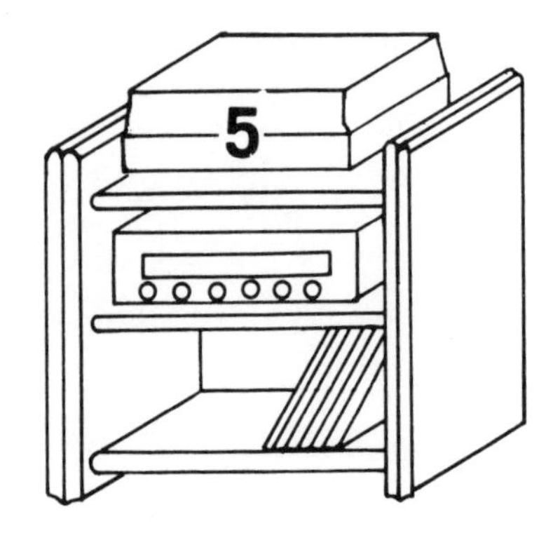

Case Wiring

Wiring a stereo system within a cabinet is much the same as wiring a system outside a cabinet. The main differences are the physical routing requirements (such as getting phono leads from the turntable to the main equipment compartment), concealment and service requirements, and a slightly greater risk of electronic interference because of proximity of both power and signal leads.

MECHANICAL CONSIDERATIONS

All wiring of any type should be installed in such a fashion that the possibilities of future testing, service, or replacement are left open. In the simpler cabinet designs, an access panel in the back of the equipment compartment satisfies this requirement. In a more complicated cabinet it is advisable to also allow a fully removable back because of the several different equipment compartments and concealed wire runs (Fig. 5-1). In any case, never, never, never glue the back on. While it is true that the back is the main sheer bracing for the cabinet, it is better to secure the back with lots of nails, staples, or screws so that disassembly is possible.

CONCEALMENT

Most people enclose their sound systems in a cabinet to hide the wire tangle. This can be accomplished both internally of externally.

The easiest way to conceal the wires from the front of the cabinet is to drill holes in the cabinet back directly behind each piece of equipment. The wires are run straight out the holes and across the back to their destination, and then reinserted.

There are three main drawbacks to this external method. First, the wires are still exposed behind the cabinet. This is an open invitation to any toddlers in the vicinity to come and yank. Secondly, the accumulation of dust on the wires is not eliminated, it is merely relocated. Lastly, this technique presupposes shelf mounting the equipment, which in turn limits ventilation and dustproofing options.

The internal method involves mounting a false back a minimum of one inch in front of the actual cabinet back and running the wires between the two. See Fig. 3-5. This offers the best mechanical protection for the wires since they are only accessible by removing the cabinet back or access panel.

The wiring duct may be used simultaneously as the main intake duct for the ventilation system (see Chapter 4), which is by nature a part of the dustproofing system (see Chapter 3).

Another major advantage of a false back concealment system is the effective doubling of the carcase sheer bracing. The finished cabinet will have a much greater resistance to racking. Fully concealed wire runs do make it more difficult to install the audio cables, but this is remedied with access panels and pull wires.

PULL WIRES

Pull wires may be installed through blind wire runs during the final assembly stage of the cabinet. These should be used only to

Fig. 5-1. With several equipment compartments, it is wise to make the entire case back removable to facilitate wiring.

simplify the initial equipment installation. In later equipment demounting, the original cables should be left in place and disconnected at the equipment.

Pull wires are not substitutes for access panels or removeable backs, and should never be used where several turns must be negotiated. If the pull is too complicated, strain will be put on the audio cable during installation. Broken, shorted, or intermittent connections are the likely result.

Make your pull wire out of smaller gauge wire than the cable to be pulled. This is your insurance against damaged signal cables. If the strain is too great, the pull wire will stretch and break before the patch cord is damaged.

Cables should be securely taped to the pull wire with plenty of electrical tape. If you are stingy, you will be rewarded with an empty pull wire end and all your patch cord ends lost somewhere behind the false back. Tape completely over a.c. plugs and audio plugs to minimize their chance of snagging at turns or entry holes.

After the wires have been successfully fished through the blind wire runs, always untape the pull wire by hand. If you get in a hurry at this stage and try to cut the tape off, you are liable to damage the patch cords as well.

CABLE LENGTH

It seems silly to mention, but the connecting wires should all be long enough. Unfortunately, the temptation to stretch innocent patch cords to their limits and beyond is quite common.

All wires should be long enough to permit pulling any component out of the case to check a connection or to change a fuse. No wire should be taut. The phono ground strap is the most likely offender. For reasons known only to them, turntable manufacturers rarely allow more than 36″ for the ground strap. The wire will barely reach the amplifier ground in a typical vertical cabinet with normal equipment placement. Do not be tempted to just leave the ground unconnected. You will get a very bothersome hum every time the turntable is used. Splice a wire of equal gauge to the groundstrap to make it long enough. Do it right with solder.

If other audio leads run short, replace the entire cord wherever possible with a good quality, but longer patch cord. Extensions should only be used where the audio cables are permanently attached to the equipment at one end as on many brands of turntables. The more connections in any given input or output lead, the higher impedance at the expense of the signal.

PATCHING OCCASIONAL EQUIPMENT

Before you build or buy your cabinet, consider how often you might like to patch additional equipment into your main system. Maybe you have a friend who likes to bring his tape deck and make copies, or you have an old open reel deck you only use twice a year and don't wish to mount permanently in the cabinet. Perhaps you continually like to compare the latest technological advances to your present equipment.

These contingencies may be easily accommodated by running an extra set of patch cords out the back of the cabinet or to an easily accessible area within the cabinet, like behind a drawer. Then, when you wish to patch the auxiliary equipment in, you merely pull out the extra set of wires, hook up, and you're in business.

A more expensive way to do the same thing involves making a special patch panel and mounting it on some appropriate surface within the cabinet. See Figs. 5-2 through 5-4.

DESIGNING OUT INTERFERENCE

The chance for electronic interference can be minimized by how you route the input and output cables and speaker leads relative to the a.c. power lines. If at all possible do not run any signal lead parallel to the power leads. If the two must be in the same vicinity, it is best to have them cross at right angles.

After mounting your equipment in the cabinet, if you have a low (60 hz.) hum that was not there before, try reversing the main power plug in the wall socket, or the component plugs in the power tap on the back of the receiver, or both. Sometimes reversing polarity is all it takes. Check radiation from the ventilation fan by unplugging it. If the hum or static then disappears, replace the fan with a better model. Cheap fans are almost always electronically noisy.

Less frequently, the hum can be caused by your particular equipment placement. Mounting a large transformer (such as the amplifier transformer) too close to the tape heads or the magnetic phono cartridge places the sensitive pickups within the magnetic field of the transformer. This is another good reason to mount the tuner or preamplifier between a large amplifier and a cassette deck.

INTERNAL POWER CONNECTIONS

Power leads for most components are plugged into the auxiliary receptacles on the back of the receiver or preamplifier

Fig. 5-2. RCA jacks and a blank switch plate.

chassis. Make sure that the sum of the component input wattages does not exceed the rated capacity of the circuit (printed next to the receptacle). It is good practice to leave twenty or thirty percent of the capacity unused as a margin of safety.

Fig. 5-3. Cover the switch plate with veneer to make a convenient and attractive patch panel.

In general, the tape decks, tuner, amplifier, lights, and fan can be plugged into the switched outlets. This way the receiver or preamplifier switch acts as a master switch for the entire system. The turntable should always be plugged into an unswitched outlet, or lacking one, the wall. This permits the turntable to complete its automatic cycle and switch itself off independently of the rest of the system.

If the switched circuits do not have enough capacity to connect all the components desired, sacrifice a tape deck to an unswitched outlet. Always leave any ventilation fans hooked up to switched outlets so that they come on with the heat producing power units.

OUTSIDE LEADS

The power line from the wall, speaker leads from externally placed speakers, and the antenna lead all need to be run from outside the cabinet in to the back of the appropriate component.

In a dust-baffled ventilation system, the air intake holes provide the most convenient pathway for bringing in the wires. Care should be taken to keep any power leads as far as possible from the speaker and antenna wires. Wire slack should not be allowed to accumulate within the air duct where it may restrict air flow. Instead it should be pulled either to the outside of the cabinet or into the equipment compartment behind the equipment panel.

Fig. 5-4. Rear view of the patch panel.

Most connections within the case utilize factory plugs and jacks. These require little attention other than making sure that the plugs are securely inserted. These friction-fit connections do need occasional inspection. They may corrode, increasing the resistance in the circuit to the detriment or discontinuance of the signal. In my experience, it is much simpler and more reliable to replace worn or corroded patch cords rather than to attempt to clean or repair them. Completely new cords cost only a fraction more than replacement plugs or jacks.

You might have to make some of your connections because of installation problems inherent in your cabinet design. Lengthening the ground strap of the turntable, hooking up the lights and fan, and installing auxiliary switches within the 115 v.a.c. lines are the most common "custom" connections.

WIRING YOUR STEREO

Flexible 18 gauge zip cord (lamp cord) is the most frequently used wire in stereo installations. It can carry enough current to be used in most main power circuits, and can do good service as speaker leads as well.

If the wire is to be used in a 115 v.a.c. circuit, total the maximum electrical load imposed on the wire and make sure it is under ten or fifteen amps (for eight to ten feet of 18 ga. wire). To do this, add the total input wattages of all the equipment to be run at the same time and divide by the house voltage (normally 115 volts). If the load exceeds the capacity of the wire, use heavier wire or run an additional power line for part of your components (Fig. 5-5). If your total power consumption is high, you may have to divide it among several separate wall circuits since most household circuits are rated either twenty or thirty amps. However, this would be an extremely rare occurence in a home sound system. Most systems use less than three amps at a time.

If the wire is used as speaker wire, the stripped ends should be tinned with solder to prevent stray strands from shorting across the terminal blocks. A direct short could blow out your amplifier.

If the speaker wires have to be longer than twenty feet, it is a good idea to check the loop impedance of each speaker with an ohmmeter to make sure it is within bounds of the amplifier specifications. Heavier gauge wire may be required.

Mechanical 115 v.a.c. Connections

Most of the time power leads will end in a "quick-connect" male plug on one end, and a female "quick-connect" receptacle or screw terminal on the other end. Different brand "quick-connect" plugs have different instructions. Follow the manufacturer's recommendations.

Wire ending at a screw terminal should be stripped about ½″ from the end, tinned, and looped around the screw with long nose

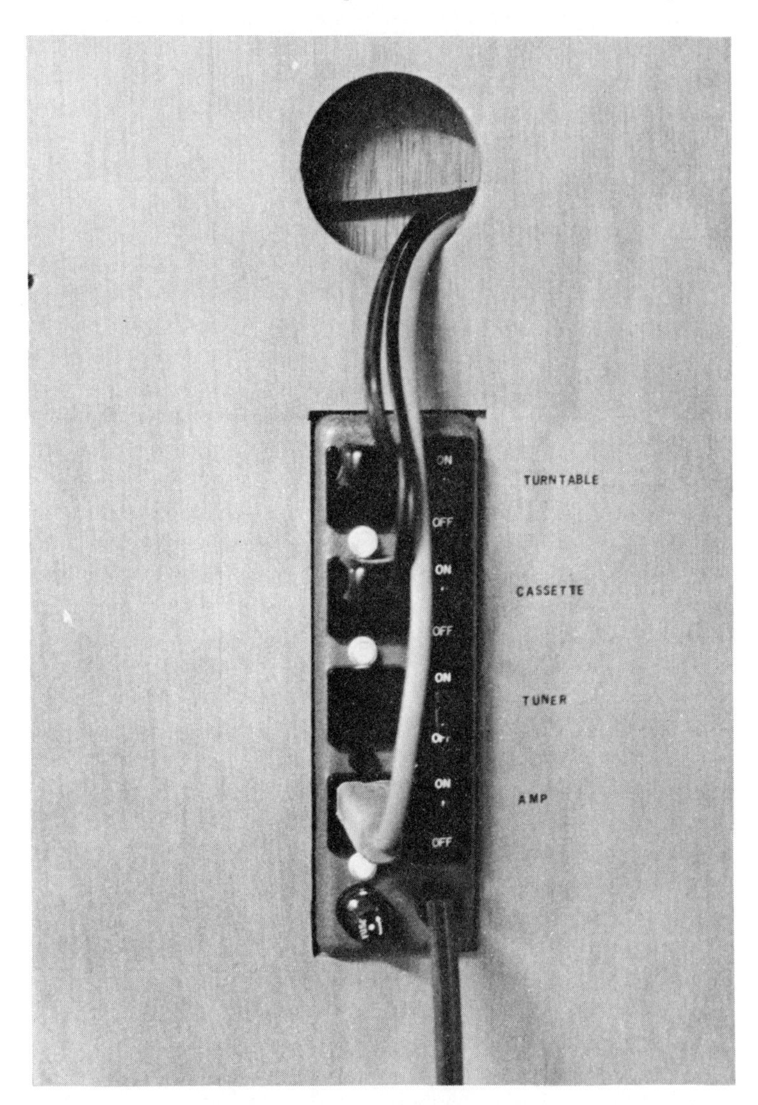

Fig. 5-5. An independently fused and switched receptacle box handles the current necessary for higher power equipment.

pliers. The loop should be tight and make a three-quarters to one complete turn around the terminal screw. Excess wire should be snipped off with diagonal cutters. After the screw has been tightened, the end of the insulation should be no closer to the screw head than 1/32″, and no further away than ⅛″.

In joining two separate wires together, strip the insulation off ½″ from the end and twist the wires neatly together. The joint should then be soldered, the excess wire snipped off, and the whole junction thoroughly wrapped with electrical tape.

Solder Connections

In stereo furniture construction, a 50 to 100 watt soldering gun or pencil is sufficient for nearly all connections encountered. The tip should be either pointed or chisel-shaped, and it should be properly tinned. To tin a new tip, apply flux and solder to the tip as it heats up. Maintain the bright, shiny surface by wiping the oxide buildup off with a damp sponge from time to time.

If the soldering gun has not been used for some time, loosen the tip mounting nuts slightly and retighten to insure a good heat transfer to the tip.

Resin core 60/40 solder (60% tin, 40% lead) is preferred for electronics work. The resin flux melts at a temperature just under the solder and flows ahead, cleaning off the solder-resisting metal oxides. *Acid core solder should never be used for electronics*. The residue from the flux will itself attack and corrode the connection.

Solder Application

Heat up the properly tinned soldering tip and apply the flat face directly to the connection being made. Almost immediately apply the end of a strip of resin core solder at the same point so that it touches both the tip and the wire being connected. As soon as the molten solder has been completely drawn into the entire area being joined, remove the solder strip, and then the gun tip.

A properly soldered connection looks shiny and smooth. There should be no dull or uneven globs. This type of connection is called a "cold" solder joint and is caused by not enough heat at the joint, or unclean joint surfaces.

If the soldered connection was heated too long or too hot, the ends of the insulation can melt or burn. The best way to prevent this while applying enough heat for a good connection is to clip an alligator clip tightly to the wire between the solder joint and the insulation. The clip will draw much of the heat off the wire and away from the insulation.

After soldering the joint and clipping off all excess wire, tape all with a good quality electrical tape.

Securing Interior Cables

Low power cables within the cabinet may be stapled to the case with round headed staples. Wires and patch cords that may be changed in the future should remain unattached. The equipment compartment may be tidied up by tying groups of patch cords neatly together with wire ties. *Never* tie a signal or speaker wire with a high power line or you will most likely introduce a hum into the system. Power lines should be clamped to the inside of the case with standard plastic wire clamps or left free.

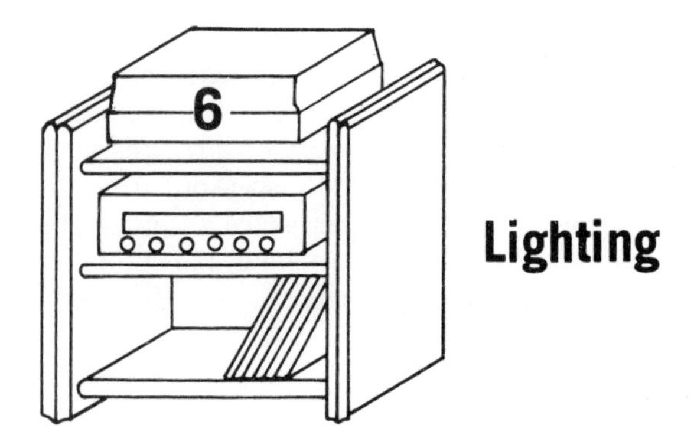

Lighting

Discretely placed lights within the cabinet greatly improve the ease of equipment operation while enhancing the overall visual effect. Lights over the turntable shelf make it easier to locate that special selection on your favorite album or to spot fingerprints and dust as you clean your records. An external master pilot light reminds you that your system is still on lest you go off to bed and forget it.

FORM FOLLOWS FUNCTION

The lighting should not be overdone. Too many lights distract from the inherent beauty of the cabinetry and equipment, waste too much input wattage, and frequently generate more heat than the equipment.

Illuminating lights should be placed only where it is necessary to see the controls and the room light is restricted. A good example of this is the turntable. Both the controls and the platter must be fully visible for normal operation. One or both of these areas is in the shadow of the next shelf up, and so the situation calls for some lighting.

On the other hand, a shelf mounted receiver has all its controls on its front panel. Under normal conditions, the room light is sufficient for proper operation.

Horizontal console designs having the equipment facing up from under a lift lid usually do not require supplemental lighting.

TURNTABLE LIGHTING

Turntable lighting should be as high above the platter and as near the front of the cabinet as the turntable compartment will allow (Fig. 6-1). The light sources should come from both sides of the compartment since not everyone in the world is right-handed.

The direct aspect of the lighting casts a mild shadow useful for locating dust particles and fingerprints. The indirect reflections off the compartment sides and top yield a softer light that makes reading the record label more pleasant.

In vertical cabinet designs having a cassette or accessory drawer immediately below the turntable, it is possible to locate the turntable light fixtures so that they illuminate the contents of the drawer as well. Then in those occasional intimate situations, you would not have to spoil the atmosphere by flipping on a room light to change a tape.

INCANDESCENT VERSUS FLUORESCENT

Fluorescent lighting, watt for watt, generates about the same heat as incandescent lighting. But since it produces about twice the light per watt, it is frequently used in stereo installations. This is a mistake. Fluorescent lighting is caused by the bulb coating glowing

Fig. 6-1. View from the inside of the equipment compartment shows the backs of the light sockets mounted on both sides, in the front of the turntable compartment. Note the equipment mounting skids and the extra power sockets.

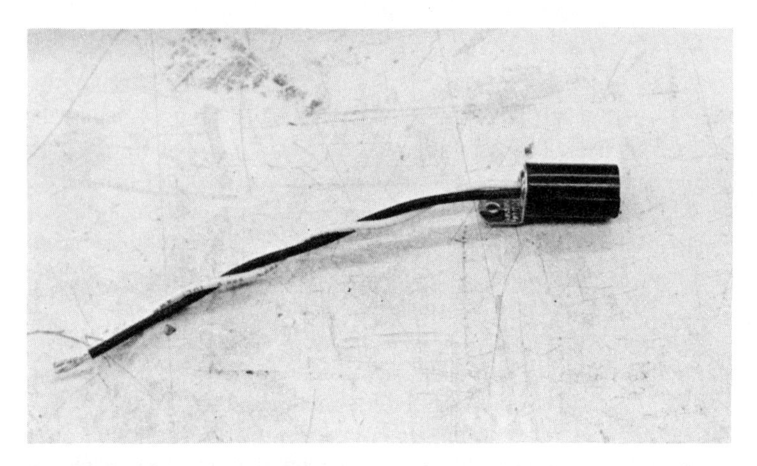

Fig. 6-2. Screw-mount type socket mounts to either the back of the light baffle or the bottom of the equipment shelf.

when it is exposed to ultraviolet light, which is generated by electron bombardment of the tube vapor. With sixty cycle house current, this means 120 separate flashes, or sparks, per minute. Each flash emits other electromagnetic radiation as well which is heard as static when picked up and amplified.

Incandescent lighting is not quite as efficient a light source, but it introduces no static into the sound system. Use the smallest lights possible to get the job done to minimize the waste heat. Night light bulbs are about the right size and are easy to replace when they burn out. Fancier "furniture" fixtures usually require more expensive, and hard to find replacement bulbs, but night light bulbs are sold in virtually every drug store, supermarket, or hardware store in the country. See Figs. 6-2 through 6-5.

ADJUSTING THE LIGHTING

With incandescent fixtures, the lighting may be increased or diminished to suit your taste. If brighter lights are desired, then wire the two socket installations in parallel. If even brighter lights are required, connect additional sockets in parallel, but keep track of the total load of the lights on your power input. Most night light bulbs are rated six or seven watts each. Two bulbs are sufficient under normal circumstances.

If the lighting is too bright, wire the two sockets in series. As a bonus, the bulbs should last longer. Avoid using a light dimmer in stereo installations. Most dimmer switches generate static.

Fig. 6-3. This type socket is usually mounted through holes bored in the equipment shelf (see Fig. 6-1).

LIGHT BAFFLES

Place the fixtures so that the bulbs are easily accessible for changing, but not directly visible from the front of the cabinet.

Fig. 6-4. Close-up of socket mounted through panel.

Fig. 6-5. Christmas-type socket mounted on the back of the light baffle with epoxy putty.

Mount a two inch wide board in front of the bulbs, or extend the equipment panel down past the front of the equipment shelf to baffle the direct light from the front.

Maintain a small clearance between the bulb and the wood baffle to facilitate bulb removal and to protect the wood from heat

Fig. 6-6. Veneered blank switch plate.

Fig. 6-7. Master switch and pilot light mounted in the switch plate.

damage. Light baffles should be mounted with screws so that they are removable. This allows greater flexibility in future revamping of the equipment section.

CONNECTING THE LIGHTS

If possible, hook the lights into the switched receptacle on the back of the receiver or preamplifier. They will then come on

Fig. 6-8. Side view of master switch and pilot light assembly.

automatically with the stereo. Before connecting, double-check to make sure that the additional twelve or fourteen watt load added to the other switched components is within the rating of the switched receptacle. A cube tap may be required to connect all the required switched components.

If there is no switched receptacle or the capacity of the receiver switch is too low, the lights may be wired through an auxiliary power switch and plugged into the wall. Mount a toggle or push button switch at one end of the light baffle. Separately switched lighting just isn't quite as dramatic.

PILOT LIGHTING

Pilot lighting is most useful in cabinets with opaque doors which conceal the turntable lighting. With the doors shut it is easy to forget that your system is on after the automatic turntable shuts itself off. Since electronics tend to have a somewhat fixed life, the practice of leaving the sytem on, intentionally or unintentionally, for days on end is wasteful.

A little neon pilot light (see Figs. 6-6, 6-7 and 6-8) can be switched through the receiver as well. The most difficult aspect of pilot lighting is placement. The light should be visible if not obvious when lit, but this is hard to do without the fixture looking like an afterthought.

A clever solution to the esthetical problem is to mount a night light socket inside the case so that the bulb shines out the vent holes in the cabinet top or back. When the room lights are on this is not visible, but it casts a noticeable corona of light when you shut off the room lights.

7 Tools and Machinery

Three key elements will determine how well your stereo cabinet will turn out: imagination, skill, and equipment. A strength in any one of these areas will go a long way towards minimizing a weakness in the other two. It is possible to construct a decent stereo rack with a hand saw, a hammer, and perhaps a screwdriver. But the better your equipment, the more room your imagination and skill have to express themselves.

In this chapter it is assumed that the reader has working knowledge of the tools he intends to use. The following is merely a catalog of the operations normally performed in stereo furniture construction and the machines or alternate machines used.

SAFETY

On the theory that safety in the shop is never stressed too often, always keep the following admonitions in mind.

■ **Keep your work area clean.** Cluttered shops are full of dangerous distractions and temptations. The same work attitude that produces a messy shop will surely produce a sloppy stereo cabinet. A clean shop saves more time in tool location, efficiency, and good attitude than it costs to clean it. Remember, "a dirty shop is a lazy shop."

■ **Keep children and any other nonparticipants completely out of the work area.** Not only can an inquisitive child hurt himself, but because of the distracting influence, he can cause you to hurt yourself.

Fig. 7-1. Commonly used safety equipment. The full face guard is both more comfortable and gives better facial protection than the goggles. Ear protectors are a must for sawing and routing.

■ **Dress for the job.** Wear comfortable clothing that you wouldn't be too upset to get a little glue or stain on. Sleeves should be short or rolled up, shirttails tucked in. Nothing should be loose

Fig. 7-2. Good quality table saw is a must for accurate cutting of equipment panels. Note sliding table for cross-cutting long panels.

or hanging. Rubber-soled shoes give both good footing and some degree of protection from electrical shock.

■ **Secure your work with a vise or clamps.** Holding the work with one hand and the tool in the other is dangerous. Clamping the work frees both hands to control the tool.

■ **Understand your equipment.** Know the normal operations and limitations of each tool, and the potential safety hazards as well. Never use a tool or an attachment to do something for which it is not intended. Compromise set-ups will always fail sooner or later, ruining material and hurting the operator.

■ **Maintain the tool properly before using it.** Make sure all cutting edges are sharp and all bearings lubricated. Dull tools have hurt more people than sharp ones. Check to see that grounded plugs are actually grounded.

■ **Use safety equipment.** Goggles and ear protectors (Fig. 7-1.) are a must. Tool guards should be checked to make sure they will operate properly, and should be replaced if damaged.

TYPES OF SAWS AND THEIR USES

Table Saw. The table saw shown in Fig. 7-2 is probably the most useful machine in the wood shop. Depending on make and model, it is safer to operate and more accurate than its cousin, the radial arm saw.

Jointing or edging lumber

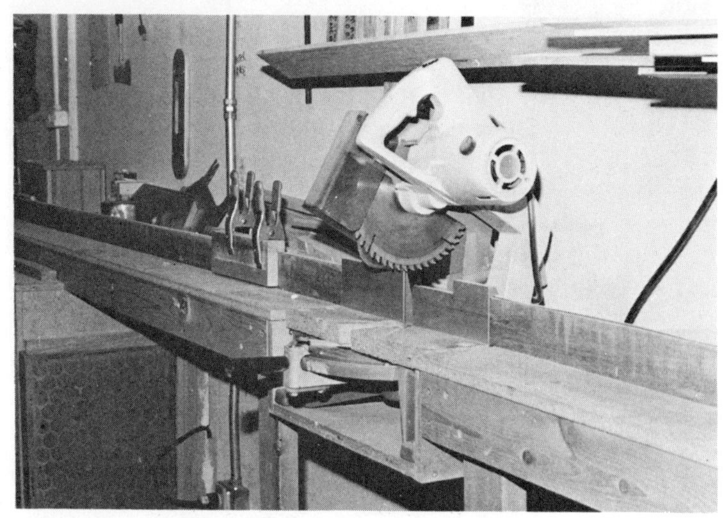

Fig. 7-3. Mounted miter saw with extension tables and stop.

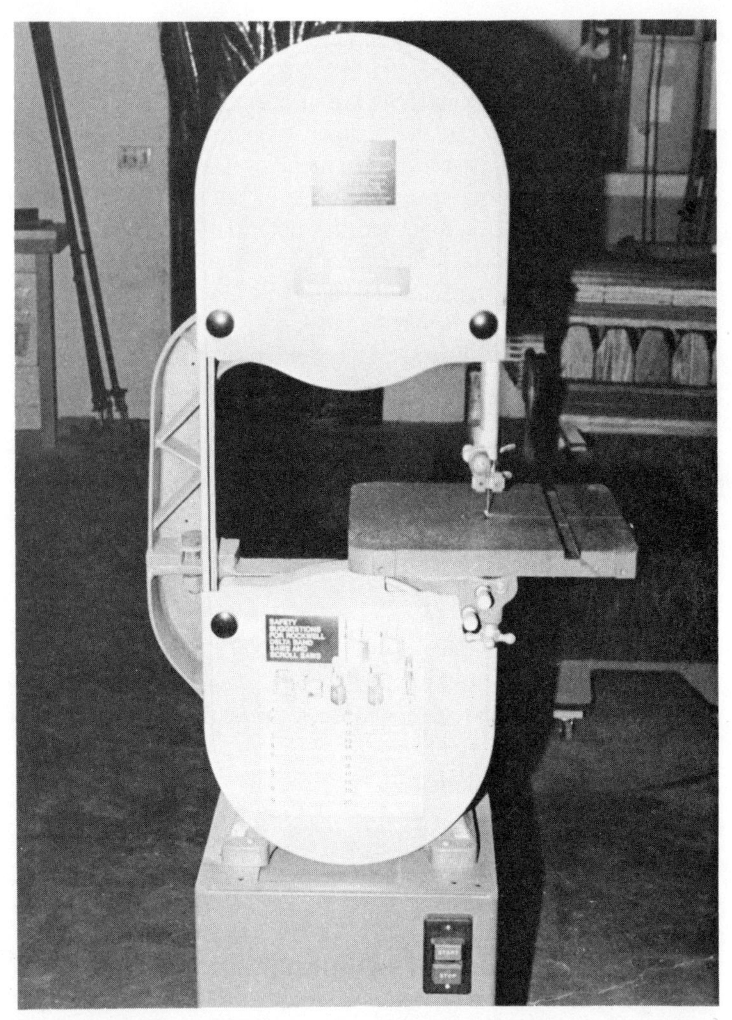

Fig. 7-4. A bandsaw is indispensable for cutting exterior curves.

Ripping lumber
Crosscutting lumber (with miter gauge)
Cutting bevels
Raising panels
Resawing lumber (up to 3″ thick)
Cutting miters
Cutting grooves and dadoes (with dado attachments)
Shaping edges (with molding attachments)
Shaping molding (with molding attachments)

Fig. 7-5. Both bayonet saw (freehand) and router (with template and flush-trimming bit) are useful for cutting interior curves.

Cutting coves
Dishing out small circles
Cutting tenons
Cutting hinge mortises
Cutting tapers
Ripping plywood
Crosscutting plywood

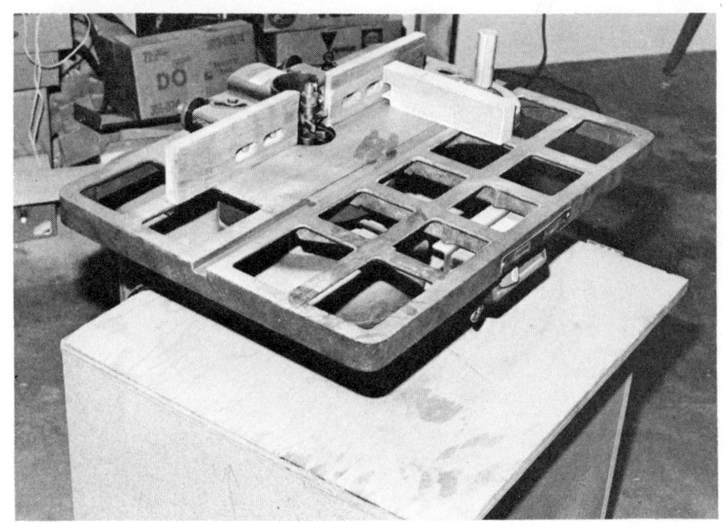

Fig. 7-6. Light shaper set up for cutting tenons.

Fig. 7-7. Router table set-up for accurate edging and cutting mortises.

Making equipment panel cutouts
Making access panel cutouts
Cutting ventilation slots in shelves
Cutting out large circles from panels
Disc sanding (with sanding attachment)

Radial Arm Saw. The radial arm saw has long been a favorite of the home craftsman. Although it is usually more

Fig. 7-8. Bottom view of router table.

cumbersome and less accurate than the table saw, it can perform nearly all the same functions, with some others besides.

Many people use the radial arm mainly as a cutoff saw. In this capacity it is superior to the table saw in convenience and accuracy. However, for widths of less than 6″ (most circumstances), a power miter saw makes much more accurate cuts.

Jointing or edging lumber
Ripping lumber
Crosscutting lumber

Fig. 7-9. The drill press can be used with hole saws for cutting tweeter holes in speaker baffles.

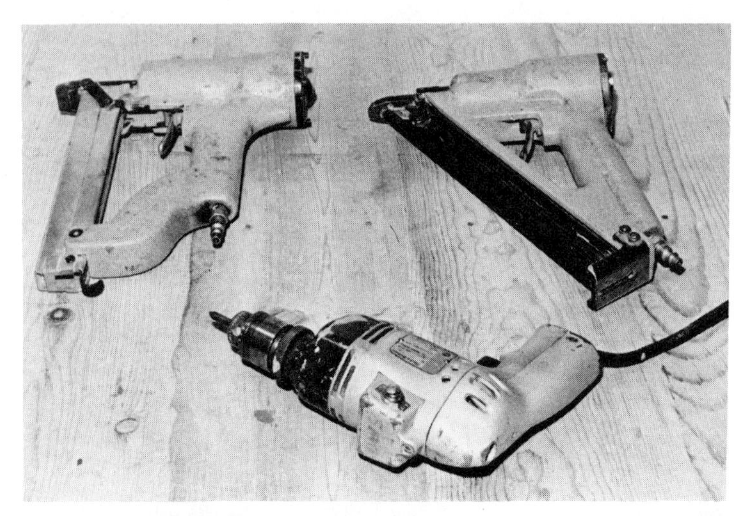

Fig. 7-10. Variable speed, reversible hand drill can not only drill all the vent holes, wire holes, and screw holes, but can drive screws as well. Power fasteners make assembly faster and more accurate.

Cutting bevels
Raising panels
Cutting miters
Cutting grooves and dadoes (with dado attachments)
Shaping edges (with molding attachments)
Shaping molding (with molding attachments)
Cutting coves
Cutting tenons
Cutting hinge mortises
Cutting tapers
Ripping plywood
Crosscutting narrower strips of plywood
Disc sanding (with sanding attachment)
Planing narrow boards (with plane attachment)
Routing grooves (with chuck attachment)
Boring holes (with chuck attachment)

Power Miter Saw. The power miter saw, by virtue of having only one pivotal point and no positional or arbor adjustments, (Fig. 7-3) is considerably more accurate than the radial arm saw for cutoff work. It is, however, limited to crosscutting widths of 6″ or less and can do little else. But with a table saw, it makes an unbeatable combination.

Cross cutting lumber (less than 6″ wide)

Crosscutting plywood (less than 6″ wide)

Cutting miters in narrow strips of plywood, lumber, or molding

Cutting aluminum tubing, aluminum shelf standards, or dowels to length

Band Saw. If you don't own a band saw as shown in Fig. 7-4 you may have to use a bayonet saw instead.

Straight cutting (with fence, limited to capacity of arm)
Cutting miters (with miter gauge)
Cutting bevels (with tilting table and miter gauge)
Freehand cutting or irregular lines and curves
Cutting cabriole legs
Resawing wide boards of lumber

Fig. 7-11. Floor belt sander can be used either vertically or horizontally. This model has a disc attachment.

Fig. 7-12. Floor belt sander with dust collection set-up.

Cutting coped molding joints
Cutting thin metal (with metal cutting blade)
Ultrathin belt sander (with special sanding belt)

Bayonet Saw. The bayonet (saber) saw (Fig. 7-5) performs many of the same functions as the band saw, but with great depth

Fig. 7-13. The secret of a good finish is superlative sanding. Hand belt sander removes knife and chatter marks, while finishing sander polishes surface free of sanding marks.

and accuracy limitations. In thick stock the blade tends to flex while cutting a curve, resulting in an out of square kerf face. Like the band saw, different functions require different blades.

Freehand cutting of irregular lines and curves in thin stock
Cutting interior cutouts (impossible on band saw)
Cutting coped molding joints
Making equipment panel cutouts
Making access panel cutouts
Cutting ventilation slots in shelves
Cutting hinges and shelf standards (with metal cutting blade)
Cutting ventilation fan mounting hole
Making speaker hole cutouts.

Portable Saws. In general, portable power saws are not accurate enough for fine case work. In the hands of an expert, they will do in a pinch, but the time and skill required for good joinery are beyond the reasonable expectations of the average craftsman.

OTHER CUTTING TOOLS

Shaper. The shaper shown in Fig. 7-6 is a very versatile woodworking tool.

Edge shaping
Cutting molding

Fig. 7-14. The larger the collection of clamps, the better the cabinet that will be produced.

Fig. 7-15. Fish tape, volt-ohmmeter (back row); 18″ hooked piece of coat hanger, round headed stapler, long nose pliers, wire strippers, bullseye level, diagonal cutters (second row); and soldering gun are necessary for a professional quality wiring job.

Jointing narrow (1″) boards
Coping standard frame joints
Cutting tenons
Cutting grooves in narrow stock

Router. The router shown in Fig. 7-5, 7-7 and 7-8 is especially useful for making internal cutouts.

Edge shaping
Cutting simple molding
Cutting curves in thin stock
Making internal cutouts
Cutting dadoes and rabbets
Cutting mortises
Cutting internal fluting, shaping, or freehand carving
Flush trimming plastic laminates or veneers
Flush trimming wood edging or case joints
Cutting dovetail joints
Cutting circles
Making speaker hole cutouts
Cutting ventilation fan mounting hole
Duplicating patterns

Carving letters

Drill Press. The drill press shown in Fig. 7-9 is much more accurate than a hand drill.

Boring screw holes
Boring wire ports (1″)
Boring ventilation holes in cabinet backs or shelves
Boring starter holes for interior routing
Boring doweling holes
Cutting out small speaker holes (with circle cutter bit)
Countersinking screw holes
Boring semicircles for wire clearance

Hand Drill. Most of the functions of the drill press can be performed with a portable electric drill, but with a great loss of efficiency and accuracy. A variable speed hand drill (Fig. 7-10) may, if equipped with a Phillips-type bit, be used as a power screwdriver.

SANDERS AND MISCELLANEOUS TOOLS

Power Staplers And Nailers. Few home shops have access to power nailers and staplers (also Fig. 7-10), which require an air compressor to operate. In most circumstances, hand nailing, screwing, and glue and clamping are sufficient. Power nailers have the advantage of saving time, and freeing up one hand to steady the work (instead of holding a nail or screw), but require an investment (including compressor) in the vicinity of $500.

Floor Belt Sander. Floor belt sanders as shown in Fig. 7-11 and 7-12 are rugged and dependable.

Sanding edge kerf marks
Sanding face mill or chatter marks
Sanding edges of larger curves
Fitting out-of-square doors

Floor Disk Sander. This floor disk sander shown in Fig. 7-11 is mounted next to a belt sander.

Smoothing rough-cut curves
Sharpening pencils

Hand Belt Sander. The hand belt sander shown in Fig. 7-13 is completely portable.

Removing kerf marks
Removing mill and chatter marks

Sanding solid glued up panels smooth
Sanding face and door frame joints smooth
Sanding face frame/carcase joint flush
Removing excess filler from joints and nail holes

Hand Finishing Sander. The small hand finishing sander shown in Fig. 7-13 is used for final touch-ups.

Finish sanding

Clamps. One of the greatest differences between an amateur and a professional craftsman is the importance placed on clamping. The amateur tries to get by with one or two makeshift clamps or weights. The professional has a major financial investment in his clamps, and uses every one, all the time wishing he had more (Fig. 7-14).

The clamp is the chief assembly tool, even more important than the nail or screw fastener itself. After all, the main purpose of fasteners is only to maintain pressure on the joint until the glue dries, not to secure the joint itself.

Glue is what holds the cabinet together. A glue joint that cures under pressure is often stronger than the surrounding wood, but a glue joint without sufficient pressure to force the glue into the exposed wood cells is not a glue joint at all.

Clamps come in such a diversity of sizes and shapes and purposes that it would be impossible to list them all here. For our purposes, the most commonly used clamps are 6″ C-clamps, and 36″ and 96″ bar or pipe clamps. The specific job will dictate just how many of each and what sizes are required, but it is safe to say that a minimum of three or four C-clamps, six or eight 36″ bar clamps, and three or four 96″ pipe clamps will be required. If you can arrange to have more than that, you will build a better cabinet.

Hand Tools. A normal selection of common hand tools is, of course, required. In addition to the faithful hammer, squares, tape measure, chisels, and so forth, a few specialty tools will be needed. See Fig. 7-15.

A **volt-ohmmeter** is useful for checking for shorts and continuity in the finished wiring. It may also be used to check the loop impedance in the speakers after installation.

A **fishtape** comes in handy for especially long wire pulls, or when pulling wires through walls, floors, and ceilings of a building. An **18″ piece of coat hanger** with a small hook bent on the end is indespensable for fishing wire the short distances in the cabinet.

A **bubble "bulls-eye" level** is a must for checking turntable level.

For the electrical work, a 100 watt **soldering gun, wire strippers, diagonal cutting pliers,** and **long nose pliers** will be required. In addition, a **round-headed staple gun** will secure any low-power wires out of the way inside the case.

The complexity of the design of the cabinet will dictate what tools are required what compromises in technique must be made, in short, the feasibility of your project. Think through the entire construction procedure before you begin, and make sure you have access to the equipment required to complete your project.

Equipment Mounting

The single most important consideration in both designing and actual construction of the cabinet is how your equipment is to be mounted. Ventilation, wire concealment, dustproofing, lighting, proportion, and construction technique all revolve around equipment placement and mounting.

There are three basic methods of mounting stereo equipment into a cabinet: shelf mounting, panel mounting, and combination mounting. Each method has several adaptations, advantages, and disadvantages. The typical cabinet will use two of the three techniques, and frequently all three are used in the same cabinet.

SHELF MOUNTING

The simplest and most frequently used method is shelf mounting (Figs. 8-1 and 8-2). This technique is just what it seems to be—mounting the component by setting it on a shelf, as is. Shelf mounting is generally used in better cabinets in conjunction with false backs. The component signal and power leads are run straight out the hole in the false back. The equipment bulk itself tends to hide the exposed portion of the wires.

Turntables are invariably mounted in this fashion (Fig. 8-1). Most modern turntables feature a rather sophisticated base designed for stability and to minimize acoustical feedback. They would be extremely difficult to disassemble for direct deck mounting (as was common some years back), and should be shelf mounted intact. Turntables by nature are very mechanical and require much more frequent servicing and replacement in comparison with the other components.

Pullout shelves or platforms are a variant of this method, but they should only be used with large, infrequently used open reel

Fig. 8-1. Both simple and angle shelf mounting in one unit.

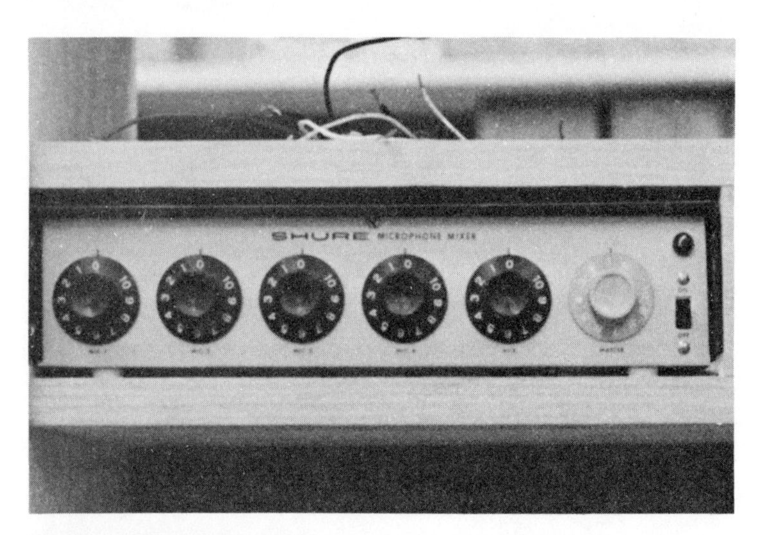

Fig. 8-2. Closely fit shelf mounting.

Fig. 8-3. Recesses are drilled on top of angle shelf to prevent unit from sliding off.

decks to save space. Pullouts should never be used with a good quality turntable. The turntable manufacturer has spent a lot of time and money in designing a stable vibration-resistant base. By mounting this technological wonder on a shelf, mounted in turn on drawer guides, at best you are condemning it to a life of nearly constant vibration and shifting level. At worst, you will shove the whole assembly into the cabinet too hard sometime with the tone arm down and leave a permanent canyon across the grooves of a valuable record. Pullout turntable shelves are a sign of obsolete or careless design.

The open reel deck is another component frequently shelf mounted for service or replacement reasons. Power equipment, tuners, cassette decks, and all the other components may also be shelf mounted if desired. The main advantage of doing so is the simplicity of changing components. If you trade in your system for the latest state-of-the-art exotics on a semi-annual basis, you probably would be better off shelf mounting everything.

Fig. 8-4. Wires and light fixtures mounted to the underside of angle shelves are not visible from the front of the cabinet.

Angle Shelves

Top loading units such as many older video and audio cassette decks can be mounted just below eye level on angled shelves (Fig. 8-1). This directs the meters at your eyes while raising the cassette deck from space competition with the turntable.

Fig. 8-5. Television shelf mounted on top of cabinet.

Fig. 8-6. Cabinet designed around console television with legs removed.

Mount the angle shelf with cleats and screws. Besides being easier to install than cutting an angle dado, most angle mounted equipment will probably be changed for horizontal mounted front control equipment in the near future. Cleat and screw shelf mounting allows for future change.

Stop cleats for feet recesses (Fig. 8-3) must be positioned on the angle shelf to keep the equipment in position. The angle used will vary with exact altitude, but most angles are steep enough to cause the equipment to slide.

Another way to angle mount equipment is to make a wedge or stand that sits on a horizontal shelf. When you eventually get front loading equipment, you simply remove the angle wedge support from the shelf and put in the new unit.

Angle shelves are very good at concealing wires (Fig. 8-4). Drill holes for the power and signal leads directly beneath where

Fig. 8-7. Shelf mounted television and angle mounted video cassette deck.

they come out of the component. If the shelf is directly over the turntable, you can mount your lights to the underside.

Televisions and Video

Televisions are normally shelf mounted (Figs. 8-5, 8-6 and 8-7). If the shelf mounted television cannot be easily viewed from all around the room it is mounted on a lazy susan turntable (Figs. 8-8 through 8-13). Frequently this turntable is itself mounted on a pullout platform to permit facing the screen towards any given point in the room, even along the same wall as the cabinet. A very heavy duty (rated for 150 lbs. or more) pair of glides is required for this.

Video accessories are also normally shelf mounted. Video cassette recorders, cable television control boxes, and other accoutrements of a well-equipped television have not at this date become standardized enough to risk panel mounting. This may change with time.

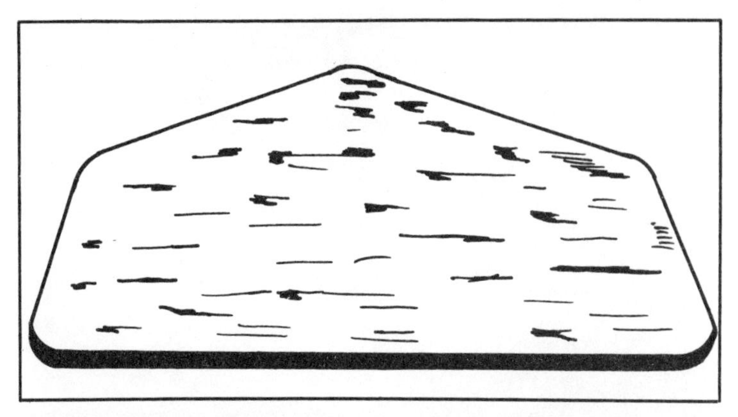

Fig. 8-8. Lazy susan platform has back corners cut away to permit swivel clearance inside case.

Fig. 8-9. Parts of lazy susan pullot platform laid out ready to assemble.

Fig. 8-10. Platform mounted in case ready to receive lazy susan.

PANEL MOUNTING

Panel mounting requires cutting a vertical wood or plastic panel to closely fit specific equipment (Figs. 8-14, 8-15 and 8-16).

Fig. 8-11. Lazy susan pullout platform complete. Recesses will be drilled in platform to receive television feet to prevent unit from sliding off platform.

Fig. 8-12. Lazy susan platform mounted with television set in place.

Fig. 8-13. Close-up view of the lazy susan platform mounted TV.

Minimal shelves or skids are required behind the panel to support the equipment weight. Properly executed panel mounting looks very professional and is intimidating to thieves. A burglar faced with panel mounted equipment usually cannot risk the time it would take to figure out how to extract it.

Panel mounting is also a principal element in the ventilation and dustproofing systems. Shelf mounted equipment rests on shelves which block air movement within the cabinet. Panel mounted equipment is suspended out into the middle of a bath of cooling air. With a properly designed intake and exhaust flow, panel mounted equipment is frequently better ventilated than the same piece of equipment placed on a table top out in the open.

Likewise, a panel closely fitted around the equipment seals off the front of the equipment compartment from dust infiltration. Shelf mounted equipment by contrast is continually accumulating all the dust coming in from the front of the cabinet.

Even though most equipment panels are made to be removable and replaceable, changing them does require expense and trouble. Panel mounting should therefore not be used where the equipment is changed frequently—say, two or three times a year.

Equipment with front mounted controls (receivers, tuners, preamplifiers, amplifiers) are prime candidates for panel mounting. Front loading cassette decks can likewise be panel mounted, but top loading cassette decks should be shelf mounted. Horizontal planes within the case are generally major structural braces and should rarely be tampered with. It is difficult to make these surfaces removable and replaceable without compromising the

Fig. 8-14. Equipment panel mounted into built-in.

Fig. 8-15. Virtually all types of equipment may be panel mounted. Note switch box and antenna at lower right.

strength of the case. Front facing vertical panels, on the other hand, are superfluous from a structural point of view. The well-constructed face frame, the back, and any false backs supply sheer strength to the cabinet. Since minimal fastening is required for a vertical panel, it is easily made removable.

Fig. 8-16. Equipment panel mounted in horizontal panel.

The first order of business for panel mounting is to remove the feet from the components that will slide in from the front. Feet are mounted several different ways. The easiest to remove are mounted with a Phillips screw (Fig. 8-17), and are taken off by simply unscrewing them.

Plastic feet that have a little plastic plug instead of a screw are removed by prying up the plug with a small screwdriver (Fig. 8-18). This allows the three notched prongs to retract in their slots, permitting removal. Expansion prong feet may have a felt pad covering over the plastic plug. Carefully peel off the felt pad to expose the plug.

Small rubber feet (Fig. 8-19) are generally glued on with a contact adhesive. Gently peel these off, and be careful not to tear

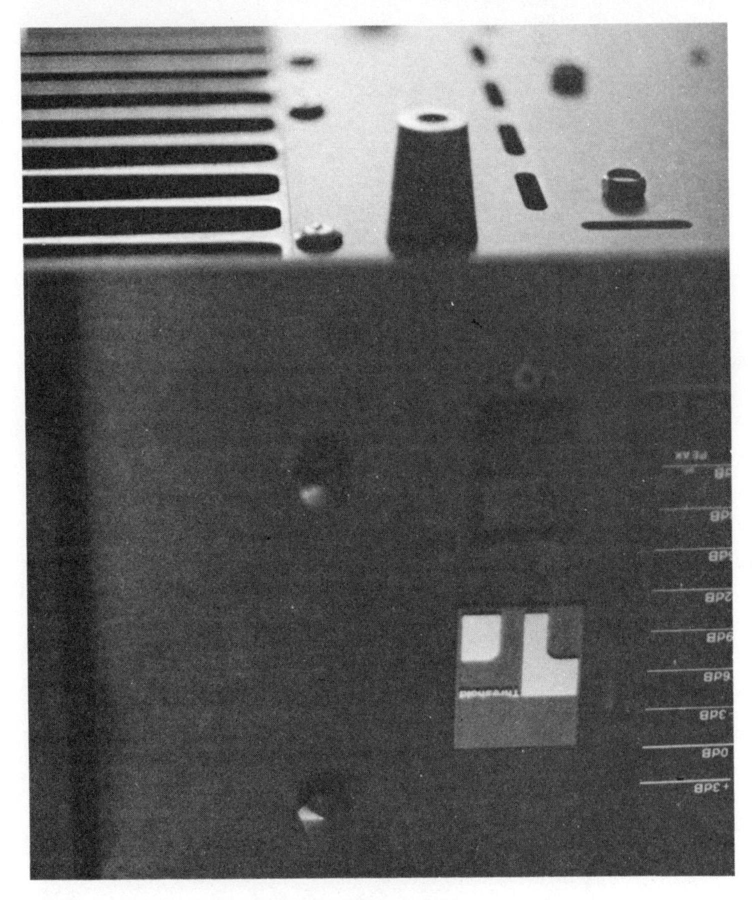

Fig. 8-17. Typical screw mounted equipment foot.

Fig. 8-18. Prong and plug mounted foot.

the rubber. In all cases the feet should be placed in an envelope, labeled for the piece of equipment they came from, and stored for the future. You never know when you might want to trade or sell your equipment.

Some components are designed to be "rack" mounted. Typically, these units have a large face plate with a wide lip all the

Fig. 8-19. Rubber stick-on feet.

Fig. 8-20. Back of face plate, rack mounted equipment.

way around the metal cover, and holes are provided on both sides of the face plate to bolt the equipment to a rack or panel (Figs. 8-20 and 8-21). The metal case should be left on these units. Measure

Fig. 8-21. Close-up of back of face plate.

Fig. 8-22. Projecting screw heads. Most metal covers have recessed screw heads.

for the panel cutout directly from the largest case dimensions (but smaller than the face plate dimensions). Be sure to allow clearance for any projecting screws (Fig. 8-22), but otherwise keep the case clearance to a minimum. This is done for two reasons. First, it is

Fig. 8-23. Rear view of a face plate showing the panel cutout overlap.

Fig. 8-24. Edge view of face plate.

easier to enlarge a cutout if it is too small than it is to reduce it. Second, the more snugly the panel fits around the equipment chassis, the less dust can infiltrate the equipment panel.

Fig. 8-25. Back of face plate.

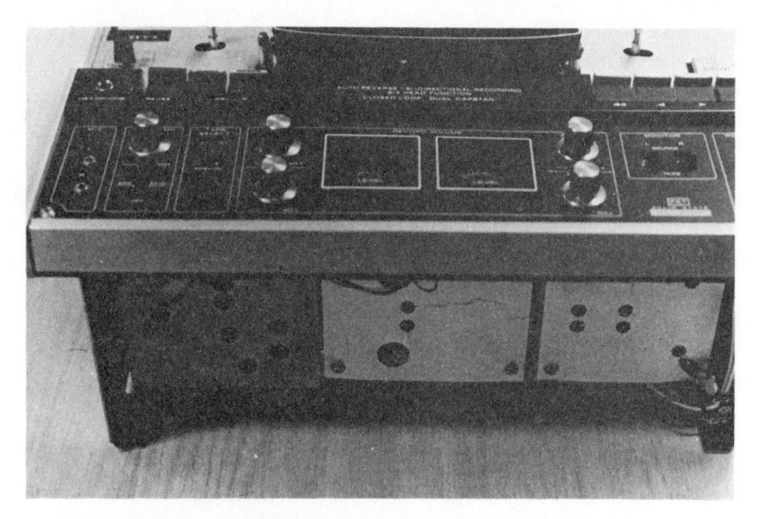

Fig. 8-26. Top view of face plate. Note the panel cutout overlap.

Other slab-type face plates may or may not require removal of the metal case depending on screw clearance and allowable hole tolerance. However, most equipment not originally intended by the manufacturer to be rack or panel mounted barely has enough tolerance even with the metal case removed (Figs. 8-23 through 8-27). Remove the case, if necessary by removing the two mounting screws on each side. Sometimes you have to take out the screws across the top immediately behind the face plate as well. At this point, check and make sure that the metal case is not an integral part of the chassis design. Normally, it is not, but it may be on the cheaper brands of equipment.

Fig. 8-27. Side view of face plate.

Fig. 8-28. Open ended face plate requiring built-up enclosure on equipment panel cutout.

WARNING! Once you find it necessary to remove the cover from the equipment, you are exposing very high voltage circuitry. You must design the equipment compartment in such a way to prevent human contact with the internal wiring. This is not just for the protection of small children, either. Even equipment that is switched off has hot electrical contacts that might now be exposed. You can suffer a bad shock or worse just by grabbing around the back chassis plate to push

Fig. 8-29. Rear view of an open ended face plate.

Fig. 8-30. Open ended face plate showing exposed inner workings.

in a patch cord. Clearly label any access with a sticker warning you and other people to unplug the entire system before opening up the equipment compartment, and use screws for any removable access panels. Easily opened trap doors court disaster. The author assumes no responsibility for people removing equipment cases.

Fig. 8-31. Building up an enclosure on equipment panel to conceal open ended face plates.

Fig. 8-32. Close-up of equipment panel enclosure.

Sometimes when the case is removed, the end of the face plate is exposed (Figs. 8-28 and 8-29). Normal panel mounting would result in a partial view into the inner workings of the component (Fig. 8-30). This is undesirable. Build up an enclosure around the equipment panel cutout for the exposed edges with matching lumber. Make sure to maintain the same depth as the unexposed face plate (Figs. 8-31 and 8-32).

COMBINATION MOUNTING

If the equipment has no lip around the chassis (Figs. 8-33 and 8-34), it must be combination mounted. This is a hybrid of shelf and

Fig. 8-33. After the decorative end panel is removed from some equipment, no lip exists to conceal the panel cutout.

Fig. 8-34. This equipment requires combination mounting.

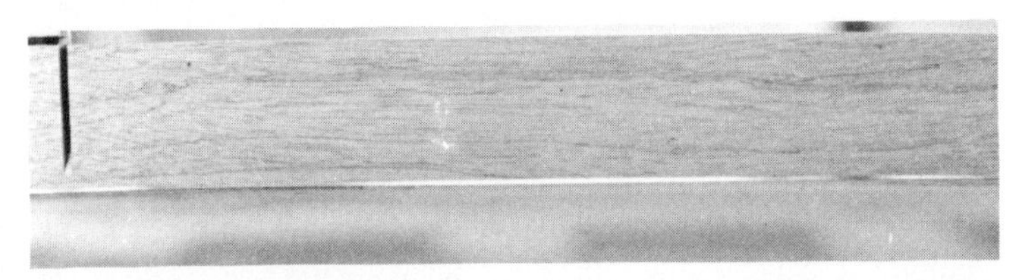

Fig. 8-35. Close-up of 1/32″ clearance for combination mounted tuner (viewed from back of equipment panel).

Fig. 8-36. Combination mounted equipment is inserted into the panel from the rear. Be sure to leave cabinet back removable to get equipment in and out.

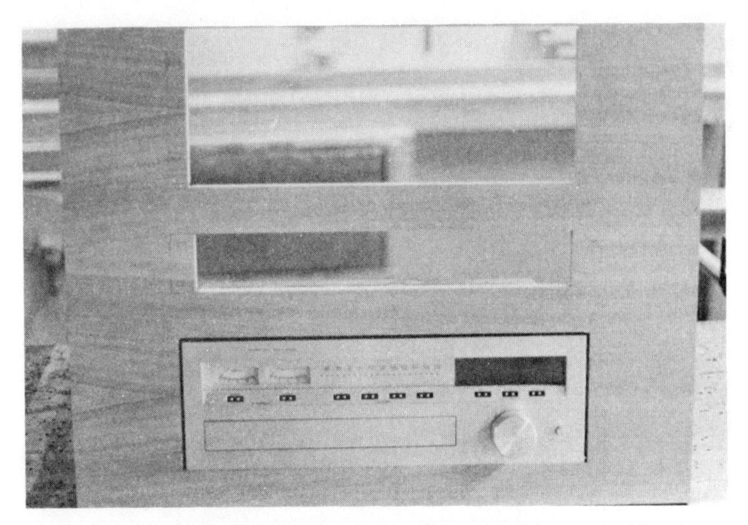

Fig. 8-37. Black metal case outlines combination mounted tuner. Inside edges of cutout should be painted flat black to keep gap from showing.

Fig. 8-38. Mark the equipment panel for cutouts. Be sure to make a notation on which side the line kerf should lie.

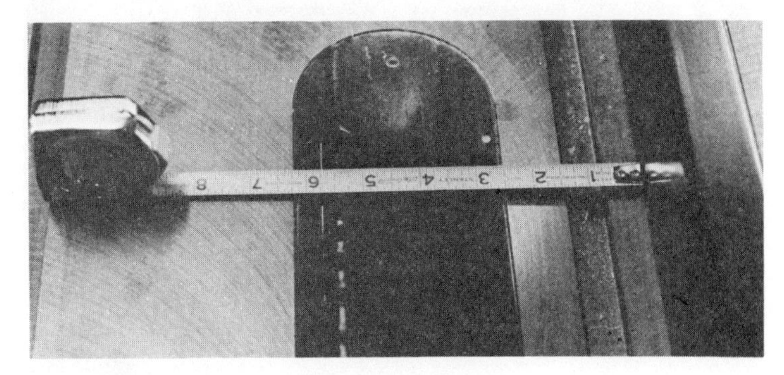

Fig. 8-39. Setting the table saw fence for panel cutouts. This setting is for either 5⅜″ "in" or 5¼″ "out."

Fig. 8-40. Making blind equipment panel cutout. Shirt sleeves should be rolled up for safe operation.

panel mounting where the panel is cut to closely fit around the very front of the component, but the cutout is not overlapped by the face plate. The cabinet is basically held in place by the shelf it rests on.

Each situation poses individual problems, but normally the cutout is made about 1/32″ larger (Fig. 8-35) than the outside measure of the face plate of the equipment. The equipment is then slid into the cutout from the rear (Fig. 8-36) and secured by blocks mounted behind the chassis to keep it from sliding back.

Since the gap may be slightly visible when viewed at certain angles, paint the inside surfaces of the cutout flat black to make them less noticeable (Fig. 8-37).

Fig. 8-41. Front of completed equipment cutout.

Fig. 8-42. Back of equipment cutout showing extended kerfs.

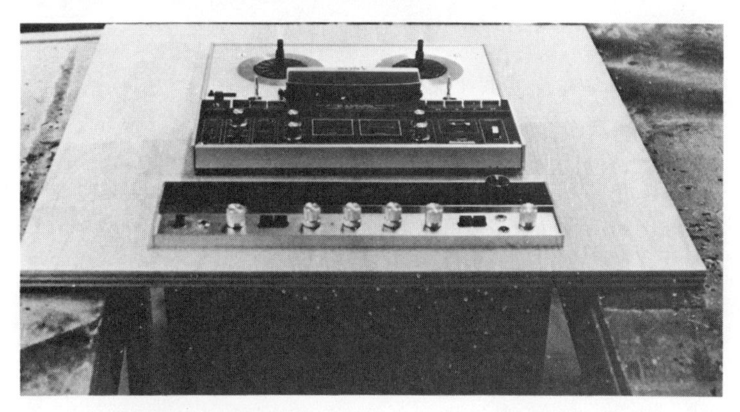

Fig. 8-43. Checking equipment fit.

Fig. 8-44. Rack mounted equipment is test fitted.

Fig. 8-45. Trace around the edge of the rack mounted equipment to locate bolt holes.

Making the Cutout

After making careful measurements of the equipment and determining the size cutouts required, lay out the cuts on the front of the raw equipment panel. If possible, allow a 2″ space between

Fig. 8-46. Mark holes with awl, and drill before sanding or finishing.

Fig. 8-47. Parts for light equipment support.

the components. This accommodates an efficient air flow and eliminates stagnant air pockets. The room is often necessary for mounting the equipment support structures.

You can cut out the openings with a bayonet or keyhole saw, but a table saw is much more accurate and leaves a cleaner kerf. If you intend to use a table saw, mark your measurements as to whether they are to be the inside (fence side) of the kerf, or the outside (Figs. 8-38 and 8-39). It is very easy to cut on the wrong side of the line.

Fig. 8-48. Equipment support assembled.

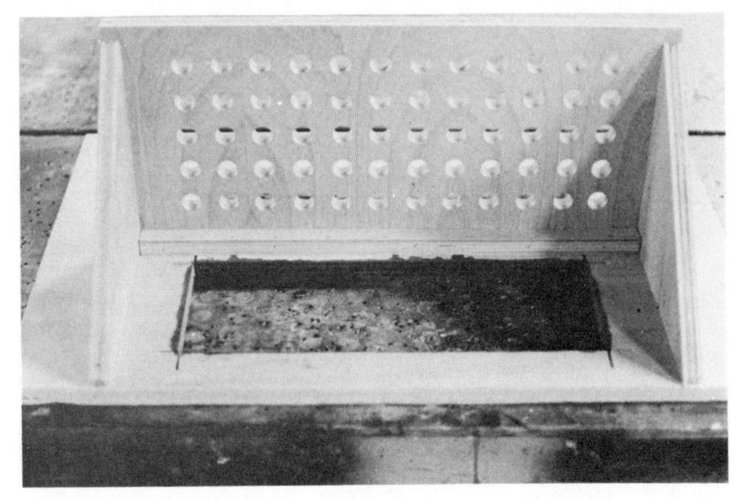

Fig. 8-49. Equipment support mounted on back of equipment panel.

Cutting an equipment panel on a table saw is a "blind" cut requiring some previously acquired skill and experience. If you do not have the skill and experience, use a bayonet saw instead. Kickbacks are very probable especially for the inexperienced.

Fig. 8-50. Spread a liberal amount of glue along the back of the equipment support corbels and back cleat. Nail or screw in position being careful not to go through front of equipment panel.

Fig. 8-51. Shelf slotted to vent a preamplifier.

Raise the blade of the table saw about ¼″ higher than the thickness of the equipment panel. Stand at the end of the saw table instead of directly behind the blade. Keeping one edge of the panel on the saw table and one end against the fence *at all times*, gently but firmly lower the panel onto the moving blade (Fig. 8-40). You must gauge the distance with your eye very carefully so that the blade breaks through the panel near the back of the cut. Cautiously move the panel back until the kerf reaches your marked line. This is where the kickback danger is the greatest. After cutting to the back line, push the board forward to the front line. Keeping one

Fig. 8-52. Another view of a slotted shelf.

Fig. 8-53. Top of equipment support designed for heavy equipment producing a lot of heat. Support is the same width as the equipment panel and has cleats at either end that screw into equipment compartment sides.

edge on the table still, and one end against the fence, lift the panel off the blade by pivoting at the edge on the saw table. Reset the saw for the next cuts and repeat the procedure.

When you are through, you should have a perfect opening on the front side of the equipment panel (Fig. 8-41), and extended kerfs on the back side (Fig. 8-42).

Fig. 8-54. Bottom of same equipment support.

Fig. 8-55. Same equipment panel with preamp support also glued into place.

Place the equipment panel across two tables (Fig. 8-43) to check the fit of the equipment. Sometimes, an additional 1/32″ or 1/16″ trim is required. Grooves or notches for previously unnoticed chassis screws are trimmed. If some of the components

Fig. 8-56. Completed equipment panel assembly. Note 1½″ wide oak brace at the back of the heavier amplifier support.

Fig. 8-57. With equipment panel in place in the cabinet, support for the heavier amplifier (top) is screwed directly into case sides.

are the "rack" type, mark the mounting slots, and bore the bolt holes at this time (Figs. 8-44, 8-45 and 8-46).

Equipment Supports

The bottom component rest on wooden skids (Fig. 6-1) mounted to the bottom shelf of the equipment compartment. This

Fig. 8-58. McIntosh Tuner/Preamplifier and Panloc brackets ready for installation.

Fig. 8-59. Future McIntosh upgrading does not necessitate panel change. Merely slide out old unit, slide in new.

allows free circulation of air under and around the equipment, and leaves plenty of clearance for light fixtures, wires, etc.

The upper components must be supported by shelves. If the component weighs less than five pounds, the shelf can merely be

Fig. 8-60. McIntosh unit at lock halfway out permits adjustment of lesser used controls on top.

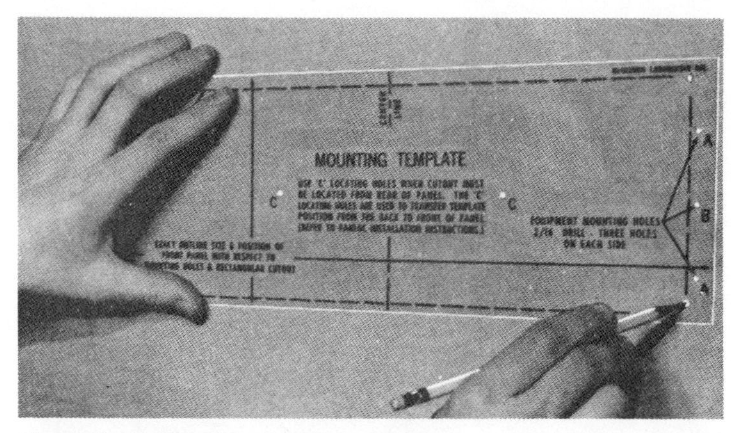

Fig. 8-61. All McIntosh equipment designed for panel mounting comes complete with Panloc brackets and mounting template (courtesy McIntosh Laboratories, Inc.).

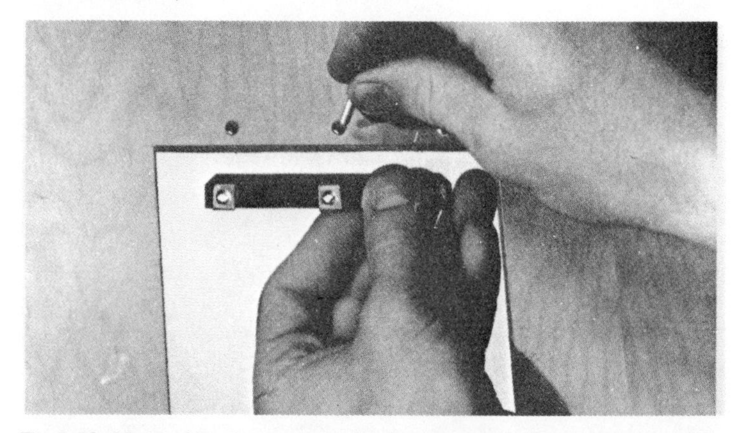

Fig. 8-62. After making the cutout and drilling the holes where indicated on the template, screw mounting strip in place (courtesy McIntosh Laboratories, Inc.).

Fig. 8-63. Brackets are set in place and screwed to the panel and mounting strip (courtesy McIntosh Laboratories, Inc.).

Fig. 8-64. Amplifiers and other heavier components should still be supported from the bottom to prevent the equipment panel from bulging out.

attached and braced to the back of the equipment panel with 45° corbels (Figs. 8-47, 8-48 and 8-49). It can be attached permanently with glue and screws (Fig. 8-50) since if you change your equipment, you will most likely also change both your panel and the shelf.

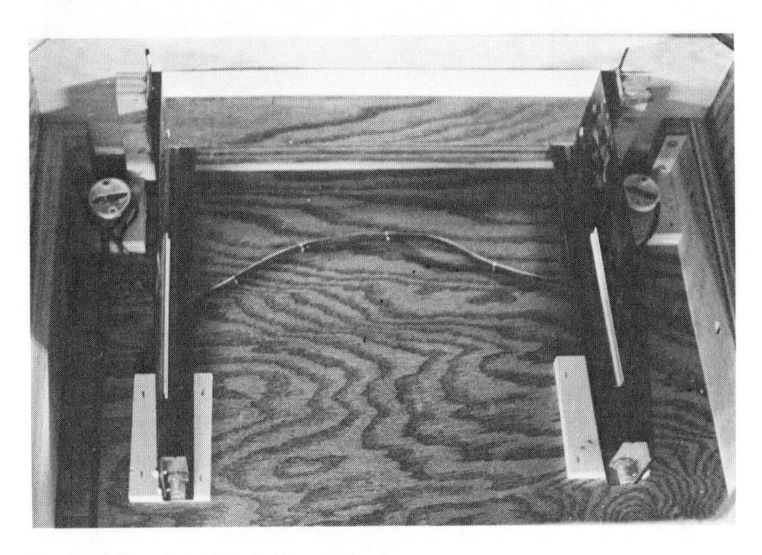

Fig. 8-65. Top view of the bottom supports.

Fig. 8-66. Drill holes in equipment shelf cleats so that clearance is left for screwdriver.

If the equipment chassis has vent holes or slots on the bottom, you must vent the shelf directly below them. This can be done with either holes (Fig. 8-47) or by cutting a slot (Figs. 8-51 and 8-52).

Fig. 8-67. Glue and nail cleats to the equipment side of the shelf. Make sure you leave enough of a setback for the equipment panel (normally 2¾").

Fig. 8-68. Temporarily support equipment shelf with precut spacers the same height as the compartment immediately below the shelf.

Fig. 8-69. Another view of temporary support spacers.

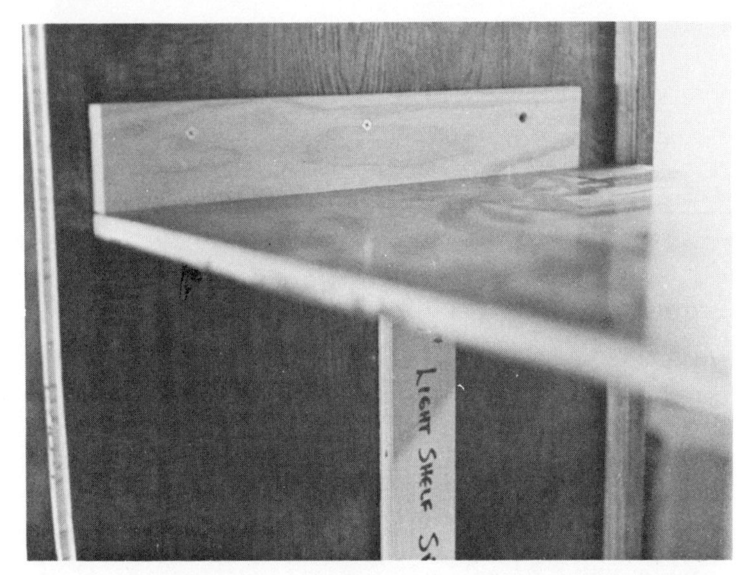

Fig. 8-70. Drive screws into the back two holes of the shelf cleat, but leave the screw out of the front hole for the time being.

Fig. 8-71. Equipment shelf mounting blocks.

Fig. 8-72. Mounting blocks in position on equipment panel.

Fig. 8-73. Positioning equipment panel into carcase.

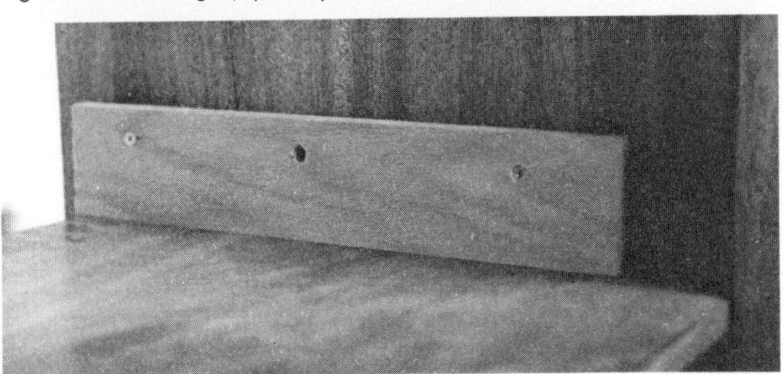

Fig. 8-74. After screwing equipment panel to top and equipment shelf, drive screw removed from middle mounting hole into front hole (equipment panel removed for clarity).

Fig. 8-75. An adaptation of the equipment panel mounting, required by the short panel and front vented top.

Equipment over five pounds must be supported by a shelf, itself screwed to the case sides. Make the shelf the same width as the equipment panel with cleats on the underside of the ends. If the equipment is much over ten pounds, brace the shelf with a piece of solid oak or other hardwood (Figs. 8-53 through 8-56). After the equipment panel is in place, level the shelf, and screw securely to the case sides with # 8 or #10 wood screws (Fig. 8-57).

Fig. 8-76. Interior view of this adaptation.

Fig. 8-77. Detailed corner view of this adaptation.

McIntosh Panlocs

Every now and then, the path of the stereo furniture maker hits a little smooth sailing. McIntosh Laboratories of Bing-

Fig. 8-78. Back of equipment shelf showing edge of false back. False back is mounted right up to the back of the equipment shelf, but is not fastened to it in any way. This permits future panel changes.

hampton, N.Y., not only makes superior quality sound equipment, but characteristically takes the details of mounting that equipment into account.

The equipment is sized and designed to fit their standardized brackets (Figs. 8-58 and 8-59). The equipment slides in and out easily, and locks both halfway in (Fig. 8-60) and all the way in. If you upgrade your McIntosh equipment to a later model, it fits the same bracket. Slide out the old one, slide in the new one. Panel mounting couldn't be easier.

Each piece of equipment comes with Panlocs and a mounting template. Make your cutout to the exact dimensions listed on the template (Fig. 8-61), drill 3/16" holes where indicated, set the mounting strips (Fig. 8-62), and screw the brackets into place (Fig. 8-63). Heavier amplifiers should be blocked up or braced from the cabinet sides to help support the weight (Figs. 8-64 and 8-65).

Mounting the Equipment Panel

Once the equipment cutouts have been made, and the equipment supports are assembled, the equipment panel is ready to be installed into the carcase.

The cabinet should be designed so that either the top or bottom of the equipment compartment is a semi-permanent shelf mounted with cleats and screws only. The cleats (Fig. 8-66) should be predrilled for the screws with at least three mounting holes which are set back far enough to allow screwdriver clearance. Glue and nail the cleats to the equipment shelf on the surface that will be inside the equipment compartment (Fig. 8-67).

Cut two spacers from scrap wood for the exact height of the compartment immediately below the equipment section. These spacers will hold up the equipment shelf on both sides while it is screwed into place (Figs. 8-68 and 8-69). Set the spacers in the middle of each side directly under the center mounting hole in the shelf cleat. Drive in a #8 or #10 wood screw in the center hole. Slide the spacers directly under the back hole in the cleat and drive a screw in these. Leave the front hole empty for the time being (Fig. 8-70).

Cut four wooden blocks (Fig. 8-71) and screw these on the back edges of the equipment panel, one to each side (Fig. 8-72).

Loosen the two back mounting screws in the equipment shelf cleats, and remove the center screws completely. This allows the equipment shelf to pivot down, giving you room to jockey the equipment panel in from the front if necessary. Work the equipment panel into place (Fig. 8-73) and pull the equipment shelf back up so that the equipment panel is snug top and bottom. After carefully checking the correct alignment and setback of the equipment panel (2″), screw the four mounting blocks tightly to the case sides, the top and the bottom of the equipment compartment. At this point, replace the screws removed from the shelf cleats in the *front* hole (Fig. 8-74).

Very short equipment panels may not allow enough clearance for this standard cleating method, so you may have to adapt the technique to suit the circumstances (Figs. 8-75 through 8-77).

If you should wish to change your stereo equipment in the future, you can remove the equipment panel and change it simply by reversing installation procedure. Be sure not to nail the false backs to the shelf that pivots (Fig. 8-78). The overlap will conceal any minor inconsistency since it is always in a shadow.

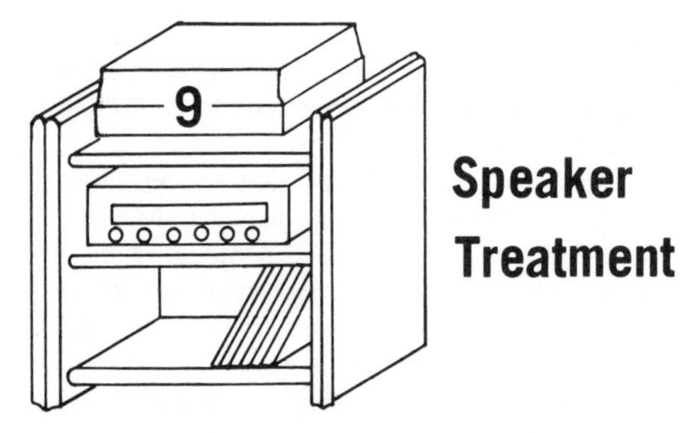

Speaker Treatment

All stereos have at least two speakers that are usually, but not necessarily, placed in the same room with the stereo components. In most cases, the speakers are independently placed in the room where they sound the best, and they do not affect the cabinet design at all. Sometimes the layout of the room dictates incorporating the speakers in the same case with the rest of the stereo components.

ESTHETICS

The main design problem with speakers is simply that they look like speakers no matter what you do with them. Of course, if you are ambitious, the best solution is to build speakers to match the stereo cabinet. Speaker cabinet construction and placement are subjects thoroughly covered in many other works, so we shall concentrate only on speakers and their interaction with stereo furniture.

Probably the best thing you can do with speakers (and certainly the most expedient thing) is to get used to them, designing your room to complement their best features. A mistake commonly made at this point is to make a large stereo cabinet to match the speakers rather than to match the rest of the furnishings in the room. The usual result of this logic is to have both the speakers and the stereo cabinet look out of place. If the choice exists, always attempt to match the style and tone of the furniture, and let the speakers be speakers.

MOUNTING THE SPEAKERS

As with the rest of the equipment, the simplest way to mount speakers in a cabinet is to shelf mount them (Fig. 9-1). If room

Fig. 9-1. Shelf mounted speaker.

Fig. 9-2. Speaker stands can improve the apparent sound coming from nearly all speakers.

Fig. 9-3. This cabinet was designed around the speakers to feature the fancy wooden speaker grills.

exists, try to place the tweeters as close to listening ear level as possible. Faced with a choice, higher is better than lower in this regard. Likewise, if you intend to place the speakers by themselves on the floor, you might want to consider making some sort of speaker stand to raise the speaker cabinets a little (Fig. 9-2). This

Fig. 9-4. Speaker compartment has its own grill cover to fit many sizes of future speakers. Speaker cabinet in this case was made to slide into the compartment.

Fig. 9-5. Permanently built-in speakers can improve cabinet esthetics, but limit future options.

simple little act can make quite an improvement in the apparent sound quality.

If you expect to keep your speakers for a long time, you may wish to consider shelf mounting them in a compartment that fits closely around them and makes them look built-in (Fig. 9-3). This

Fig. 9-6. Grill frames, marked and partially cut out.

Fig. 9-7. If the edges of the frame will be exposed, a bevel can make their appearance more interesting.

Fig. 9-8. Corner of a completed grill frame ready to receive the grill cloth.

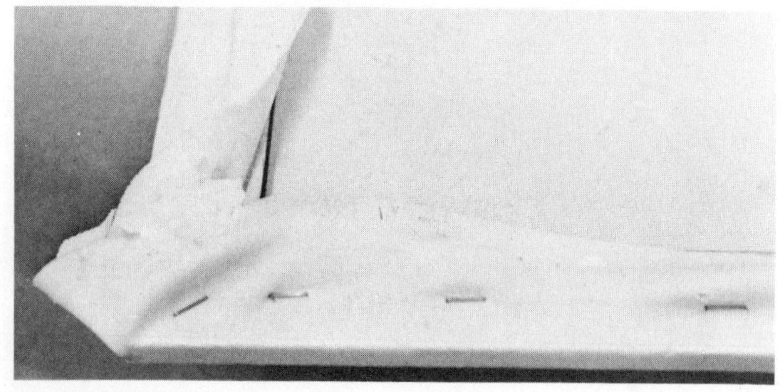

Fig. 9-9. Grill fabric correctly stretched and stapled.

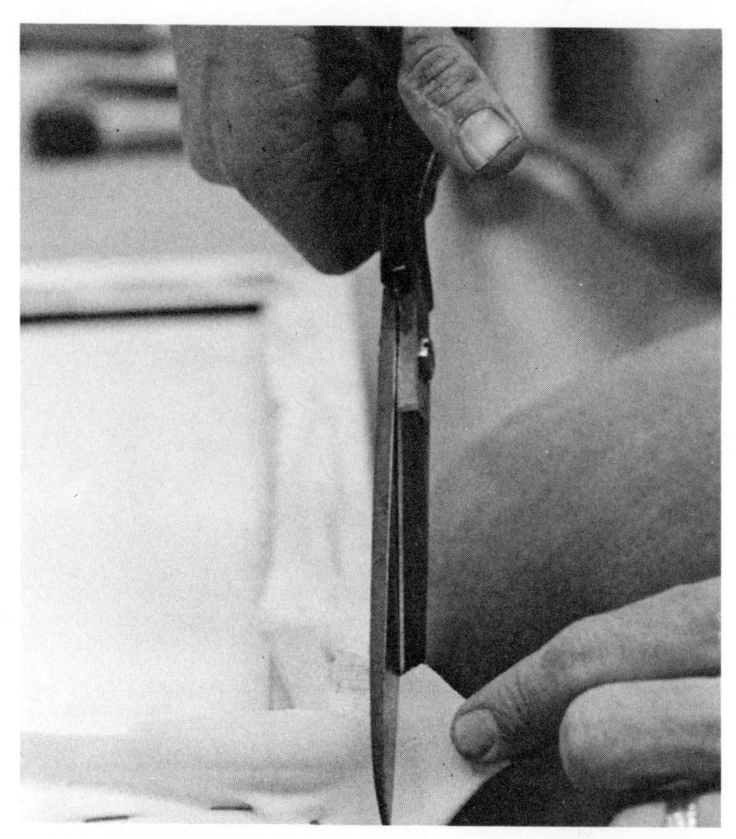
Fig. 9-10. Making the first corner cut.

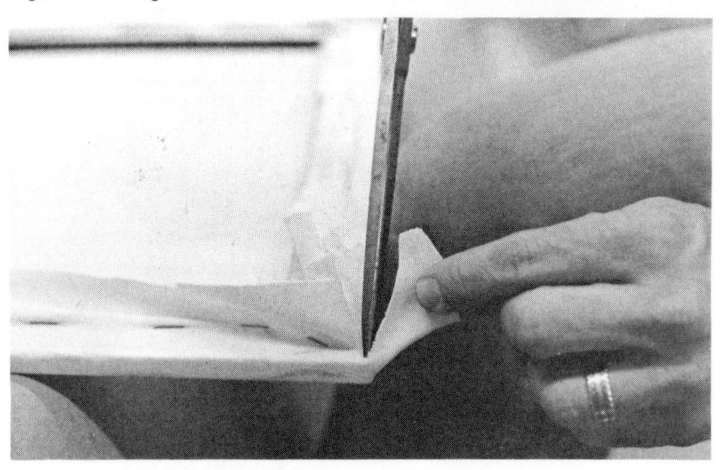
Fig. 9-11. Finishing the first corner cut.

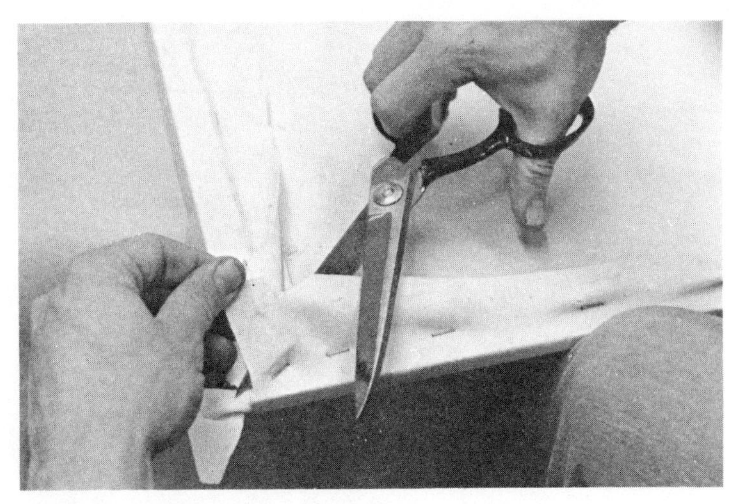

Fig. 9-12. Making the second corner cuts.

saves you from having to make a grill cover. A variation of this is to make your own speaker cabinets just the right size to slide into the compartment allowed for them.

If the possibility exists that you might upgrade your speakers in the future, make the speaker compartment oversize to allow a larger future speaker to fit. The margin around the speaker will not

Fig. 9-13. Note margin left on waste side of staples.

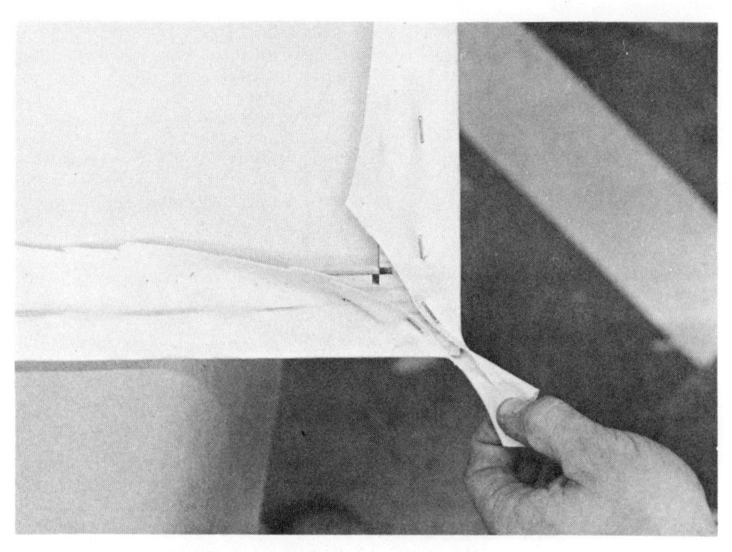

Fig. 9-14. Stretching the corner flap.

show as in Fig. 9-4 since it will be covered with a custom made grill cloth. When in doubt, keep your options open.

Speakers should be built permanently into the cabinet (Fig. 9-5) only if you expect to keep the same speakers forever. The reasons for doing so are mainly space and design considerations; that is, if the form of the cabinet is more important than the

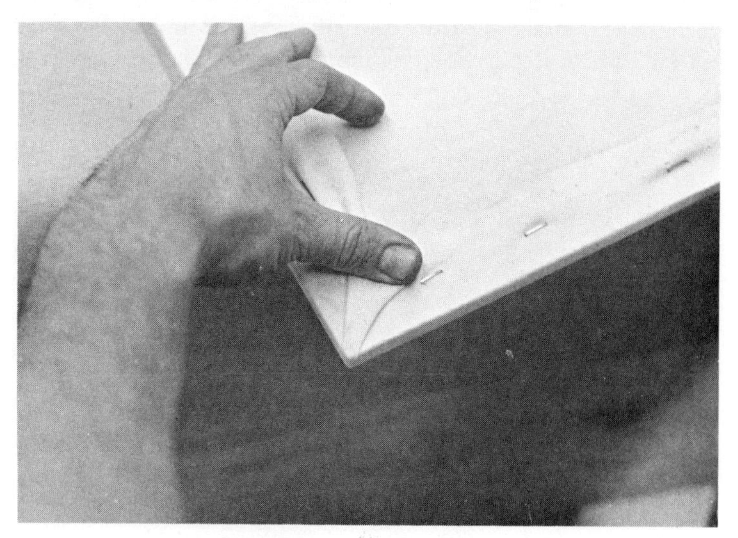

Fig. 9-15. Hold the flap down firmly.

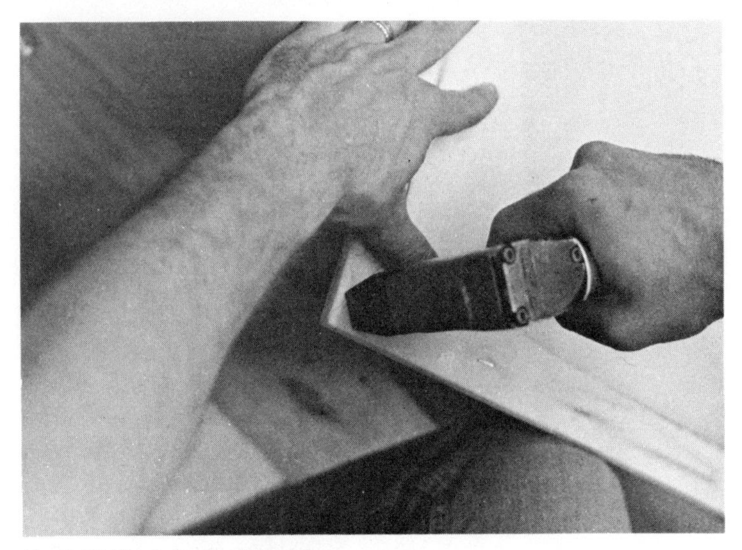

Fig. 9-16. Staple the flap into place.

function, and if you cannot allow much room for the speakers in the design.

ACOUSTICAL FEEDBACK

Mounting speakers in the same case as your turntable is tantamount to asking for acoustical feedback. The vibration generated from the speakers can be transmitted through the wood of the cabinet itself, vibrating the turntable enough to cause an

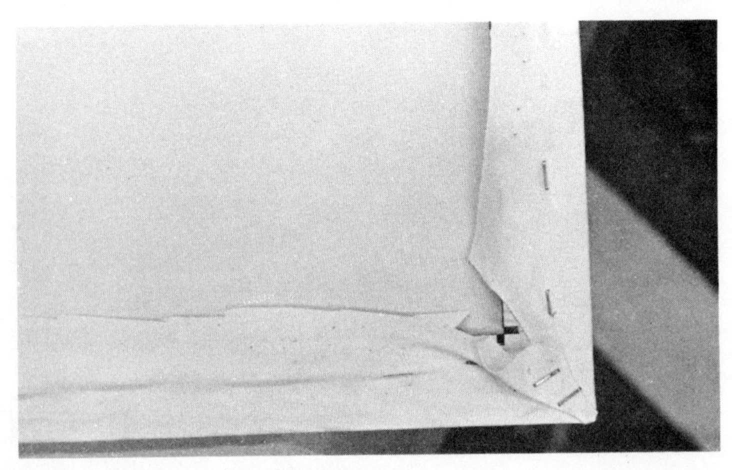

Fig. 9-17. Back view of the completed corner.

Fig. 9-18. Front view of completed corner. Compare this to Fig. 9-8.

awful howling sound. Speakers permanently mounted in the cabinet, because of their intimacy with the basic structure, can be twice the problem. This was one of the major flaws of the old console designs.

Shelf mounted speakers can be damped with felt, rubber, or carpet pads between the speaker cabinet and the stereo cabinet. Permanently mounted speakers must be lined with particle board,

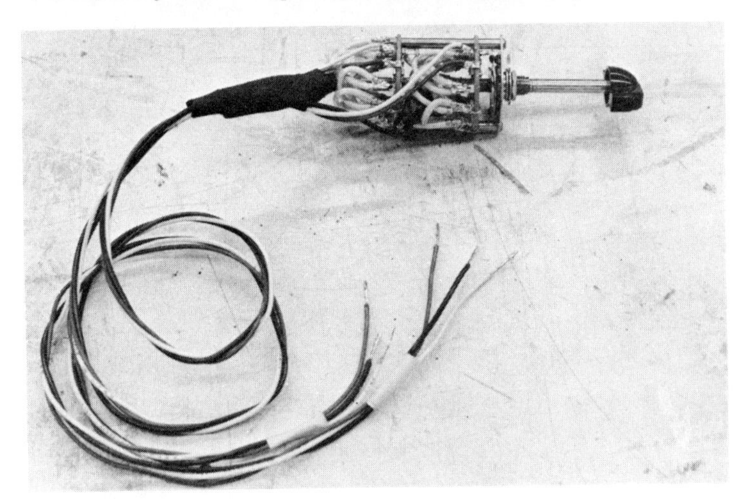

Fig. 9-19. Constant impedance stereo volume control.

Fig. 9-20. Volume controls mounted along a special control panel.

the thicker the better. If your taste runs towards loud, bass-intensive music (rock-and-roll, 1812 Overture, etc.), it is better to forget mounting the speakers and turntable in the same cabinet altogether.

Fig. 9-21. Front view of volume controls mounted on one side of equipment panel.

MAKING GRILL FRAMES

Use AC or AD fir plywood at least 3/8″ thick to make the grill frames. Thinner plywood will bow too much under the tension of the taut grill cloth.

Cut the board about 1/16″ less than the dimensions of the speaker compartment opening. You may have to allow more clearance if you use thicker fabric. With a marking gauge or combination square, scribe a line a minimum 1¼″ inside the edge of the board (Fig. 9-6).

Using a jar lid or handy roll of masking tape, trace a radius inside the marked corners tangent to the scribed lines. This radius is actually optional, but it acts as a corner block to strengthen the frame and give greater resistance to bowing under fabric tension.

Set up the table saw (or bayonet saw) for a blind cut on the inside of the scribed lines, and make the cut right up to the point of tangency (Fig. 9-6). Cut the radii with a bayonet or coping saw.

After cutting out the center of the frame, ease the inside edges slightly with sandpaper to prevent a sharp frame line from showing through the fabric. If the frame is to be mounted externally with velcro, improve the appearance of the exposed edges with a pleasing looking bevel (Fig. 9-7).

After cutting the frame, paint the front and edge surfaces flat black to prevent the wood of the frame from showing through the fabric (Fig. 9-8).

MOUNTING GRILL CLOTH

The hardest part of making a speaker grill is choosing the color and pattern of the material. Any material that light can readily be seen through should be acoustically transparent enough for grill cloth purposes. This includes dyed burlap, most double knits, and many other fabrics. Simpler patterns and stretchy material work better. A very dominant linear pattern can heighten mounting mistakes by not ending up perfectly parallel to the frame.

Cut the fabric about eight or ten inches larger than the grill frame in both dimensions. This should leave you with enough material to attain good leverage for stretching. Center the frame face down on the back side of the spread out fabric, and evenly fold one flap of the cloth over a frame end.

With an all-purpose stapler, fasten the material to the back of the one end of the frame. Then pull the fabric tight around the opposite end and staple. Repeat with the two sides, being careful to keep the warp and woof of the fabric parallel to the frame sides. Pull the material tight in both directions at the corners and fasten each side of the corner fold with a staple 45° to the other staples (Fig. 9-9). Each corner should end up with two such staples (Fig. 9-13).

Pull the corner material up and make a vertical cut with shears about ¼″ inside the corner right to the frame (Figs. 9-10 and 9-11). Now clip off the extra material on the waste side of the diagonal staples, leaving about a ¼″ margin so that the staples don't unravel the material and pull through (Figs. 9-12 and 9-13). Stretch out the remaining corner flap (Fig. 9-14), and keeping it under tension, fold it back over the diagonal staples on the frame (Fig. 9-15). Secure this flap in place with two or three staples at right angles to the diagonal staples (Figs. 9-16 and 9-17). Trim off the excess material all around the back side of the grill frame. You should end up with a neat, wrinkle-free grill cover (Fig. 9-18).

SPEAKER VOLUME CONTROLS

Home sound systems frequently have several sets of speakers in various parts of the house. If all the speakers are driven by one amplifier with one volume control, every speaker will have the same relative loudness. This is not always desirable.

Various loudness controls are made to be used with each pair of speakers to give everyone in the house a little better control over their environment. These volume controls should be a "constant impedance" type and can either serve both (Fig. 9-19) or just one channel of a pair of speakers. They can be mounted in a special control panel (Fig. 9-20), or directly on the face of the equipment panel (Fig. 9-21).

Accessory Storage

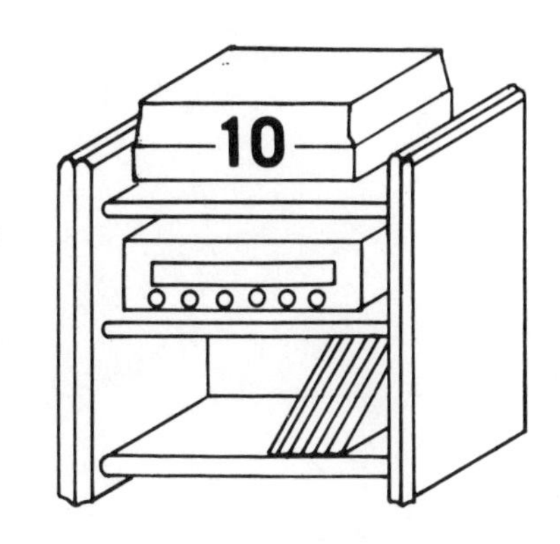

With the exception of the tuner input, a sound system is just so many knobs, switches, and wires. The signal input must be provided by records, cassette tapes, 8-track tapes, and open reel tapes. Both the components and the signal sources need maintenance with record cleaners and demagnetizers. Complete listening pleasure may additionally require headphones, additional patch cords, and many other minor accessories, some of which may be developed in the future.

Storing all these accessories with or near the electronic equipment is quite convenient. Accessory storage may even be the main motivation for acquiring a stereo cabinet, especially if you have ever fished around the back of a dark, dusty closet in search of a lost tape or patch cord.

SHELF STORAGE

Appropriately spaced shelving is the most expedient type of storage, and is commonly used for housing records, open reel tapes, and miscellaneous minor accessories. Records require shelf spacing a minimum 13″ deep and 13″ high, with about 1″ in length for every five or six single albums.

Records must be stored vertically to keep them from warping. This is accomplished with record dividers (Fig. 10-1) which act as "bookends" for a given section of records. These should be scalloped in the front so that the albums may be easily grasped. Division increments will vary depending on available room in the

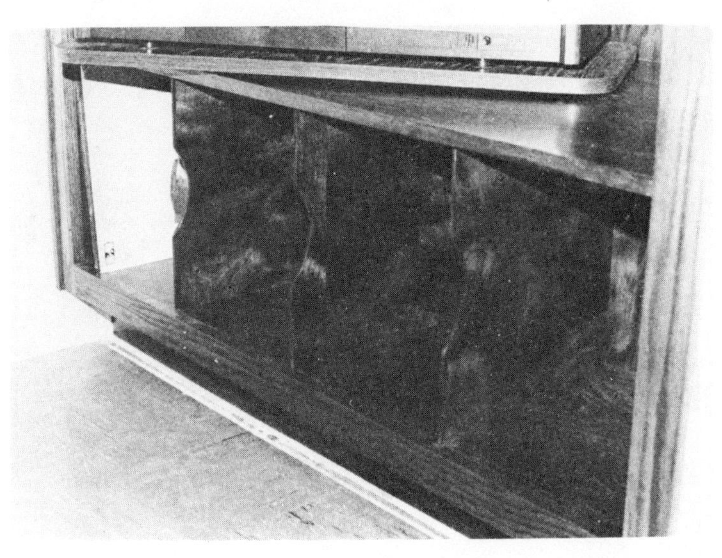

Fig. 10-1. Record dividers permanently fixed in cabinet. Divider spacing should not exceed ten inches.

design and the number of albums, but it is a good idea to keep the divider spacings under ten inches (about sixty records worth).

Dividers are commonly made a permanent part of the cabinet by dadoing the partitions into the shelves (Fig. 10-2). However,

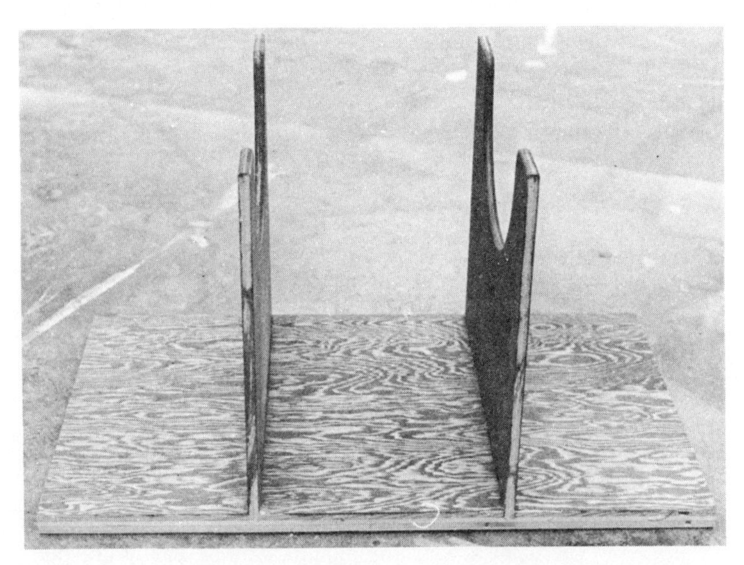

Fig. 10-2. Removable record dividers give more future equipment options.

Fig. 10-3. Turntable compartment may be subdivided to store a dozen or more frequently used albums.

removable dividers preserve a greater degree of future cabinet flexibility. Should you buy additional components for your system in the future, you can take out the records and dividers in one section and convert it to equipment housing. This keeps all your stereo gear together in one cabinet, and looks better than setting the new component on the top of the cabinet for lack of room inside.

Fig. 10-4. Shallow cassette storage shelves. Amplifier is set behind false back.

Fig. 10-5. Components and television are raised as high as possible in horizontal format cabinet. Cassette and accessory storage drawers keep space from being wasted.

Fig. 10-6. Close-up of interior drawer. Magnetic touch-latch holds drawer in and eliminates need for drawer pull.

Fig. 10-7. File-type record storage drawers.

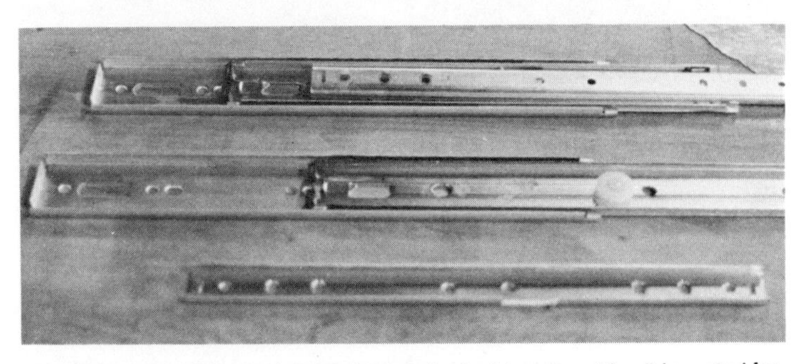

Fig. 10-8. A pair of good quality full extension ball bearing side slides rated for 100 pounds are needed for record drawers.

Fig. 10-9. Tape storage pullout. Top two rows can hold up to twenty cassettes, bottom row holds about 15 seven inch reels.

The back panel of removable dividers is made of plywood at least ⅜″ thick, and is cut ¼″ less than the inside dimensions of the shelf. The divider panels themselves extend thirteen inches in front of the back panel. When installed, there is usually four or five inches of space behind the removable section where you can hide

Fig. 10-10. Console format cabinet with tape pullouts makes good use of deep case. Note open reel deck pullout on right.

Fig. 10-11. Slab type drawer front. These can be solid pieces of lumber, or edged veneer plywood.

Fig. 10-12. One slab type drawer front and three different frame and panel fronts show variety of designs.

Fig. 10-13. Bottom view of drawer sides. Note ¼″ wide groove for drawer bottom.

Fig. 10-14. Back of drawer is joined to side with butt joint. Note how bottom will overlap the drawer front (or front-back).

unused patch cords, record cleaning solution, and gold bullion. Your infrequently used accessories remain close at hand, but stay clean and out of sight.

In wider cabinets, records may be stored on the turntable shelf. The turntable compartment can even be subdivided (Fig. 10-3) for your most frequently used albums.

Fig. 10-15. Glue the front-back liberally in a serpentine pattern.

Fig. 10-16. Drawer ready to receive bottom.

Open reels are treated in the same manner as records. Smaller reels require a shelf 8″ high and 8″ deep, professional size reels take a shelf 11″ high and deep.

Shelf storage is less convenient for compact cassettes and 8-tracks. If you use the full depth of the case, you awkwardly have to dig through all the cassettes just to get one. If you make shallow

Fig. 10-17. Sliding the bottom in.

Fig. 10-18. Drawer bottom in place.

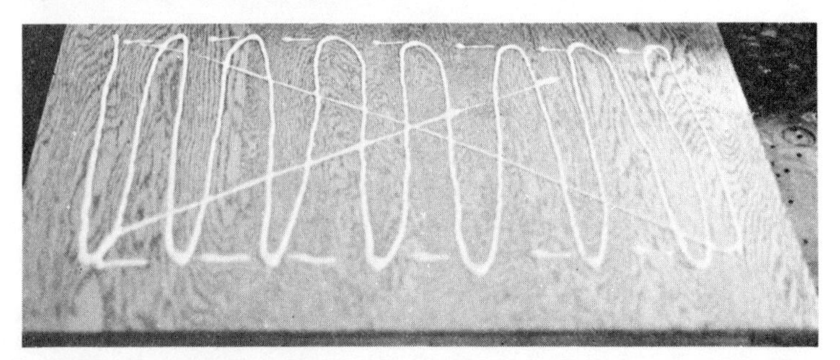

Fig. 10-19. Alternate method of assembly: preparing the drawer front to receive completely assembled drawer box.

Fig. 10-20. Positioning assembled drawer box.

Fig. 10-21. Lower front corner of completed drawer.

shelves for convenience, you end up wasting much of the cabinet volume.

One exception to this is if you have a piece of equipment which does not necessarily have to be exposed to the front of the cabinet (e.g., a power amplifier with no knobs, a rarely adjusted equalizer).

Fig. 10-22. Top view of completed drawer.

Fig. 10-23. Record and cassette drawers ready to be mounted in vertical format cabinet.

Shallow shelves may be built for the tapes with a false back (Fig. 10-4). The equipment is set behind the false back resulting in efficiently used case volume.

DRAWER STORAGE

Cassettes and 8-tracks are usually stored in drawers. Although the drawer mechanisms take up some space, the advan-

Fig. 10-24. Mounting hardware to drawer.

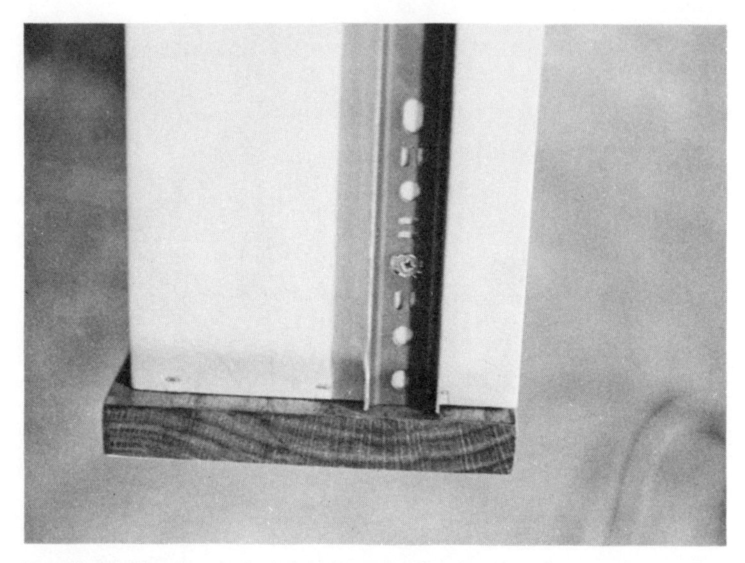

Fig. 10-25. Use the vertical slots for mounting hardware on the drawer, the horizontal slots for mounting the hardware in the case.

tages of top access and use of the full depth of the carcase make up for the loss. Drawer storage is more expensive than shelf storage, but it has much greater convenience.

Fig. 10-26. Spacing cleats bring drawer hardware mounting surface flush with inside of face frame.

Fig. 10-27. Always mount the top cleat first. The bottom cleat will otherwise get in the way of the spacers.

Drawers actually improve case volume utilization by making otherwise blanked and wasted space useable. Built-in speakers raised to improve sound can be built with a drawer designed below (Fig. 1-30). Television and turntables can similarly be raised to a comfortable height (Fig. 10-5). This can be achieved either beneath the external cabinet doors (Fig. 1-8), or behind them (Figs. 10-5 and 10-6).

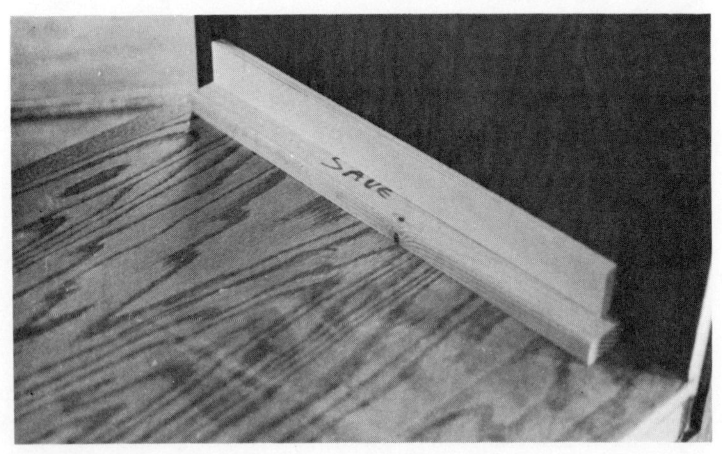

Fig. 10-28. Spacing the bottom drawer cleat (if required).

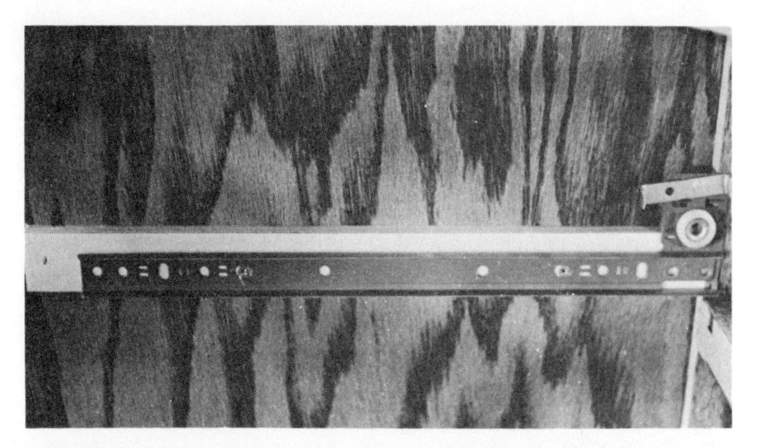

Fig. 10-29. Drawer hardware mounted in case. Notice that the horizontal slots were used.

Separate file drawers, both standard (Fig. 10-7) and lateral (Fig. 1-24), are the best way to store records. The jacket spine can be visually scanned from the top without having to crouch on the floor as with shelves. Since a fully loaded drawer of records can frequently weigh 70 or 80 pounds, use heavy duty ball bearing side slides (Fig. 10-8) rated for 100 pounds or more to carry the weight. Full extension slides give greater useable drawer space.

Storage Pullouts

Pullout trays and racks can be used for cassette tapes, 8-tracks, or open reel tapes. Pullouts can also be used to store

Fig. 10-30. Record drawer hardware ready to be set.

various infrequently used accessories, but may cost more than they are worth for this purpose. Well thought out pullouts use space very efficiently. It is possible to store two courses of cassette tapes over the top of one row of 7″ open reel storage, and maintain full access for both (Figs. 10-9 and 10-10). Allow 4½″ × 3″ × ¾″ for each cassette case, 8″ × 8″ × 1″ for each 7″ reel, and 11″ × 11″ × 1″ for 10″ reels.

Drawer Planning

The drawer front must be designed to complement the rest of the cabinet design. Interior drawer fronts are normally slab type (Fig. 10-11), while exterior drawer fronts should match the door style and may either be slab type or frame-and-panel (Fig. 10-12). The front should be completely assembled, machined, and finished on the front side and edges prior to drawer assembly. Do not at this point finish the back of the drawer front. A clean, white wood surface on the back insures a good glue bond between the drawer front and the drawer box.

In planning the drawers, make sure that the top of the highest drawer opening is lower than eye level, otherwise you will have a hard time seeing the contents in actual use. Likewise, a higher drawer with a heavier load can cause a major shift in the center of gravity as it is opened. A fully loaded stereo cabinet can weigh over

Fig. 10-31. Lateral record file drawer should pull all the way out of case to allow the albums to be easily removed.

Fig. 10-32. Record drawer hardware is a size longer than the drawer to insure that drawer pulls out far enough.

three hundred pounds and will do a lot of damage to itself, and possibly you, should it fall over.

The drawer front, which will be seen, should be made out of the same type of exterior material that the rest of the case is built of. Leave a minimum 1/16″ clearance on both sides and top of the drawer front, and ⅛″ at the bottom to allow room for a little settling under load.

Before cutting out the rest of the drawer parts, carefully read the instructions supplied with the drawer glides you intend to use. Each manufacturer can require a different clearance at the drawer sides, top and bottom.

The interior height of the drawer will vary with the intended contents. Record drawers should be 13″ high by 13″ deep inside measure, cassette tapes and accessory drawers should be 5″ high inside measure.

The drawer box parts can be made of most any smooth, strong, cheaper material. Yellow poplar, red oak, and lauan lumber are common, as is Baltic birch plywood. The material should be a minimum 3/8″ thick, ideally a maximum of 1/2″. The thickness can be greater, but only at the expense of interior storage room.

Particle board may be used as well, but it usually performs less satisfactorily over the left of the cabinet. Since the board has no fibrous integrity to securely hold the drawer hardware screw threads, and since the weight of the loaded record drawer will subject the screw mount to much abuse, it is best to avoid particle board altogether.

Drawer Construction

The material chosen will determine the inner dimensions of the drawer. Cut drawer sides the full front-to-back depth of the drawer, and one-half inch taller than the expected interior height of the drawer. Make sure you have allowed an additional one-half inch above the drawer sides in the case opening for adjustment and removal clearance. Run a 1/4″ wide groove 1/4″ from the bottom edge the length of the drawer side on the inside to receive the drawer bottom panel (Fig. 10-13).

Cut the drawer back and the front-back one-half inch narrower than the drawer sides (Fig. 10-14) to allow the bottom to slide in. With glue and nails, assemble the drawer sides, backs, and the front-backs into the basic box using butt joints. If skill, equipment, and time permit, you can get as fancy as you want here and use traditional dovetail joinery. However, considering the minimal strain on the drawer sides, the strength and stability of modern materials (particularly Baltic birch plywood), and the holding power of modern glues and fasteners, the structural gain of fancy joinery is negligible compared to the cost. But it will look impressive.

Spread plenty of glue on the outside of the front-back in a serpentine fashion (Fig. 10-15). This binds parallel fibers of the surface grain together so that they support each other and makes a stronger joint. If a glue line runs straight with the grain and starts to fail, it will probably fail the whole length rather rapidly.

Locate the drawer box on the back of the drawer front (Fig. 1-16), and secure with either power staples or screws. Nails do not work very well in this situation. There is no room to swing a hammer, and the vibration from hammering one nail home will loosen all the others. The glue joint under compression provides a strong joint, but since the drawer front will be continually pulled and pushed, it doesn't hurt to have the mechanical backup of screws.

While the glue in all the drawer joints is still soft, slide the ¼″ plywood or hardboard bottom into the grooves in the drawer side (Figs. 10-17 and 10-18), check all the box diagonals for squareness, and adjust if necessary. This is very important. One or two minutes spent making sure the drawer is perfectly square can save an hour or two of makeshift adjustment later. Once the drawer is perfectly square, nail the bottom to the drawer back and the front-back with plenty of nails. Since the bottoms of the drawers are not seen, use box nails instead of finish nails. The smaller finish nail head can pull through the bottom panel under the weight of the records.

An alternate way of fastening the drawer front to the drawer box is to assemble, square, and put the bottom in the drawer box first, then glue on the drawer front (Figs. 10-19 and 10-20). This method is more suitable for power staplers and screw guns since assembly access from the bottom side of the drawer is precluded by the already installed bottom panel.

The completed drawer (Figs. 10-21 through 10-23) should be strong and neat and provide scores of years of service.

Mounting the Drawers

Buy good quality ball bearing side slides for long life of the cabinet. Plastic glides or tri-roller kitchen cabinet slides just cannot hold up to the constant usage and weight of the records. Slides rated for fifty pounds are adequate for the cassette and accessory drawers, but use slides rated for at least one hundred pounds for any record drawer. Make sure that the record drawer can be pulled completely out of the case so that record access is unrestricted. This usually is accomplished by using a slide one size larger than the actual drawer requirements (e.eg., using a 16″ full extension slide for a lateral 14″ deep drawer).

Mount the hardware to the drawers as specified by the hardware manufacturer (Fig. 10-24). If the drawer half of the hardware has both vertical and horizontal slots, use the vertical slots only for the time being (Fig. 10-25).

If the carcase does not have a face frame, mount the case half of the drawer slides directly to the interior case sides. If the carcase has a face frame, you will have to install cleats the same thickness as the overlap of the face frame (Fig. 10-26). Use hardwood or hardwood core plywood (such as Baltic birch) for the cleats. The screws will pull out of softwood and softwood plywood too easily.

Mount the cleats for the top drawers first using only the horizontal slots. Cut spacers to hold the cleats evenly off the cabinet floor (Figs. 10-27 and 10-28). Use the same spacers in the same order to mount the case half of the slides. Once the slides are in place (Figs. 10-29 and 10-30) slide the drawers into the carcase and check the fit. If the drawer needs to go up or down a little, loosen and adjust the screws in the vertical slots of the drawer hardware. If the drawer needs to come out or go into the case a little, adjust the screws in the horizontal slots in the case hardware. Bias the drawers towards the upper part of the drawer opening to allow for a little settling from the weight of the contents.

Well adjusted drawers should roll very easily and should not bind against the face frame or the bottom of the case. Again, make sure that lateral record file drawers pull all the way out of the case to permit record removal (Figs. 10-31 and 10-32).

Once the alignment and clearances of the drawers have been completely adjusted, very carefully remove the drawers and set the hardware permanently with the extra screws in the round hardware mounting holes.

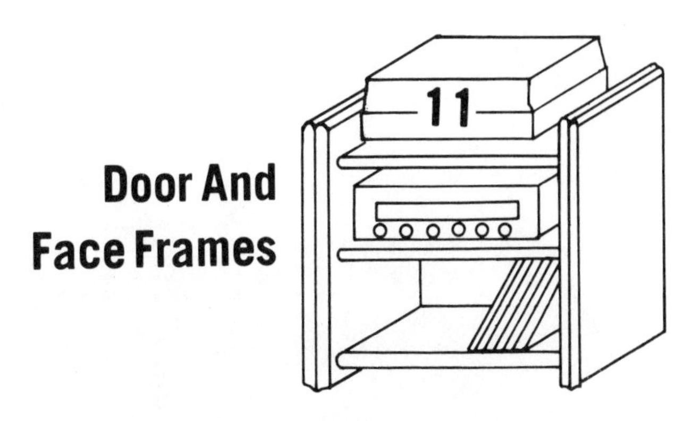

Door And Face Frames

Frame construction technique must be adequately mastered in order to build a lasting and good looking piece of furniture. Hardwood lumber stile and rail face frames and doors are the first and foremost esthetic noticed when viewing the finished piece. The quality of the frame construction (read joinery) has a lot to do with how much everyday use and climatic changes the cabinet can take before disassembling itself into a pile of parts on the floor.

FACE FRAME AND DOOR ADVANTAGES

The advent of good quality plywood simplified case construction considerably by eliminating the structural necessity of building the entire carcase around a frame. But a well constructed solid wood face frame enhances the rack-resistance of the case by lending more rigidity to the open face. While face frames do not provide anywhere near the sheer bracing effect of the cabinet back, together they tend to better stabilize the front.

It is safe to assume that fully inset doors with dust inhibiting minimum clearance are impossible without a face frame. If the floor is just a little uneven under the cabinet base, the carcase will rack under its own weight and cause the doors to bind. With a face frame there is increased resistance to case racking. Depending on the shrinkage and expansion factors of the door frame lumber, it is possible to have fully inset doors with 1/16″ clearance on four sides that never bind.

Face and door frames take eighty or ninety percent of the physical abuse the cabinet will undergo. Fingernails, jewelry,

Fig. 11-1. Frame stiles marked and ready for mortising.

shoes, stereo equipment, various stereo accessories, vacuum cleaners, and many other hard objects with corners will bang into the cabinet during its lifetime (100 - 200 years if well made and maintained). Whereas a veneer edged panel can tear, look

Fig. 11-2. Cutting a third shoulder on a tenon. This covers the radius at the end of a router cut mortise.

unsightly, and be difficult to repair, a lumber frame merely dents a little. Dents are less noticeable than veneer punctures and can be repaired easily.

Large and respectable furniture manufacturers spend thousands of dollars either putting dents in brand new furniture (called distressing), or spatter painting "distress marks" in the finish just to achieve some appearance of aging. Actually, the painted on distressing is an industry-wide method of disguising

Fig. 11-3. Frame members machined and ready to assemble.

Fig. 11-4. Cutting a hinge mortise in a door stile prior to assembly.

declining veneer quality. In any event, solid framing can take the bumps much longer and harder than veneered edges which are less than 1/28″ thick.

Doors get opened and shut, smacked into walls and other furniture, and are handled each and every time the stereo is used. Since the doors are moving parts, racking and warping affects their performance directly. A frame and panel designed door can not only handle the certain physical abuse, but can also handle dimensional changes in the wood itself.

FRAME AND PANEL CONSTRUCTION

Frame and panel construction is more or less dimensionally stable. Wood, as a medium, shrinks across the width of the grain,

Fig. 11-5. Door frame glued and ready to assemble.

Fig. 11-6. Proper glue application on a tenon. Notice the third shoulder.

but doesn't shrink much along its length. The frame and panel design puts the low shrinking long grain along the significant frame dimensions. The vertical members (stiles) have the grain running up and down, the horizontal members (rails) have the grain running across the width of the frame. The net result is that the frame tend to retain the same outside dimensions regardless of wood expansion and shrinkage. By contrast, a solid or glued up lumber slab

Fig. 11-7. Assembling the joint.

Fig. 11-8. Excess glue either collects in the extra 1/16″ depth of the tenon or squeezes out the top as shown here.

type panel can shrink or expand as much as ¼″ depending on the type wood and the width of the slab.

Any tendency for one frame member to bow or twist is usually counterbalanced by the other frame members. This reduces the

Fig. 11-9. Clamping the joint. By adjusting the placement of the clamp contact points, a door can be adjusted to perfect squareness. Note ¼″ plywood clamp pads.

Fig. 11-10. Tap the rails flush with the stile ends.

likelihood of the doors twisting or warping out of line. For our purposes, we shall examine mortise and tenon joined frames and dimensionally stable plywood panels.

Mortise and Tenon Joints

Frames may be assembled using any number of joining methods. Clinching staples, corrugated nails, splined miters, half

Fig. 11-11. Checking the frame for squareness. The diagonals should be equal.

Fig. 11-12. Pinning the tenons with power staples.

laps, and doweling might be used, but none of these methods have the simplicity to strength factor of the mortise and tenon joint. Doweling comes the closest, but dowel joints require machinery and supplies specifically designed for the joining task. A good supply of dowels is necessary. The dowels must be grooved to

Fig. 11-13. Excess glue may be wiped away while still soft with a damp rag.

Fig. 11-14. The ends of the door joints are filled prior to door sanding.

allow glue relief. The wherewithal is required to bore a series of extremely accurate doweling holes. Dowel joints depend on a good glue joint in two places: where the dowel meets the stile, and where the dowel meets the rail.

Fig. 11-15. Hardwood veneer plywood panel may be glued directly on the back of door and drawer front frames. Note the overlap that will be sanded off later. This allows for slight variations in size on the part of the frame.

Fig. 11-16. Inset door panels set in rabbet in back of door frame.

On the other hand, mortise and tenon joints may be cut accurately on the same equipment that is required to cut out the rest of the cabinet parts. Mortise and tenon joints need no other supply than the glue and wood of the stiles and rails themselves. Also, mortise and tenon joints depend on only one glue interface: that directly between the cheeks of the stile and rail.

Since the glue joints in casework are the most corruptible element used in construction, minimizing the total number of glue joints will increase the durability and life of the case as a whole. That mortise and tenon joints can stand up to the task is evidenced by an Egyptian sarcophagus now residing in a British museum—

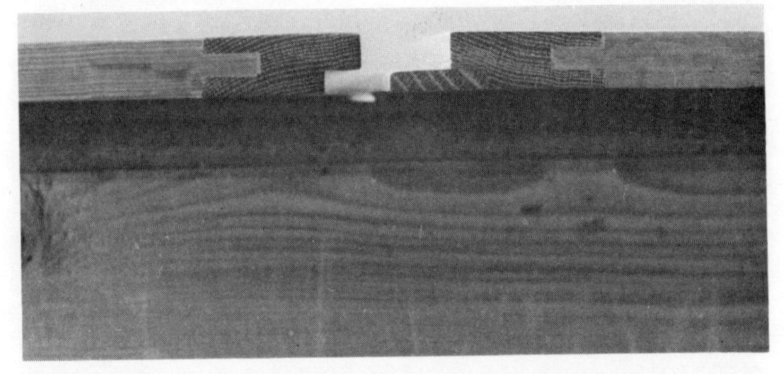

Fig. 11-17. Door frames with inset panels must be milled for dust lip.

Fig. 11-18. Door frame with glued on panel. Notice how the panel is recessed to receive male dust lip on other door.

still together five thousand years after its mortise and tenon construction.

Cutting the Mortise

Mortises may be chiseled out by hand, drilled with a drill and cleaned up by hand with a chisel, or cut with a router. The router method is more accurate for most framing mortises.

The frame material is usually ¾ to 1 inch thick, and works well with a ¼″ wide mortise. Mount a one horsepower or larger router

Fig. 11-19. Hinge mortising cut on a table saw prior to gluing on door panel.

Fig. 11-20. Hinge mortising cut with router.

under a router table (See Fig. 7-7) with an adjustable fence. Use a ¼″ straight cutting router bit with a plunge cutting tip, and set the fence ¼″ away from the inside edge of the bit. This will cut a mortise exactly in the center of the ¾″ thick stile.

The depth of the cut will depend on the wood used, strength required, width of the stiles, the power of the router, and the sharpness of the bit. Stereo furniture requires a tenon ½″ to ⅝″ long. Set the depth of the mortise 1/16″ deeper than the tenon to allow space for glue relief.

One the fence has been set, the depth of cut determined, and all adjustments tightened, test the set-up with scrap pieces of lumber. In a set-up as shown in Fig. 7-7, the material will be fed on edge from right to left. Like a saw blade, a router bit has a right and

Fig. 11-21. Hinge mortising cut on table saw after door panel is glued on. Note square corners.

Fig. 11-22. Screw tends to wander pulling the hinge out of position unless a guide hole is punched in the center with an awl.

a wrong direction of feed. Go the right direction and the cutting action of the bit will hold the material against the fence. Go the wrong way and part will be forced away from the fence. At best, cutting the wrong way produces an inaccurate joint. At worst, you can lose big hunks of your hand.

Mark the router fence with lines 1/16″ outside the actual cutting limits of the bit. On the outside edge of the stiles, mark the exact rail placement (Fig. 11-1). This way the cut will end up 1/16″ short of the actual rail size. Since the router bit leaves a radius at

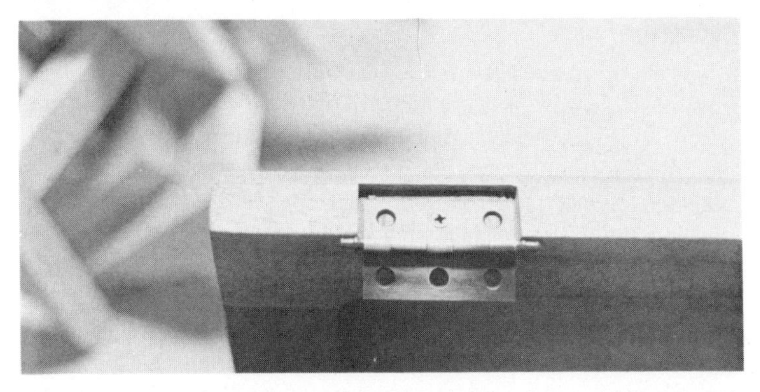

Fig. 11-23. Center screw holds hinge in place until door is completely mounted and lined up.

Fig. 11-24. Spacer pads hold door above the face frame the same height as the hinge knuckle is thick. The spacer stick supports the door freeing both hands for mounting.

each end of the slot, cutting the slot slightly short permits covering up the radius with a third shoulder on the tenon. The alternative is laborious squaring of the slot ends with a hand chisel or rounding the tenons with a file—cut-and-try at its worst.

Always mortise the stiles with the good face against the router fence. The distance between the fence and the router bit is a constant and should not change. The exact thickness of the lumber may. This way, when the joints are assembled, the front side of the frame will be smooth jointed and require little sanding.

Cutting the Tenon

Tenons may be cut either on the table saw or on a light shaper. For the table saw, set the blade height the exact length of the tenon. Space the fence from the blade so that the distance from the outside of the kerf to the fence is exactly the same as the distance between the mortising router bit and router fence. It is a good idea to set up the router for mortising at the same time the table saw or shaper is set up for tenoning. Then the set-ups can be tested with pieces of scrap lumber until the joints mate perfectly.

With the face of the rail against the fence, cut the first cheek of the tenon. If your set-up is symmetrical, the other cheek may be cut using the same saw setting. After cutting all the cheeks of the rails and interior stiles, reset the table saw to cut off the shoulders.

Take great pains in setting the depth of the shoulder cut. While the cheek cut can be made slightly deep without weakening the tenon, the shoulder cut must be on the money to insure maximum joint strength.

After the shoulders have been trimmed, reset the saw to the height of the tenon length. Clamp a scrap piece of lumber or plywood to the saw fence and move the fence right next to the blade. The blade must rotate freely, but should scrape the wooden fence pad slightly. Using this set-up, cut a slight (⅛″ deep) third shoulder (Figs. 11-1 and 11-2) which will conceal the mortise radii.

To cut tenons on the shaper, use two flat cutting shaper cutters with a steel ¼″ washer between the cutter heads (Fig. 7-6). Both the cheeks and the shoulders will then be cut in one step. Use

Fig. 11-25. Door mounted with piano hinge.

Fig. 11-26. Magnetic latches mounted in position for double doors.

appropriate caution and equipment such as hold-downs, sandpaper liner miter gauge, and sharp blades. *This method is dangerous and should only be performed by people with a lot of shaper experience.* The third shoulder is cut on a table saw as previously described.

Mortising for Hinge Mounting

After all the cabinet frame parts have been mortised and tenoned (Fig. 11-3), and shouldered, the hinge mortises can easily be cut into the door stiles. This operation may also be done after complete door frame assembly, but is easier and more accurate if done now.

Set the table saw blade 1/16″ lower than the width of the hinge knuckle. Set the fence 1/16″ less than the distance from the end of the stile to the hinge, and cut a kerf across the back edge of the stile. Reset the fence and make another kerf slightly further in on the stile than the hinge plate is wide (Fig. 11-4). The idea is to outline the hinge mortise leaving 1/16″ clearance on either side of the hinge plate to prevent binding.

With the miter gauge, slide the hinged edge of the stile back over the moving saw blade in the second kerf. Center the stile directly over the saw arbor and slowly and carefully slide the stile towards the first kerf. This will evenly waste the whole mortise in one step. If the mortise is any deeper than ⅛″, make two passes with different settings to avoid a kickback.

An alternate and slightly safer way to accomplish the same end is to make a series of kerfs between the two outline kerfs. This is slow and laborious, but nearly impossible to kick back.

ASSEMBLING THE FRAME

Prior to permanent assembly, test assemble all the frames to make sure the mortises are long enough or the shoulders deep enough. Nothing is more frustrating than to have a rather complicated frame halfway glued up, and only then discover that a mortise was miscut, or the length of an internal stile or rail

Fig. 11-27. Overlay type door with self-closing spring loaded hinges.

Fig. 11-28. Finger pull bevel in back of door frame eliminates the need for door pull hardware.

misfigured. Test fitting the frame dry will flush out mistakes of this type without ruining all the parts.

The joints should fit snugly, but should need nothing other than hand pressure to assemble. If you have to hammer the joints together, they will eventually split under the additional pressure from the glue or hygroscopic swelling of the wood tissue.

Lay the test-fit frame face down on a large, level table and gently pull the frame members apart. Apply a bead of glue to the top inside cheeks of the mortises, and also to the end edge of the tenons (Figs. 11-5 and 11-6). When the tenon is slipped into the mortise (Fig. 11-7), the glue will be smeared evenly and thoroughly across the faces of the adjoining cheeks. The excess glue will accumulate in the 1/16″ extra depth of the mortise or squeeze out the top of the joint (Fig. 11-8).

Clamp the frame together with just enough pressure to snug up the joint (Fig. 11-9). With a hammer, gently but firmly tap the rails until their outside edges line up perfectly with the ends of the stiles (Fig. 11-10). Check the diagonals for squareness (Fig. 11-11). Slightly out of square doors are usually caused by out of square tenon cuts or asymmetrically placed clamps. This can be adjusted out by realigning the clamps to pull out the discrepancy.

After the frame is squared up, tighten the clamps some more and check the diagonals again for shift. If the frame remains square,

Fig. 11-29. Door with slightly crowned rails. Note vee-grooves evenly spaced in panels.

Fig. 11-30. Close-up of drawer front designed to look like a pair of doors.

pin the tenons with staples (Fig. 11-12) or short brads. Two or three per joint is sufficient. Actually, the purpose of the pins is only to hold the joint until the glue dries. If you have more clamps or time, you can leave the frames clamped for two or three hours and unclamp without pining.

After removing the clamps, wipe or sand off the excess glue (Fig. 11-13). At this point face frames are ready for assembly with the case. Door frames must be shaped with a router, filled (Fig. 11-14), and finish sanded to make them ready to receive the panel.

DOOR PANELS

Hardwood veneer plywood panels are dimensionally stable and may be glued directly on the back of the door or drawer front frame (Fig. 11-15) or rabbeted into the back of the frame (Fig. 11-16).

If the panel is to be glued directly on the back, cut it a little oversize (Fig. 11-15) and flush-trim or sand it down after the glue has dried.

Solid lumber panels require a more complicated mounting system, and generally need special equipment and set-ups. The principles of solid panel mounting, and dealing with expansion problems are well known and covered in many other books on general cabinet making.

Dust Lips

Larger cabinet openings such as are common in most stereo furniture designs are normally enclosed with a pair of doors rather than a single door. This is usually more esthetically pleasing, and

Fig. 11-31. French curves can make a frame and panel door look more interesting.

Fig. 11-32. Fancier door frame molding can complicate joinery. Here the beading requires either compound coped or mitered joints.

functionally more shrewd. The larger the dimensions of the door, the greater the likelihood of its warping or twisting. A pair of doors will pose far fewer alignment problems than a large single door.

In a cabinet that is two or more sections wide, making simultaneous adjustments on equipment in two different sections (e.g., setting recording levels on the tape player while recording off a turntable) can be awkward if you have to work around a wide door swing.

The main problem with door pairs is the gap in the middle between them. An overlapping dust lip (Fig. 11-17) will block out fifty or sixty percent of the dust infiltration from the front of the cabinet. These must be machined into the hardwood lumber frame for doors with inset panels (Fig. 11-17), or incorporated with panel recess (Fig. 11-18) and overlap if the panel is glued directly on the back of the frame.

Hinging the Doors

If the hinge mortises have not been previously cut in the door stiles, now is the time to do them. These may be cut on the table saw as previously described (Fig. 11-19), or cut with a jig and a straight-cutting router bit (Fig. 11-20). Cutting the mortise with a router bit will leave rounded corners which may require additional hinge clearance. With table sawn mortises the corners will be square and professional looking when the panel is glued on (Fig. 11-21).

Pad and clamp the completed door with the hinged edge up. Hold the hinge in place with one hand and mark the center of the middle screw hole with an awl (Fig. 11-22). This insures that the mounting screw starts centered and doesn't pull the hinge out of position.

With one screw only, set the hinge in the door mortise (Fig. 11-23). Save the remaining holes for possible adjustment later. Cut spacer pads (Fig. 11-24) the same thickness as the hinge knuckle. Cut a spacer stick the same length as the case is deep. Lay the cabinet on its back and position the spacer pads on the face frame approximately next to where the hinges will be mounted.

Lay the door in opened position by propping with the spacer stick. With both spacer pads and stick in place (Fig. 11-24) both hands are left free for the careful attachment of the inside edge of the face frame. Again use only one screw per hinge.

After all the doors have been positioned and mounted into place, set the cabinet back upright and check the swing of the doors for binding and alignment. By loosening the single screws, you can "walk" the door into proper alignment with the balance of the mounting screws. After alignment is perfect, set the hinges by installing any additional screws.

Piano hinges (Fig. 11-25) can be installed using similar technique. Use one screw at each end of the hinge (and one in the middle if the door is long) to position the door and check the fit. After aligning the doors, use alternating holes in the piano hinge plates to prevent screw heads from hitting.

LATCHES AND HARDWARE

All types of cabinet door latches are available on the market. Avoid the mechanical type. These require very precise door alignment to work properly and cannot tolerate slight changes in alignment that will occur with time. They also wear out much faster than magnetic latches.

Magnetic latches (Fig. 11-26) are quite impossible to wear out and allow more variation in the door alignment while working properly. If the cabinet is moved or shifted so that the doors rack slightly, magnetic catches will continue to perform without adjustment. For the same reason, magnetic touchlatches are better to use than mechanical touchlatches, especially if no external door hardware is desirable.

If self-closing spring loaded hinges (Fig. 11-27) are used (as with overlay ⅜″ inset doors), a finger pull bevel can be cut in the back edge of the door frame to eliminate the need for external door pull hardware (Fig. 11-28). Figures 11-29 through 11-32 show examples of different styles of doors.

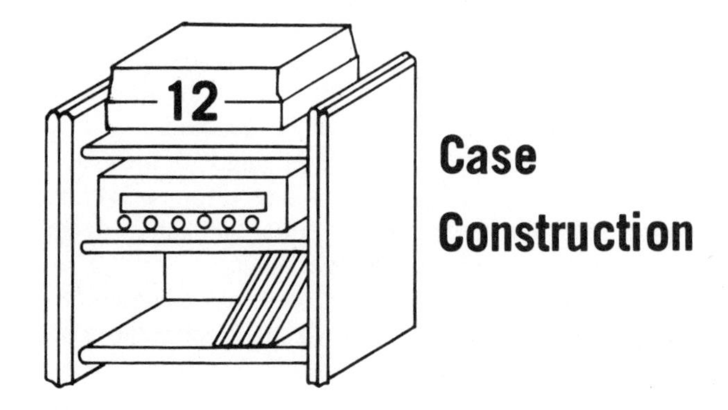

Case Construction

There is no single method of case construction. Technique varies from one cabinetmaker to another depending on skill and experience, machine availability, time and money constraints, material characteristics, and a host of other variables. Virtually every book on cabinetmaking has yet another suggestion or trick to simplify or strengthen or economize. The following techniques are just another few composite methods adequate for a durable, long lasting piece of furniture, but are general enough to be performed with typical woodshop equipment.

METHODS OF CASE CONSTRUCTION

Carcase construction has been greatly simplified with modern hardwood veneer plywood. Once upon a time, one had to either find a board that was wide enough for the case sides and top, or resort to some variation of the frame and panel method of building up planar surfaces. Now, since very few of the larger trees are left, even fewer wider boards exist. This leaves three remaining options for spanning wider areas:

First, either that rare wider board must be located, or it may be fabricated by laminating several narrower boards together. The lamination method is quite common, but itself poses several problems. It raises the total number of potentially corruptible glue joints in the cabinet. Using solid lumber increases the weight of the piece, which amplifies the risk of damage when the piece is moved. Quite a bit of expensive labor and equipment is used to join and surface the laminated boards. Since solid lumber changes size with

Fig. 12-1. Back edge of end panel. Dado is for fixed shelf, rabbet along edge is for cabinet back.

humidity changes, more complicated construction and expensive joints are required to accommodate the wood movement. The net result is a more expensive cabinet requiring more specialized equipment.

Second, narrower boards of lumber can be assembled into that old stand-by, frame and panel construction. This solves the

Fig. 12-2. Blind dado is cut most of the way with dado blade on the table saw. The end of the cut is squared up by hand with a chisel.

Fig. 12-3. Bottom view of semi-permanent shelf over turntable compartment. Shelf is cleated and screwed into case to allow future changes.

dimensional stability problem and simplifies some of the joinery, but the labor and weight problems remain.

Third, a modern hardwood veneer plywood can be used. The crossband layers within the plywood core stabilize the hygroscopic

Fig. 12-4. Fully machined partition panel. Grooves have been cut to inset the brass plated shelf standards.

Fig. 12-5. Vertical format end panel. Center dado is for fixed turntable shelf. Start assembly by laying panel on a pad on the floor as shown.

dimensional changes. This allows simpler joints for the same strength and requires less equipment and labor. Hardwood veneer plywood panels come four feet wide and up to twelve feet long, which rather thoroughly solves the available size problem. The

Fig. 12-6. Insert the fixed shelves without any glue at this stage.

Fig. 12-7. Glue the top edge of the fixed shelves and position the other end panel.

Fig. 12-8. Tap the edges flush and pin the joint with two or three finishing nails.

Fig. 12-9. Completed panel assembly ready to receive face frame.

Fig. 12-10. Horizontal format cabinet panels laid out ready to assemble.

core veneers in the plywood are usually made from lighter woods, frequently softwoods, thus reducing the weight damage problems.

The main disadvantage of hardwood veneer plywood is that any exposed edge must be covered with veneer or a hardwood strip to keep the core layers from showing. Also, since the top veneer is only 1/28″ or less, construction must be more precise. Excess sanding can go right through the veneer. Repairs are more difficult for the deeper mars and scratches for this reason. This may not be too much of a problem if the cabinet has solid lumber face and door frames since the front and corners take most of the abuse. Sometimes a handwiped finish will turn out uneven because of microscopic knife checks on one face of a book matched veneer.

Fig. 12-11. After gluing the partition, glue the end panels to the bottom and pin with finishing nails.

Fig. 12-12. Installing the rest of the fixed shelves.

In dollars per square foot, lumber and hardwood veneer plywood cost very nearly the same. But, by greatly reducing preparatory labor and equipment expenses, the advantages of plywood far outweigh the disadvantages for a small scale operation. All further case discussion will assume panel construction.

DADOED PANEL CONSTRUCTION

Butt joints are often used for panel constructed cases because they are simple. But since they depend completely on fasteners and glue joints for their strength, the joints will fail in time.

Fig. 12-13. Partition should be notched out to receive back cleat.

Fig. 12-14. Back cleat in place. Ends are nailed from outside end panels.

Dadoed structural parts (Figs. 12-1 and 12-2) not only have the strength of glue and fasteners holding them in place, but also rest in a groove which transfers the shelf load directly down the end panels to the floor. Dadoed joints will not pull out even if the glue joint fails.

VERTICAL DESIGNS

In vertical stereo furniture designs, the cabinet top and bottom and any permanent shelves are dadoed into the end panel. Semi-permanent shelves such as the equipment shelf are mounted into place with cleats and screws (Fig. 12-3). The end panel should

Fig. 12-15. Taping a mitered case corner joint with masking tape.

Fig. 12-16. Flip the taped pieces over and apply glue in the mitered edges.

be grooved or bored to receive the adjustable shelf standards at the same time it is dadoed for the shelves (Fig. 12-4). The cabinet back is set into a rabbet on the back edge of the panel (Fig. 12-1).

After all the parts have been cut out and all the exposed shelves have been edged, carefully measure and mark where the dadoes and rabbets will go on the end panel. It is much better to

Fig. 12-17. The masking tape acts as a hinge when the case is folded together. The last joint is glued and taped at this point.

Fig. 12-18. Check the assembly for squareness.

check your layout from pencil lines which can be changed than from possibly wrong cuts.

After checking the layout, cut the top, bottom, and shelf dadoes and the back rabbets in the end panels. If any inset shelf standards are part of the design, lay out and cut the grooves for

Fig. 12-19. A rabbeted mitered corner joint shows how nicely the machining goes around the corner.

Fig. 12-20. Turntable shelves need 1½″ spacer block to keep from being knocked out of line when the face frame is nailed on.

these at this time. After all the machining is done, sand all the surfaces that will be in the interior of the cabinet while the parts are still unassembled. It is much easier to sand a stack of parts laying flat on the table than it would be to fight with the interior corners and awkward positions later. You can also pre-stain the interior surfaces at this time, but make sure not to get any stain on any surface that will be glued. Keep the stain out of the dadoes.

Fig. 12-21. Case bottoms and unexposed tops need ¼″ spacer block.

Fig. 12-22. This tall bookcase will have the top four inches of its back sloping in at a 45° angle to deflect air from a wall register over the top of the cabinet. Specially sized spacer blocks are required for face frame mounting.

Lay one end panel on a pad on the floor (Fig. 12-5). Insert the fixed shelves, top and bottom in the appropriate dadoes without gluing them (Fig. 12-6). Spread a bead of glue along the top edge of

Fig. 12-23. A glue bead is laid along the front edges of the panel assembly just before the face frame is set into place.

Fig. 12-24. Bottom corner of panel assembly. Notice how glue bead is biased towards the bottom side of the cabinet floor. Excess glue should only bleed out where it is not seen.

the now upright shelves (Fig. 12-7). Bias the bead of glue towards the side of the shelf that will not be seen since some glue will probably bleed out of the joint later.

Take the other end panel and carefully fit the glued end of the shelves into their respective dadoes as you lay the panel on the case. This is easier if you start at one end and work your way towards the other end. It should smear some glue on an exposed surface, wipe it away immediately with a damp rag.

Tap all the front shelf edges flush with the face edge of the end panel and set the shelves with two or three finish nails each (Fig.

Fig. 12-25. Face frame is set in place and nailed. Compare with Fig. 12-9.

12-8). Roll the case over on the padding, remove the unglued end panel, and repeat the gluing steps. The panel assembly is then rolled on its back (Fig. 12-9) and the diagonals are checked for squareness. Once the case is square, it is ready for immediate installation of the cabinet face frame.

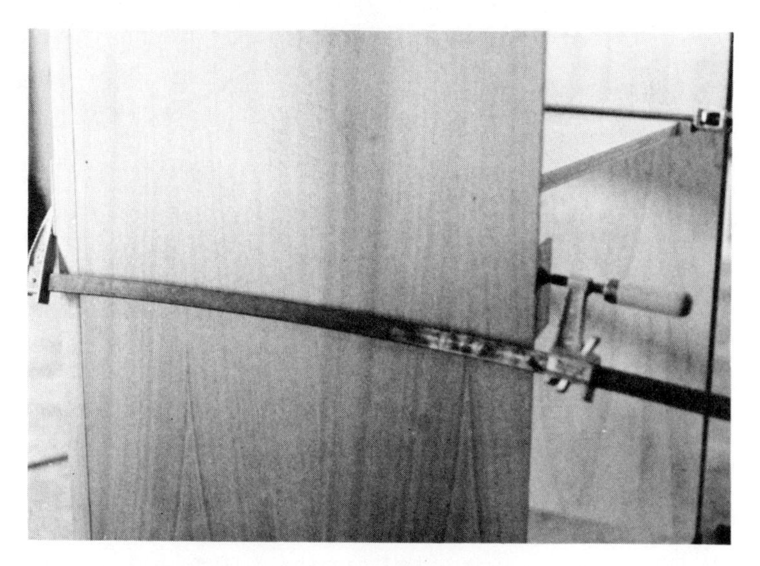

Fig. 12-26. Bar clamps snug up the face frame joint. Note the 1/4" plywood clamp pads.

Fig. 12-27. Sometimes a web clamp is useful. Here clamp holds face frame, end panels, and case top all under pressure at the same time.

HORIZONTAL DESIGNS

Horizontal cases are assembled very much like vertical cases. The main differences are the presence of one or more interior partitions and the absence of a fixed top panel. Because of these

Fig. 12-28. Face of this base cabinet lined up so well that only four clamps were required to snug up the joint. Normally two or three times as many clamps are required.

Fig. 12-29. Glue block cut out of 5/4 oak scrap lumber. Rule gives approximate size.

Fig. 12-30. Normal depth stereo case requires two blocks at the bottom along each end panel.

Fig. 12-31. Back of face frame should have a minimum of three blocks.

Fig. 12-32. Glue should bleed out from all around a properly glued and installed block.

Fig. 12-33. The sides of this plywood corner block are about four inches long each. Back corner is notched slightly to allow glue clearance and to prevent corner of block from acting as a lever fulcrum, forcing cabinet joint apart as the case is moved.

Fig. 12-34. Corner block at a top front corner on a horizontal format cabinet.

Fig. 12-35. Corner block at top front corner of case that will have end panel frames. Notice how the corner block is mounted on top of an inset metal shelf standard.

complications, the main case is assembled on its back rather than on its side.

After the parts have been fully machined and prepared for assembly, lay them out flat on the floor in their relative positions

Fig. 12-36. Corner block at a top back corner.

Fig. 12-37. Case ready to receive end panel frames. Face frame is wider than panel assembly by the width of the two end panel frames.

(Fig. 12-10). Spread a bead of glue in each of the dadoes in the bottom panel, and lift up the bottom with its back edge to the floor. Insert the bottom edge of the partitions and tap all the front edges flush. Shims or spacers may be required to hold the partitions the proper height off the floor (Fig. 12-21).

After nailing the partitions in place, run a bead of glue in the bottom dado of both end panels, lift them into positions, tap the edges flush, and nail the joints (Fig. 12-11).

Fig. 12-38. Close-up of end view of face frame. End panel frame will set in the rabbet.

Fig. 12-39. Test fitting the end panel frames to the face frame.

Roll the partial assembly into an upright position and glue and nail the other fixed shelves into place (Fig. 12-12). Apply glue to the receiving notches in the upper back of the partitions (Fig. 12-13) and on either end of the back cleat, and nail the back cleat into place (Fig. 12-14). The cabinet is then placed on its back to immediately receive the completed face frame.

Fig. 12-40. Close-up of end panel/face frame joint during test fitting.

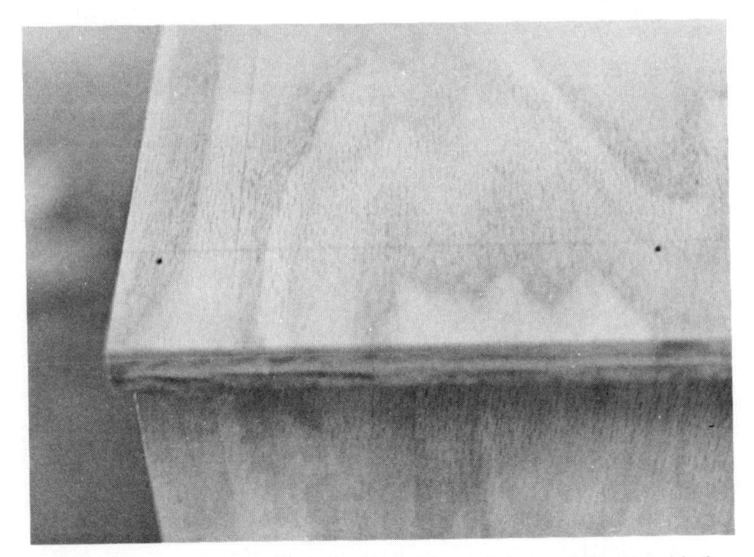

Fig. 12-41. Pencil lines mark ¼″ less than end panel frame position to guide the glue bead. No glue should be allowed to bleed to the inside of the end panel frame.

MITERED PANEL ASSEMBLY

Speaker boxes and some stereo cabinet designs may have mitered corners where the panels join. This technique is frequently used in Scandinavian and other contemporary styled pieces. The main advantage is a simpler appearance where the two

Fig. 12-42. Glue bead for end panel frame.

Fig. 12-43. The end panel frame is set into position.

planes join and are visible at the same time. This occurs most frequently in lower or horizontal formats. The main disadvantage of the joint is its weakness unless it is splined.

The miters are carefully cut on the ends of the panels slightly less than 45°. This insures that the visible portion of the joint will be perfect.

With the exposed faces up, the panels are lined up on a long table and the edges are carefully mated. The joint is then taped with

Fig. 12-44. End panel frame is clamped tightly to face frame. Downward pressure is applied while frame is stapled to the end panel from the inside of the case.

Fig. 12-45. Back edge of mounted end panel frame.

a wide strip of masking tape which acts as a temporary hinge (Fig. 12-15). The whole row of parts is then carefully flipped over with the tape side down (Fig. 12-16). A liberal amount of glue is spread in each joint, and the case is folded into final position where the last joint is taped (Fig. 12-17). The assembly is then laid on its back and the diagonals are checked for squareness (Fig. 12-18). After the glue is thoroughly dry, the masking tape is removed revealing what

Fig. 12-46. View during test fitting shows inside the extended stile legs.

Fig. 12-47. Make sure the leg portion of the joint is clamped tightly as well.

should be perfect joints. Any dadoes and rabbets should be cut in the panel prior to assembly, and will line up nicely on completion (Fig. 12-19).

FACE FRAME INSTALLATION

Any of the fixed shelves that have been recessed to receive a false back (Fig. 12-20), the cabinet back (Fig. 12-21), or some other consideration (Fig. 12-22) must be blocked up with a spacer.

Fig. 12-48. Clamping the filler block for the back corner of the leg.

Fig. 12-49. Joint between face frame and panel is filled and ready for finish sanding.

Otherwise, the act of nailing on the face frame wil sheer the glue joint in the shelf dadoes and knock them out of line.

Spread a bead of glue along all the front edges of the panels (Fig. 12-23). Bias the glue bead slightly towards the face of the panels that will not be seen. Any glue bleed out will then be on the side of the joint requiring less finish work. On the end panels run the bead slightly to the inside edge where the excess will be hidden by the face frame overlap. On the bottom and shelves, run the bead slightly towards the bottom faces of the panels (Fig. 12-24). The cabinet top panel is glued slightly towards the top edge.

Lay the face frame in position on the panel assembly, and line up the diametrically opposite corners. The face frame should be the exact height of the carcase and should overhang each end panel slightly. If the case is slightly wider than the face frame, you cannot sand the panels down without sanding right through the veneer. The easy way to insure proper tolerance is to cut the stiles 1/16″ wider than necessary, and plan to trim off the excess with a hand belt sander or router after assembly.

Starting with the top, line up the inside edge of the face frame flush with the adjacent panel, and nail the frame to the shelf about every eight to ten inches with finsihing nails. Line up the shelves and cabinet bottom next, and nail the end panels last.

After the face is pinned into place with nails (Fig. 12-25), stand the case upright and clamp the joints with as many clamps as it takes to close the gaps snugly (Figs. 12-26 through 12-28). Let

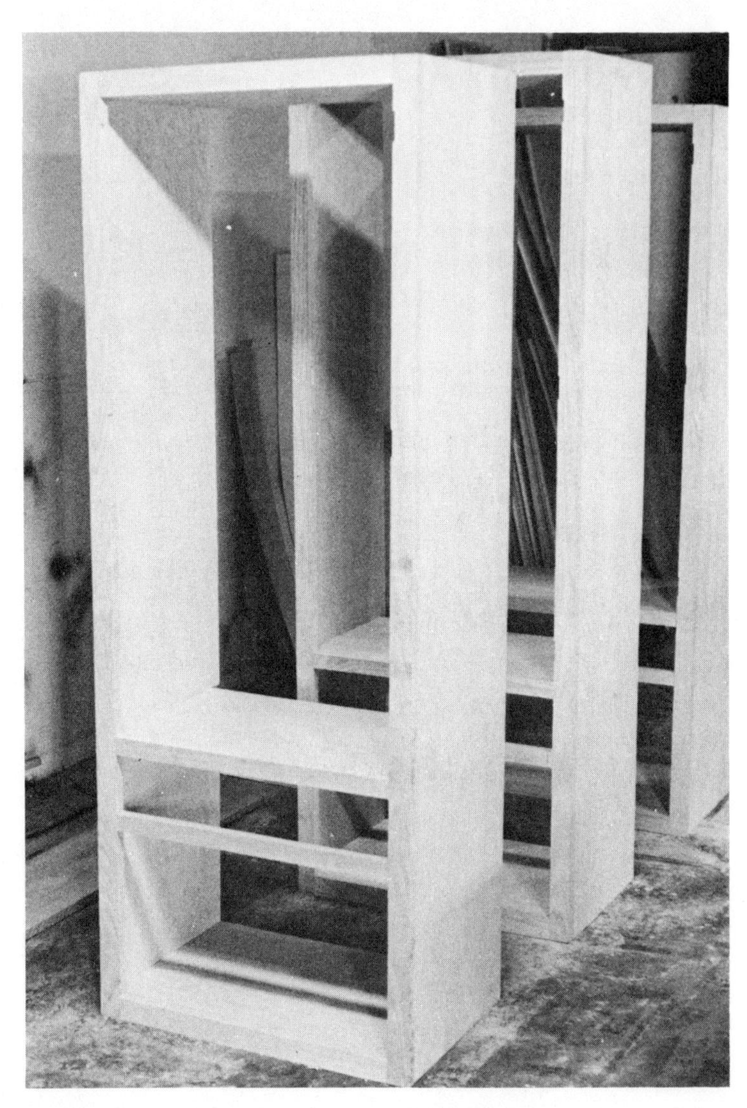

Fig. 12-50. Completed vertical format carcases ready for finishing and final assembly.

Fig. 12-51. Back view of horizontal carcase shows fixed shelves recessed for false backs. Concealed area will be used for wire runs and ventilation ducts.

the carcase assembly stand undisturbed overnight to allow the glue to cure thoroughly before the next step.

BRACING THE CASE

After the case has been assembled, the key joints are reinforced with either glue blocks or corner blocks.

Fig. 12-52. Cassette storage shelf is very shallow with a false back to be mounted behind it. Amplifier will be concealed in the hidden compartment.

Fig. 12-53. False back and mounting cleats ready to assemble.

Glue blocks are short pieces of scrap lumber (Fig. 12-29) that are glued every few inches on the back side of a major panel joint (Figs. 12-30 and 12-31) to increase racking resistance in the case, and to supply additional glue surface area for a proportionally stronger joint. Glue should be applied very liberally to the faces of the glue blocks. After the block is in position it should be pressed into the corner rather hard and slid sideways slightly. This "wringing" action squeezes all the air out and insures a good joint

Fig. 12-54. Front view of cleated false back.

Fig. 12-55. Back view of cleated false back.

without the necessity of clamps. Glue should bleed out all around the blocks (Fig. 12-32).

Corner blocks are triangular pieces of plywood (Fig. 12-33) that are liberally glued, and nailed or screwed into the corners of an open top case (Fig. 12-34 through 12-36). Any glue that bleeds out around the corner blocks should be wiped with a damp rag since the interior panels may be exposed to view.

Fig. 12-56. Back view of false back installed in cabinet. Compare to Fig. 12-51.

FRAMED END PANELS

Traditional end panels were constructed with the frame and panel design in order to allow the use of narrower lumber stock or to eliminate the complications caused by lateral hygroscopic expansion. With plywood panel construction the framed end panels are mostly used for appearance.

Stereo cabinet carcases must be a minimum of 18″ deep, and frequently 20″ deep to accommodate the stereo equipment, false backs,and ventilation ducting. From an esthetical point of view, the

Fig. 12-57. Back view of false back from concealed amplifier compartment. Compare to Fig. 12-52.

Fig. 12-58. Corner blocks in horizontal cabinet drilled for mounting finished top. Always drill at least two holes in case top needs some minor adjusting during installation.

deeper the cabinet, the worse the cabinet looks. One way of minimizing the boxy appearance of such a deep end panel is to frame it. The additional vertical lines and slightly narrower panel proportion will then make the cabinet look a little shallower than it actually is.

With framed end panels the case is put together as previously described, but the face frame has to be wider to allow for the thickness of the end panel frames (Figs. 12-37 and 12-38). The end

Fig. 12-59. Test fitting plinth to carcase.

panel frame can merely butt into the back of the face frame or be rabbeted (Fig. 12-38). In either case allow an extra 1/16″ in width of the outside face stiles to be sanded down flush with the end panel frame after assembly. Be sure to reduce the width of the front end panel still by the thickness of the face frame so that both end panel stiles appear the same width after assembly.

Fig. 12-60. Test fitting base to upper.

Fig. 12-61. Test fitting waist molding between upper and base cabinets.

Make the face and end panel frames before assembling the carcase and pre-fit the frame assembly to check all your dimensions (Figs. 12-39 and 12-40). After assembling the carcase, completely sand the end panel since it will be difficult to sand after mounting the frame. Scribe a pencil line ¼″ closer to the panel edge than the inside of the frame will be positioned (Fig. 12-41). This line will be used as a guide for spreading the glue. With a serpentine pattern, lay a bead of glue on the edge side of the line (Fig. 12-42). Use plenty of glue, but not so much that it would bleed to the inside of the frame. Should that happen it will leave an ugly glue spot that will show up when the carcase is stained. If the glue bleeds to the outside no harm is done.

Position the end panel frame (Fig. 12-43) and clamp tightly at the joint with the face frame (Fig. 12-44). Screw or staple the frame to the end panel from inside the carcase while applying pressure around the glue joint. Set the assembly aside and let it cure for an hour or more before mounting the frame to the other side of the

Fig. 12-62. Test fitting door and drawer fronts.

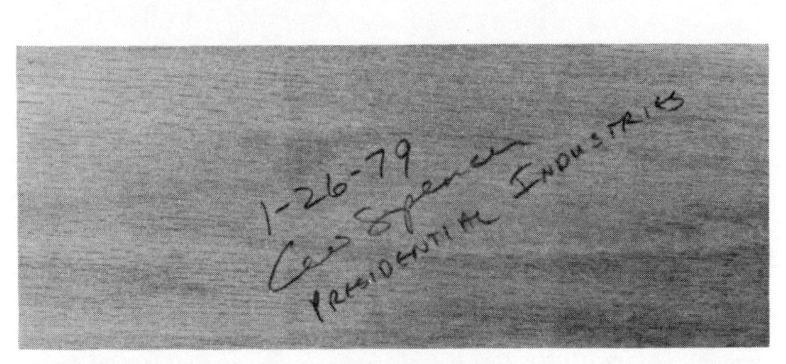

Fig. 12-63. Always sign and date the completed case in pencil on the bottom. Make it easy for future antique dealers.

case. The completed assembly will be very strong with the hardwood frame providing additional protection for the veneer (Fig. 12-45).

STRUCTURAL LEGS

The face and end panel stiles can be extended below the bottom of the case to form legs if desired (Fig. 12-46). While installing the end panel frames, make sure that the leg portion of the stiles is tightly clamped together (Fig. 12-47). The back of the leg is filled in with another block of lumber cut to fit, glued, and clamped into place (Fig. 12-48). The assembled leg is then belt sanded smooth and finish sanded with the rest of the case. Legs that are an extension of the stiles are very, very sturdy compared to legs that are made separately and screwed into place later.

FINISHING PREPARATIONS

Set all the nail heads in the face frame and end panels with a nail set, and fill the holes with a stainable filler that matches the raw wood. Fill all the exposed joints between the face frame and the panel surfaces (Fig. 12-49).

Chuck a flush-trimming bit into a router and trim off all the excess wood from the face frame stiles. Pad the floor and lay the case on one side. With a hand belt sander, lightly and carefully sand the face frame joint until it is flush with the end panel. Then sand down all the unstained exterior surfaces with a finishing sander, slightly easing all the exposed edges with the finest grade of sandpaper used. Set the case aside (Fig. 12-50) to await final assembly.

FALSE BACKS

A false back is a ¼″ hardwood veneer plywood panel that is mounted inside the case to conceal the wiring, ventillation ducting, and sometimes a piece of equipment. Any shelves in front of the false back should have clearance on the back side for the wire run gap as well as the thickness of the backs (Figs. 12-51 and 12-52).

The false back itself is cut the inside width of the carcase section and overlaps the back edge of the interior shelves. The back side is cleated with ¾″ by ¾″ cleats cut out of plywood scrap (Figs. 12-53 through 12-55). Edges that overlap the back of the shelves do not need any cleating since they may be nailed directly to the back of the shelf. The exception to this is the equipment shelf (see Fig. 8-78) which should remain unattached. The equipment shelf must pivot freely to allow the equipment panel to be changed in the future.

After the false back subassembly has been completed, its nails are set and filled (Figs. 12-56 and 12-57), the panel is sanded, stained, and finished before mounting in the cabinet. When both the carcase and false back are ready for final assembly, slip the false back into place and fasten with either power staples or screws.

PREPARING FOR FINAL ASSEMBLY

If the case has a finished top, drill mounting holes in the appropriate corner blocks (Fig. 12-58). Also check the fit of the plinth (Fig. 12-59) and cornice, the base of the upper (Fig. 12-60), the waist molding (Fig. 12-61), and the doors and drawers (Fig. 12-62). Any other subassembly should be test fit prior to finishing. Check the equipment panel and shelf fit, the false backs, and the cabinet back.

When you are satisfied with the fit of all the subassemblies, take everything apart for finishing. This is also a good time to sign and date the bottom of the carcase in pencil (Fig. 12-63). This records for future reference the date of manufacture of the piece, who was responsible, and perhaps the location. If you end up making many stereo cabinets, you begin to forget the vital statistics of each cabinet servicing. Also, if several people have worked on the cabinet, by having all sign the case you not only give them pride in their work, but if something goes wrong you know who to blame.

And then, who knows? Maybe you will become another Stradivarius. Your signature might make the cabinet worth a fortune!

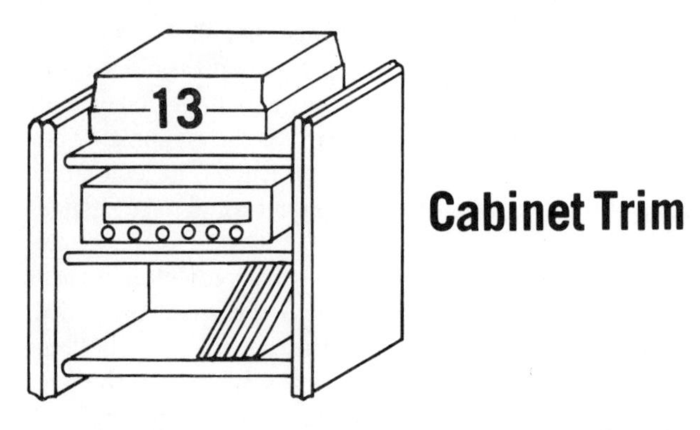

Cabinet Trim

A case without trim is just another box. Not only does the molding smooth out the lines, it also places solid lumber out in front of the veneered surfaces for their protection.

The top edging and waist molding break up the vertical lines and add a considerable amount of interest to the design. In vertical cabinets, the cornice molding can also enclose the area used for dust baffling (see Fig. 3-5). The plinth molding absorbs the beating from a vacuum cleaner wand and helps to raise the drawers high enough off the floor for more convenient use.

TOP EDGING

If the top plane of the cabinet will be visible, a finished top is required. Sometimes finished tops are used even above eye level instead of cornice molding. This permits placing other decorative objects on the top of the cabinet (framed photographs, plants, etc.) and gives a different appearance than a heavier cornice molding.

A finished top is made of hardwood veneer plywood to cover the bulk of the area, with a solid lumber edging splined to the front and two sides. The edges may be left square as they are, but more commonly they are given some sort of treatment or shaping to enhance the overall appearance of the cabinet.

The width of the molding is not very critical as long as it allows for the depth of any shaping cut on the edge. If the edges are simply to be left square a ¼″ or ⅛″ strip will suffice to conceal the core veneers of the plywood. Even so, the molding is usually eased

Fig. 13-1. Before the splining grooves are cut, the top edging is fit to the top panel. The front piece must be a perfect fit.

slightly with the final sandpaper to knock down the sharp edges. Not only is it possible to cut your hand by running it along a very sharp wood edge, it is also very easy to lift up a splinter which can stick into your palm or snag the dusting rag later.

Fig. 13-2. Checking the miter joint. The angle must be exactly 90°.

Fig. 13-3. The shaper set up to cut the splining groove. The flat cutter cuts a ¼″ wide groove in the edge of the panel and molding. Depth of the cut depends on the width of the molding. Here the groove will be ⅝″ deep in 2″ molding. The limitation in this instance is the length of the tooth.

Fig. 13-4. The top panel is grooved for splining on the front edge and the two sides.

Fig. 13-5. Close-up of splining groove in the top panel.

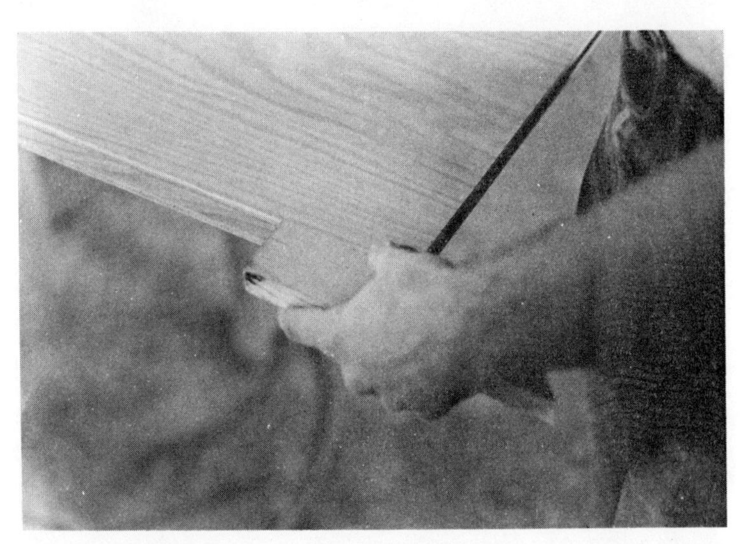

Fig. 13-6. The alignment of the grooves is checked with a piece of scrap before final assembly with glue.

For edges to be shaped, the width of the edging usually runs from ¾″ to 2″ depending on the depth of the shaping cut or what pre-ripped stock lumber is available. If you plan to use 2″ wide ripped stock for the door framing, you may as well use a 2″ wide

Fig. 13-7. The side pieces should be cut too long. This permits enough length for trimming and adjusting the miter fit. After the glue has dried the excess is cut off flush with the back of the top.

Fig. 13-8. View of splined joint after assembly and trimming.

lumber edge for the finished top. By reducing the number of different lumber rippings you increase the efficiency of time and lumber useage. Since the lumber may all be ripped at the same time with the same saw setting, the accuracy (or at least the consistency) of the rail/stile/edging stock is greatly improved.

Fig. 13-9. The corners of the top edging have been marked for rounding. Diameter must be appropriate for the size and style cabinet. Here a nickel was used.

Fig. 13-10. Cut the radius with a bayonet saw.

Fastening Top Edge Molding

Cut the top panel to size. Be sure to remember to add one times the overhang from front to back, and two times the overhang from side to side; and to subtract one times the width of the edging from front to back, and two times the edging width from side to side.

Fig. 13-11. Shaping the rounded corner.

Fig. 13-12. Corner as it appears after all shaping, sanding, and finishing are complete.

Take the best available strip of stock for the front molding and cut a miter at both ends. The measure of the edge to be joined (the shortest edge) should be *exactly* the length of the plywood top panel (Fig. 13-1), or perhaps up to 1/64" longer. It is better to cut this piece slightly too long and shave off a little from one end until the fit

Fig. 13-13. A common edge treatment (see text). This can be used alone with the proper overhang on lower and smaller cabinets.

Fig. 13-14. The same edge with a cove molding is frequently used on taller and more massive cabinets to balance the appearance.

is perfect than it is to try to get the exact size on the first try. If you accidently cut it too short, you have to start all over again.

Next cut the side pieces and miter only one end of each. These do not need to be exactly the right length. In fact, plan to cut them long and trim off the excess after fastening the edging to the top panel.

Fig. 13-15. The first step after cutting the waist molding parts is to glue and clamp the back cleat into place.

Fig. 13-16. The miters are carefully aligned as the front piece is glued on and clamped.

Test fit the miter joint between the front and side pieces of molding (Fig. 13-2). The angle should be perfectly 90°. If the angle is not quite right, adjust it by trimming the miter on the overlong side pieces until the angle and fit are perfect.

Fig. 13-17. Pin the front corners from the side edge with a finishing nail to reinforce the weak miter joint.

Fig. 13-18. Completed waist molding with corner blocks attached.

Set up the shaper (Fig. 13-3) or router table to cut the splining grooves. On the shaper use a flat cutter ¼″ wide, with the fences set for the depth of the cut. The cutter head should be raised 1/3 of the thickness of the panel (¼″ for a ¾″ panel top) above the shaper table. Or, for the router use a brand new ¼″ straight cutting bit set for the depth of the cut. The router table fence is clamped down ¼″ from the closest point of bit rotation. Notice that this is the same

Fig. 13-19. Detail showing profile of the standard edge shaping used both for the top edging and the waist molding.

Fig. 13-20. Waist molding is attached to base section by screwing the corner blocks together. Notice how side piece has been notched to allow clearance for the cleats that will hold the base and upper carcases together.

router table set-up described in Chapter 11 for mortising door stiles.

Machine both the plywood panel and the lumber edging with the good face towards the table (for a shaper) or the fence (for the

Fig. 13-21. Simple planar recessed plinth molding. Left front miter joint still has masking tape on it left from the assembly procedure.

Fig. 13-22. (Left to right) Cross section of a completed plinth or cornice molding, the detail molding unattached, and the backing cove molding unattached.

router table). It is imperative that no irregularities be present on the fence or table surface or the veneer will be scratched. By machining the good face towards the table or fence, the distance from the good face to the edge of the splining groove will be constant regardless of discrepancies between the thickness of the plywood panel and the lumber edging.

Fig. 13-23. Top view of the same molding.

Fig. 13-24. Cross section of a fancier plinth or cornice molding made with a Sears or Rockwell molding head attachment on a table saw. Notice that the side molding pieces will have a definite left and right when cut.

Cut the splining grooves in the lumber edging first. The lumber is harder than the plywood panel and more sensitive to dull blades. It will split or chip if conditions are not nearly perfect. Once the lumber edging has been machined, groove the appropriate edges of the panel (Figs. 13-4 and 13-5).

Fig. 13-25. The 5/16″ difference between the width of the backing cleat and the molding is split as follows: 1/16″ on the case mounting edge (right) and ¼″ on the floor edge (left).

·Fig. 13-26. The backing cleats for the front pieces are cut 1/16″ short at each end to prevent the backing from interfering with the miter joint alignment.

If the splining groove is ¼″ wide as previously described, ¼″ thick plywood can be used as the splining material. Cut a strip ⅛″ less than twice the depth of each groove. This allows a 1/16″ glue relief at the bottom of each groove when the edge is assembled.

Quarter inch plywood is normally three veneer construction with the grain of the thicker center veneer running at right angles to the surface veneers. If the surface veneer of the spline runs the length of the strip then the grain of the core will predominantly run

Fig. 13-27. Not only will you need to allow a small clearance for joint fitting, but on the side pieces be sure to allow room for the front and rear backing cleats.

Fig. 13-28. Plinth or cornice parts laid in position ready to assemble.

the short way. This is as it should be. The idea is to run the graining of the spline core at right angles to the joint itself. Likewise, if you plan to use a lumber spline, always run the grain the short way across the joint. Should you run it the long way the joint would shear the first time any stress is put on it.

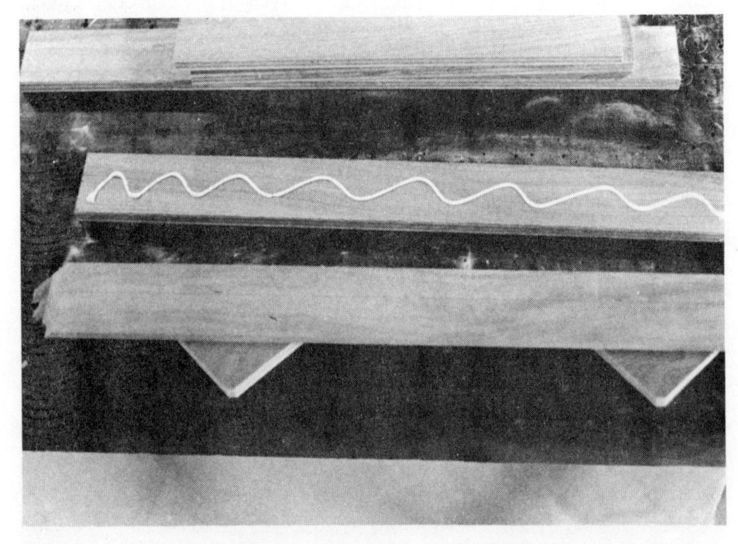

Fig. 13-29. Glue the backing cleats to the molding, allowing the clearances previously described.

Fig. 13-30. Test fit the subassemblies.

Cut the splines to length and test fit the molding to the panels (Fig. 13-6). The spline should slip onto the grooves of both the panel and the lumber edging with just a little friction. If the spline falls in with no friction at all, the glue joint will eventually fail under

Fig. 13-31. Apply glue to the miter cheeks, backing edge, and molding back face on the side piece. Glue and line up the joints one at a time.

Fig. 13-32. Align the joints and pin with one or two finishing nails or power staples.

stress. If more than hand pressure is required to insert the spline, the wood will split due either to mechanical pressure from the clamps or hygroscopic swelling of the wood from the glue.

After fitting, glue the splines into the lumber edging first. Apply glue on both cheeks of the groove and along the edge of both sides of the spline itself. As the spline is pushed into place the glue will evenly spread across the cheeks. The excess glue will end up in the extra 1/16″ glue relief space you previously allowed.

Fig. 13-33. Prop the molding upright.

Fig. 13-34. Apply glue and nail or staple the back cleat to the ends of the side piece backings.

The molding with the spline inserted is then joined to the grooved panel in a similar fashion. Notice that this is exactly like the mortise and tenon joint of the frames, only much, much wider. Clamp the molding in place with half the clamps on the bottom of the panel, half across the top side. By alternating the clamps in this way, you make sure that the molding is held squarely on the panel.

After the glue is dry trim off the extra lengths of the side molding (Fig. 13-7) to reveal what should be a neatly splined joint (Fig. 13-8). Now the top is ready for edge treatment.

Fig. 13-35. Close-up of Fig. 13-34.

Fig. 13-36. Check the frame for squareness before mounting the corner bracing.

Shaping the Top Edge

The available equipment and your imagination are the only constraints for the edge molding pattern. Several passes with different cutters can achieve quite a variety of edges, or a single router or shaper cutter may be used.

You may wish to round the corners of the top slightly depending on the edge pattern you have chosen or the amount of overhang required by your design. This softens the appearance a

Fig. 13-37. The top corner blocks are mounted first.

Fig. 13-38. Turn the plinth over.

little as viewed from the top. More importantly, this reduces the chances of some pint-sized person banging their head directly on a sharp corner.

Mark the attached edge molding with the proper radius (Fig. 13-9) for the corner. Here I have used a handy template—a nickel—which requires only a small capital investment. The curve is then cut with a bayonet saw (Fig. 13-10) and the whole lumber

Fig. 13-39. Mount the corner blocks on the bottom side. This photo clearly shows the ¼″ recess for the dust panel if this were to be used as a cornice molding.

Fig. 13-40. Rechecking the assembly for squareness. If the parts were cut right, the diagonals should be within 1/64″ of perfect square.

edged is shaped with a router (Fig. 13-11). After shaping, sanding, and finishing (Fig. 13-12), the corner becomes one of those little touches of thoughtfulness and good appearance which will increase the value of your finished cabinet.

One edge shape that appears frequently in commercial furniture can be easily simulated with a ½″ radius beading bit and a 3/16″ radius beading bit. Assuming that the material is ¾″ panel with a ¾″ thick solid wood edging at least ¾″ wide, set up the

Fig. 13-41. Completed plinth or cornice molding. Notice the 1/16″ gaps at the ends of the backing cleats.

Fig. 13-42. Front corner views of completed plinth or cornice molding assembly.

router with the ½″ radius beading bit with a 1/16″ lip to shape the top edge of the molding. Change the router to the 3/16″ beading bit and set so that the arc of the cut is tangent to the router table surface. Use this set-up to shape the bottom edge of the molding. The resulting edge treatment may be used alone on smaller and lower cabinets (Fig. 13-13) or can be used in conjunction with a cove molding made with a ½″ radius coving bit. The cove molding helps to balance the appearance of taller, more massive cabinets (Fig. 13-14).

WAIST MOLDING

The main purpose of waist molding is to break up the boxy appearance of a larger cabinet by adding horizontal lines. It also

Fig. 13-43. Back view of Fig. 13-42.

Fig. 13-44. Extra wide plinths must have cross bracing every thirty inches or less.

helps to disguise the joint between the upper carcase and the base carcase. Two piece construction makes a lot of sense in taller and wider cabinets. A completely assembled cabinet can weigh anywhere from two to six hundred pounds, and is more fun to move in smaller portions.

The waist molding is made very much like the top edging, only without the panel. It is usually an open frame joined in the front corners with miter joints and supported in the back with a cleat. Normally corner blocks are fastened inside the frame for strength.

Fig. 13-45. L-shaped plinth molding frame required for cabinet shown in Fig. 1-33.

Cut the front piece and miter both ends as with the top. The side pieces should again be cut slightly too long and mitered on one end. This time, after the miter joints have been checked for the proper angle, the side pieces are cut to the exact length prior to assembly.

Cut the back cleat out of plywood scrap, the exact length of the inside edge of the front molding. Apply glue at both ends and clamp between the two side pieces as shown (Fig. 13-15). Apply glue to the miter joint cheeks and clamp the front piece in place (Fig. 13-16). A little clamp adjusting may be necessary to square up the frame and to perfectly mate the joints. After the clamps have been properly adjusted and tightened, drive one finishing nail into each joint from the side piece edge to reinforce the corner (Fig. 13-17). The nail head should be set very deeply and filled prior to edge

Fig. 13-46. L-shaped plinth and cornice molding before assembly. Notice the top and bottom shapes are reversed.

Fig. 13-47. Early American plinth molding has cutouts requiring the backing cleats to be narrower. Otherwise, construction is similar to the fully enclosed plinths.

shaping to eliminate the possibility of hitting the nail with the router bit.

An alternate and slightly better way to reinforce the basically weak miter joint is to groove the miter cheeks for splines. But this

Fig. 13-48. Drilling the mounting holes through the backing cleats.

is more trouble and really isn't necessary since the waist (which is effectively clamped between the two carcases) receives little or no stress at the joint.

Corner blocks (Fig. 13-18) greatly reinforce the joints, much more than splining would. Corner blocks are amply glued and stapled or screwed into place while the frame is still clamped. Allow the whole assembly to thoroughly dry before taking the clamps off. After removing the clamps, shape the edge as desired (Fig. 13-19). This may be the same pattern as the top edging, or a different but complementary one.

Fig. 13-49. From the bottom side the holes are counterbored for the screwheads. The depth of the counterboring is determined by the length of the screw and the width of the molding.

Fig. 13-50. Plinth molding positioned and ready to screw to the case bottom. Use soap to lubricate the screw threads that will be driven into solid lumber.

After filling, sanding, and finishing, the completed waist molding is positioned on the base section and screwed through the corner blocks to the corner blocks of the base unit. If the upper carcase is to be cleated to the base carcase from the inside (the usual procedure), notches should be cut in the inside edge of the molding to allow the cleats to clear (Fig. 13-20).

PLINTHS AND CORNICES

A plinth is an enclosed base (as opposed to legs) beneath the base carcase. The cornice is the molding above the upper carcase.

Fig. 13-51. Simple cornice molding installed.

Fig. 13-52. Detail of L-shaped cornice installed.

The principle function of the plinth and cornice is to improve the lines and proportion of the finished cabinet.

In stereo cabinets the cornice can double in purpose to conceal part of the ventilation and dust proofing systems. A projecting heavy molded lumber plinth will protect the veneer of the base end panel by taking on itself the abuse of feet, chair legs, and vacuum cleaner attachments. A recessed plinth holds the cabinet off the ground enough to avoid the shoe marks, and is recessed deeply enough not to be marked itself.

A simple planar recessed plinth is the easiest to make (Fig. 13-21). It is assembled from hardwood veneer plywood using miter

Fig. 13-53. Projecting part of a built-in has its own cornice molding.

Fig. 13-54. Cabinet was built with removable side molding pieces. This allows several identical sections to be slid together into a larger wall unit without gaps caused by projecting molding.

joints as demonstrated in Chapter Twelve (see Fig. 12-15 through 12-18). After the basic frame has been put together, corner blocks are screwed or stapled into place in the same manner as will be shown later for the heavy molded plinths. The finished plinth is then screwed through the corner blocks to the underneath side of the carcase.

The set-back of a recessed plinth can vary a little, but is usually about 3″. This is deep enough to avoid shoe marks but not so deep as to endanger the stability of most cases. Avoid recessed plinths altogether with cabinets much over five feet tall. The smaller the footprint of the plinth in area, the less stable the completed cabinet.

Heavy Molding

Fancier plinth molding can be bought, but the stores carrying the pattern you want in the wood species desired are few and far between. Since the typical stereo cabinet requires less than ten or fifteen lineal feet of plinth and cornice molding combined, you will find it very difficult, or very expensive, to try to talk a lumber mill into making such a small quantity. If you can, hooray for you. But most circumstances require making your own plinth and cornice molding.

A simple but attractive molding can easily be made with a molding head on your table saw, or with a router, or both. The molding will be given depth by making it in two pieces (Fig. 13-22 and 13-23). The backing piece is made out of 4/4 lumber milled to approximately ¾″ thick, and ripped 3¼″ to 4″ wide depending on the exact pattern of the molding to be cut. The width of this board will determine the height of the plinth or cornice.

Fig. 13-55. Wooden door or drawer pulls can be fashioned to match the style of the cabinet. Here a walnut drawer pull has been notched into a Modern looking slab drawer front.

Cut a large cove along one edge of the face. A Sears or Rockwell ⅝" cove knife set does this very nicely. Make sure the knives are very, very sharp each time you get ready to use them. On wood harder than pine (most of them), cut the cove with several passes for safety. *This is quite dangerous and a good way to lose a finger or two, so work very slowly and carefully, following any of the molding head manufacturer's instructions to the letter*. Better yet, talk someone else into doing this part of the operation.

The fancier detail of the molding comes from a narrower strip, one and a half to two inches wide, laminated to the bottom of the face of the backing cove molding. The easiest way to guarantee that the plinth and cornice molding match the door and drawer front molding is to use the same pattern for this front piece you used for the door molding. It pays to experiment with the appearance by making a couple test tuns of molding out of scrap before making a very time consuming large run of expensive lumber that might not even look right when you are through.

It is not necessary to have the cornice molding exactly the same as the plinth molding, but it is certainly faster. If your moldings differ at all, always use the heavier, more massive version for the plinth, and the lighter appearing version for the cornice. This keeps the finished cabinet from looking top heavy. Many furniture manufacturers use the same detail molding for both the plinth and the cornice, with the cove backing for the cornice, and a large beading or ogee pattern backing for the plinth.

Cut the patterns for both the backing and the detail molding before assembly, then laminate them together with plenty of glue and pressure. Power staples are especially handy for lining up the two pieces during assembly, but wood screws from the back will do a superior job if lined up properly.

This method is very useful for generating a lot of good-looking patterns with limited equipment (Fig. 13-24).

Molded Plinth Assembly

Allow the molding laminations to completely cure before cutting the parts for the plinth. In pencil, lay out exactly where the straight and mitered saw cuts will be. If a staple or screw fastener ends up where a saw kerf must go, extract it before going any further. If the saw blade were to hit a staple or screw, at best it would be ruined. At worst, pieces of saw blade, staple or screw will fly out like shrapnel.

The cabinet should lap over the top edge of the molding by at least ⅛". This provides a direct bearing of the cabinet weight through the molding to the floor and allows for any discrepancy in the width of the case. Be sure to allow for this overlap when cutting the plinth and cornice parts.

Miter both ends of the front piece, and one each of the side pieces. Most plinth and cornice molding will have an asymmetrical pattern, with a definite left side piece, and a definite right side piece. All too often first timers cut either two lefts or two rights before they know what they have done. This is especially maddening if the molding was difficult pattern to make and just enough was cut. The odds are remote that the patterns from two different runs of molding will line up perfectly at the miter joint. Should you make this mistake you will pay a penalty of having to spend quite a bit more time fitting and trimming the corner joints by hand.

Next, the backing cleats are cut. These are pieces of scrap, usually ¾″ plywood, that have been ripped 5/16″ less in width than the cove backing molding. This 5/16″ difference will be split on the back of the plinth and cornice molding as follows (Fig. 13-25): a 1/16″ gap along the case edge of the molding, a ¼″ gap along the floor edge. The 1/16″ gap eliminates the chance of the backing cleats rising higher than the molding, forcing gaps between the molding and the case. The ¼″ gap forms the recess in the top of the cornice molding to receive the dust panel. Although this quarter inch gap is not really necessary for the plinth molding, it leaves you the option of later picking out the best of the two molding assemblies for the cornice, which is closer to eye level.

Cut two backing cleats ⅛″ shorter than the back face of the plinth front piece. One of these will back the molding itself, the other will serve as the back member of the frame. The 1/16″ clearance at both ends (Fig. 13-26) prevents the backing from interfering with the joint alignment.

Cut two more backing cleats for the side molding pieces. Again allow 1/16″ clearance at both ends plus the thicknesses of both the front and back members of the backing frame (Fig. 13-27). Using the scrap left over from the backing stock, cut two corner blocks for each frame corner. Lay out all the pieces in position and check the plinth frame for fit and size (Fig. 13-28).

If everything appears to be in order, glue and staple or screw the backings to the respective molding (Fig. 13-29). Again recheck the fit of the pieces (Fig. 13-30). One joint at a time, apply glue to the cheeks of the miter, the end of the backing, and the exposed back face of the side pieces (Fig. 13-31). Line up the joint and pin it in place with a couple of finish nails or staples. Then do the other joint the same way (Fig. 13-32).

Swing the molding up on its front pieces and prop it so that it stands by itself (Fig. 13-33). Apply glue to the back end of the side piece backing cleats and set, nail or staple the back frame member into position (Figs. 13-34 and 13-25). Lay the frame back down on the table and check it for squareness (Fig. 13-36). The acceptable degree of tolerance at this point is about 1/32″.

From the top side of the assembled frame, glue and staple or screw the top corner blocks into place. Use plenty of glue and set the corner blocks about 1/16″ down from the top edge of the backing cleats (Fig. 13-37). Turn the frame over (Fig. 13-38) and install the bottom set of corner blocks the same way (Fig. 13-39).

While the glue is still wet, check the completed assembly for squareness (Fig. 13-40). Provided that the miter joints and corner blocks were all cut accurately, the frame should be within 1/64″ of perfect square, but 1/32″ is acceptable. If the frame is further out of square than this, use a clamp to gently pull the frame square. Leave it overnight for the glue to cure hard before removing the clamp.

After the glue has cured, set the pinning nails or staples, fill the holes, and sand the surfaces smooth. Because of the variety of curves used for plinth and cornice molding, it is usually much faster and safer to sand the whole frame by hand. An electric vibrator or finishing sander will sand the wider flat surfaces fine, but tends to sand off or weaken the sharper features of the molding. The finished molding should be neat and have distinct detail (Figs. 13-41 through 13-43).

Plinths and cornices for larger cabinets are constructed the same way. Wider cabinets require additional cross-bracing in the backing frame to restrict the tendency for the solid lumber molding to bow (Fig. 13-44). Use cross-bracing every thirty inches or less.

Plinths and cornices may also be shapes other than quadrilaterals, depending on the design of the cabinet. A cabinet with two sections of different depths placed side by side requires an L-shaped molding frame (Fig. 13-45). The cornice molding must be the exact reverse of the plinth molding (Fig. 13-46).

Early American style furniture utilizes a simpler molding with a band sawn cutout (Fig. 13-47). This is made the same way as the heavier molded plinths except that the backing frame must be narrower in order not to show through the cutout. Also, the floor end of the corner joints are blocked with smaller lumber glue blocks rather than the larger plywood corner blocks.

Since the grain of the glue block will run perpendicular to the grain of the molding, it can restrict the hygroscopic movement across the width of the molding grain. The stress can actually build up to the point of splitting the molding. To get around this problem, use a series of one inch long glue blocks rather than a single longer block. The joints between the shorter glue blocks give some expansion and shrinkage relief to the plinth molding and reduce the chance of it splitting.

Mounting the Plinth and Cornice

Drill a series of 7/32″ holes through the backing frame from the case side to the bottom side (Fig. 13-48). Flip the plinth molding over and counterbore a ½″ diameter hole far enough down to give the mounting screw at least a ½″ bite in the case sides (Fig. 13-49). For example, if you have used 3″ wide backing cleat stock behind the plinth molding and you plan to use #10 × 2½″ wood screws, counterbore the bottom side of the hole 1″ deep.

Lay the carcase on its back on the floor. Carefully line up the plinth and cornice molding (Fig. 13-50) and screw them in place. Where the screw enters the plywood end panels, snug up the screws gently to prevent the wood from splitting or stripping out. Soap or wax the screw threads that will go into the bottom of the solid lumber face frame. This lubricates them and makes them less likely to twist in two as they are tightened.

The completely installed cornice (Figs. 13-51 through 13-53) looks striking and can take a lot of knocking and banging around.

MOLDING FOR MODULAR UNITS

Modular units can pose a problem with fancier plinth and cornice molding. Modular units are designed to be butted up side by side depending on the arrangement of the room. This causes a problem since the projecting plinth and cornice moldings of the two cabinets would interfer with each other and leave some wall exposed between the two. Instead of the two cases looking like one larger case, they would look like two cases placed too close together.

There are two solutions to this problem. The simplest is to stick with a simple design (e.g., Danish modern or kitchen cabinetesque) which could use a flush or recessed plinth base. The cases could then be butted together without any projecting molding interference whatsoever.

More traditional looking pieces which need the heavier, fancier molding can also be made adaptable, but not as easily (Fig.

13-54). Instead of cutting miter joints at the corners, the side pieces are butted to the ends of the front piece and merely screwed into place from the inside of the case or backing frame. The exposed ends of the side pieces are cut to the same profile as the front molding. Wooden pulls can be made to match (see Fig. 13-55).

Then, if two cases must be set side by side, the respective side pieces can be removed by taking out their mounting screws, and the cases then slid together. The main disadvantage of this is that the exposed end grain of the side pieces usually takes the stain a little darker than the front piece of molding. This can be minimized by sizing the end grain prior to staining the cabinet.

Built-Ins

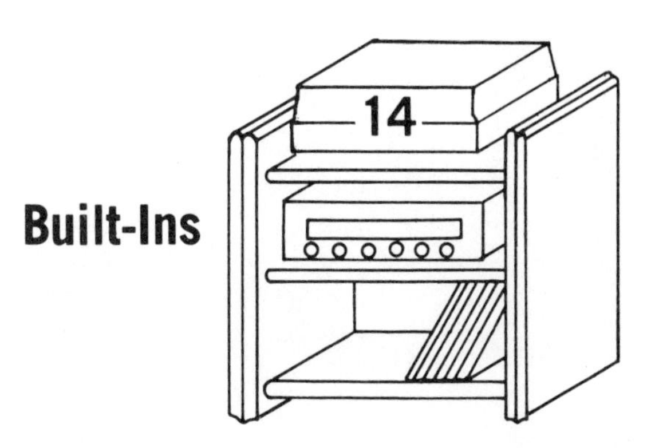

The flexibility of free-standing stereo furniture is nice but sometimes a built-in is exactly what a room needs to look good, and you plan to leave it that way forever. A good quality, good looking permanent cabinet can not only be very convenient, but can significantly raise the value and sales appeal of your home.

Built-in stereo cabinets are very similar to free-standing furniture. The main problems that arise with built-ins are ventilation, wire accessibility, retaining the option of future equipment changes, and method of attachment to the house.

TYPES OF BUILT-INS

Built-ins are generally one of two distinct types. One is actually recessed into the wall (Figs. 14-1 and 14-2). This is accomplished by cutting a hole in the wall the size of the cabinet, reinforcing the framework of the house, and setting the cabinet into place. The planning for a cabinet to be mounted in such a way can get quite complicated. You must first make sure that the very act of cutting the hole in the wall doesn't weaken the house itself. It is well worth paying a licensed contractor to inspect the wall and perhaps even make the necessary cuts and reinforcement. It really wouldn't matter how impressive your stereo cabinet looked if the house fell down around it.

You must also be aware of what is on the other side of the wall. If it is a utility room or a garage, no problem. But you wouldn't really want the ugly back end of a box sticking out through a wall of the den into a formal dining room. If the case projects into an unheated part of the house, be sure to insulate it to the R-value of

Fig. 14-1. Sound equipment cabinet set in hole in wall.

the rest of the wall. Otherwise in the winter you will pay a premium for your stereo in the lost heat.

There are some construction advantages to a recessed cabinet. Since the carcase exterior will not be seen, construction technique does not need to be any more complicated than butt joints and screws for the basic case. You can use hardwood veneer plywood with only one good face at a definite cost advantage over the A-2 plywood needed for free-standing pieces.

The other type of built-in would probably be more properly called a "built-on". The case is constructed in the shop and merely screwed in place to the wall (Fig. 14-3). The design can range from

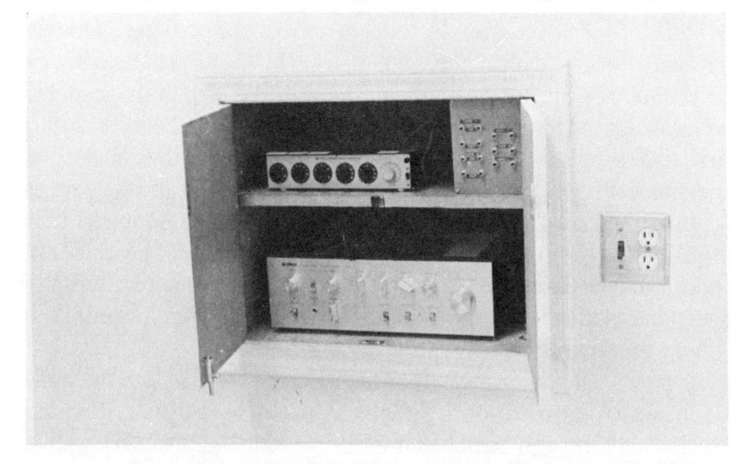

Fig. 14-2. Same cabinet showing interior. The other side of the wall is a utility closet.

a simple box with shelves to a series of fairly ornate cases complete with fancy cornice and plinth molding. Since the exterior of the carcase is exposed to view, it must be as well made as any free-standing piece. This is the type of built-in most frequently used, and the one which we shall discuss.

CASE CONSTRUCTION

Most built-ins are variations on the vertical cabinet theme, and are constructed accordingly (Figs. 14-4 through 14-8). If inset brass shelf standards are to be used for the adjustable shelves (Fig. 14-9), the grooves to receive the hardware should be cut in the appropriate end panels and partitions before assembly. In fact, the best time to mount the standards in the grooves is after all the panels have been sanded and stained on the interior sides, but are

Fig. 14-3. "Built-in" cabinet mounted by fastening the completed cabinet to the wall.

Fig. 14-4. Built-in cases are constructed the same way as free standing vertical cases. Here an end panel is laid on the floor with the top and bottom panels dropped into their dadoes.

still laying flat on the table awaiting assembly. If you wait until the case is put together, you will have problems swinging a hammer or lining up a power staple gun to set the standard ends.

A solid lumber cleat should be glued and screwed to the inside of the cabinet floor and top panel so that the cabinet may be easily screwed to the wall studs (Fig. 14-10). The strength of the joint between this cleat and the case itself is the sum total of what keeps the cabinet from falling off the wall, so spare no glue there.

PRE-FITTING MODULAR SECTIONS

If the finished built-in will have more than one case section, pre-fitting these sections together in the shop will save an awful lot of headaches at the installation site. Minor adjustments can be made with all the proper tools and room available in the shop

Fig. 14-5. The end of the top and bottom panels are glued and the other end panel is set in place and pinned with finishing nails.

Fig. 14-6. After both end panels are glued and pinned, the face frame is glued and nailed on the box.

without making a mess that is difficult to clean up. All the cases and molding subassemblies should be finished before pre-fitting.

Lay the cases on their backs on the floor oriented in their relative positions. With small bar or pipe clamps, clamp the face

Fig. 14-7. Larger built-in carcases may have several fixed shelves. Notice that one of the shelf edges has been reinforced with 1½″ wide lumber edging to hold the weight of the television.

Fig. 14-8. Detail of reinforced shelf edge. Notice that the edging stops ⅜″ short: the exact depth of the dado.

frames tightly together (Fig. 14-11). Work slowly and carefully to make sure the face frames all line up on the same plane. After setting each clamp check all the previous clamps to make sure the

Fig. 14-9. Brass plated metal shelf standards have been recessed into the end panel. Notice the mounting cleat screwed to the stud in the wall.

Fig. 14-10. Detail of mounting cleat. Cleat should be glued amply and screwed onto the cabinet floor or top panel from the unexposed side of the panel.

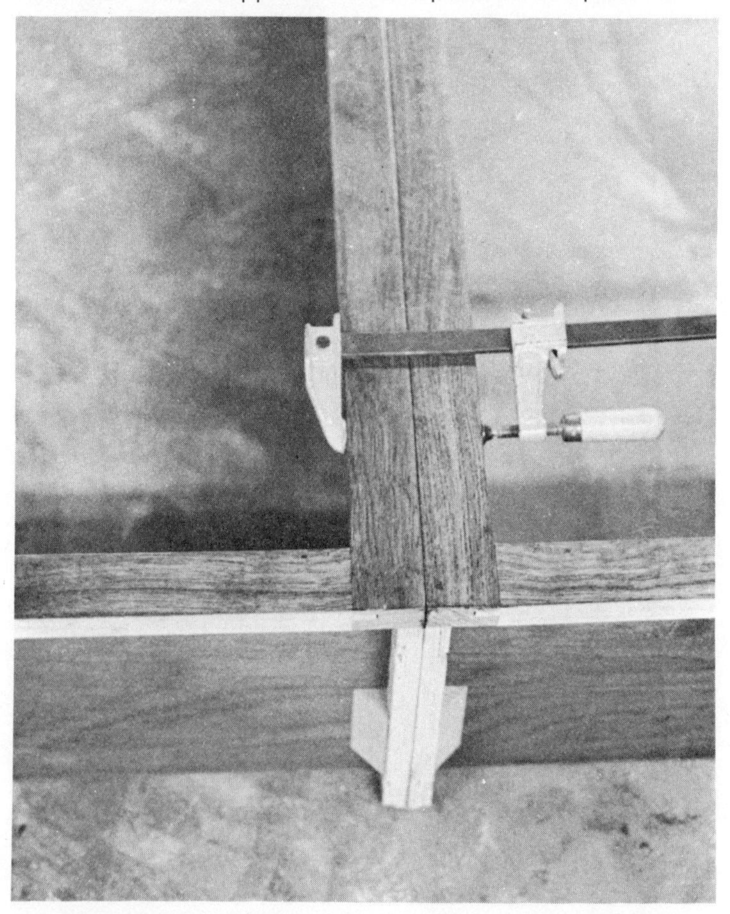

Fig. 14-11. Module face frames are carefully lined up and clamped together.

Fig. 14-12. While cabinets are still clamped together, the pilot holes for the screws are drilled. Cabinets are separated and the shank hole is enlarged and countersunk.

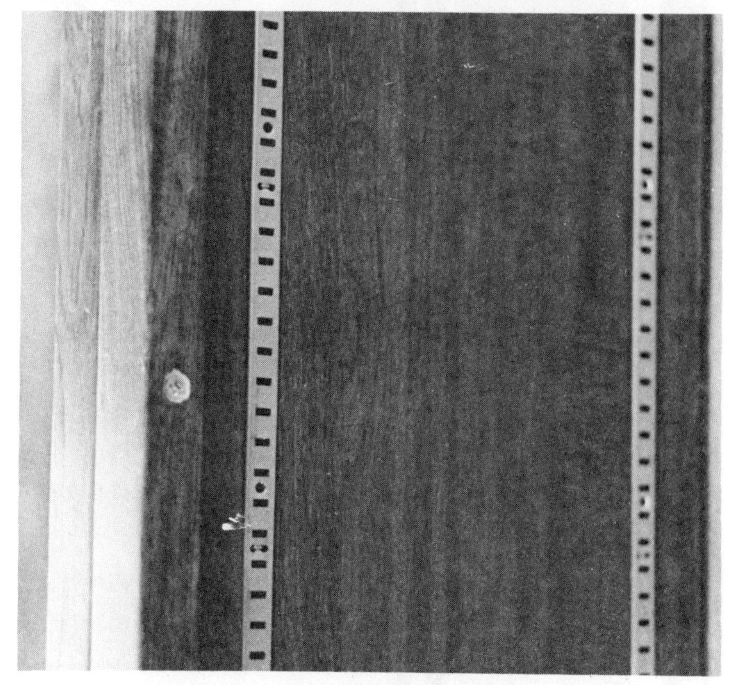

Fig. 14-13. Module face frames screwed together.

Fig. 14-14. Module units completely pre-fit. Unit will now be disassembled and moved to installation site.

Fig. 14-15. Fancy built-in top. Edging has been splined on and shaped. Notice how the edging laps the face frames of the adjoining cabinets.

Fig. 14-16. Joints in a counter top. Match up the grain as much as possible, and locate the joint over an end panel or other support.

Fig. 14-17. One inch wide scribe molding conceals half inch gap left to simplify cabinet placement and accommodate discrepancies in the ceiling and floor. Hot-melt glue saves the frustration of nailing at such an awkward height.

face frames are still lined up. One adjustment can frequently move another out of line.

The pilot hole is drilled at this time to guarantee that the screw holes will line up perfectly when the cabinet sections are mounted in position in the room. Drill the pilot hole through the edge of the face frame that will be least visible when the cabinets are installed. The hole should extend completely through one stile and partly through the mating face frame stile. The hole should be drilled ten to fifteen percent smaller than the inside diameter of the screw threads so that the screw will bite nicely.

After all the pilot holes have been drilled, unclamp the face frames and separate the modules a little way from each other. Select a drill bit equal to the diameter of the intended screw shank and enlarge the holes that go clear through the one face frame. Countersink the screw entry hole for the flat screw head (Fig. 14-12) and soap the screw threads. The two frames should snug up very tightly (Fig. 14-13), sometimes without even leaving a visible joint.

Any plinth or cornice molding should be pre-fit at this time (Fig. 14-14). The procedure is the same as for mounting a plinth on a freestanding cabinet. However, if the built-in cabinet will go all the way to the ceiling, the cornice molding must be attached to the cabinet after the case has been tilted up into standing position. Since there is no top access at this point the standard procedure is to custom fit the cornice molding (without any backing cleats) around the top of the cabinet, nailing it into place. If this must be done, plan the height of the face frame so that the molding will lap it by about ½″. This insures enough surface to hold the nail. Since

Fig. 14-18. Cabinet has been designed around a window frame that is very close to the room corner.

Fig. 14-19. Bookcase is in front of a critical wall register. The height of the cabinet is several inches less than the ceiling, and the top portion of the cabinet back has been angled to deflect the air over the cabinet and into the room.

some of the cornice moldings can be quite heavy, it may not be a bad idea to use a "panel bond" type adhesive along with the nails and actually glue the molding to the cabinet and the ceiling as well.

Needless to say, the miter joints of any molding added after hanging the carcase on the wall will have to be touched up after the molding is fixed in position.

COUNTER TOP SURFACES

Horizontal surfaces that are low enough to serve as a counter top are generally constructed out of hardwood veneer plywood with a solid lumber edging. These may also be made of formica or other plastic covering laminated to particle board. Formica is undeniably tough stuff, but it has one basic problem: any wood grain pattern formica fades or changes color over a long period of time in a different direction than the fading and color change in finished natural wood. A perfect match today will probably be an obvious mismatch within two years. This is, of course, no problem if you plan to contrast the natural wood with a black or other solid color top.

Real lumber and veneer tops somehow seem more suitable for a family room or living room. These are made much the same way as tops are made for free-standing pieces.

Two edge treatments are common. The simplest is merely a 1½″ skirting ripped out of ¾″ lumber and nailed to any exposed edge of the counter top panel. The corner joints are mitered and the edges may or may not receive any shaping. Tops made this way tend to look like most kitchen counters, only made from real wood.

The fanciest way is to spline a ¾″ thick edge like the tops on free-standing furniture and give it a furniture edge treatment (Fig. 14-15). Depending on the desired overall effect of the finished cabinets you can even add a cove molding for a heavier, more ornate look.

One of the main problems with built-ins is that they frequently run too long to make the top out of only one length of plywood. A

Fig. 14-20. Cabinet with open back elsewhere still has false back to conceal where the wires run.

Fig. 14-21. The stereo compartment. Receiver is concealed in a base cabinet at end table height beside one end of the sofa.

Fig. 14-22. Wires run up the inside of the face frame stiles and behind the cornice molding of the upper over the sofa. Speakers above opposite ends of the sofa give pretty good stereo separation for the sofa and the easy chair across the room.

good hardwood lumber yard can special order plywood up to twelve feet long, but most circumstances will require you to make a few joints in the top (Fig. 14-16).

Fig. 14-23. Cabinet drops down to a desk set up for viewing X-rays and writing. Notice that the fancy cornice molding continues all the way around the room at the ceiling.

Fig. 14-24. Desk area features four filing drawers (letter one way, legal the other), three desk storage drawers, and plenty of shelving for reference books. Desk top wraps around for lots of room to organize papers.

Fig. 14-25. Detail of light table. One half inch translucent plexiglas is set into the surface of the table and lighted from below with a single fluorescent tube.

Any joints should be matched for grain if at all possible to make them look less obvious. Twisting stress would be highly unusual for tops so the joints need not be any fancier than butt joints. Where some stress may be a consideration (e.g., a joint very near a knee

Fig. 14-26. Box under light table has conveniently located push button switch, and is drilled on the bottom side for ventilation intake holes, the top of the box side for ventilation exhaust. Box must be eight inches deep for proper ventilation.

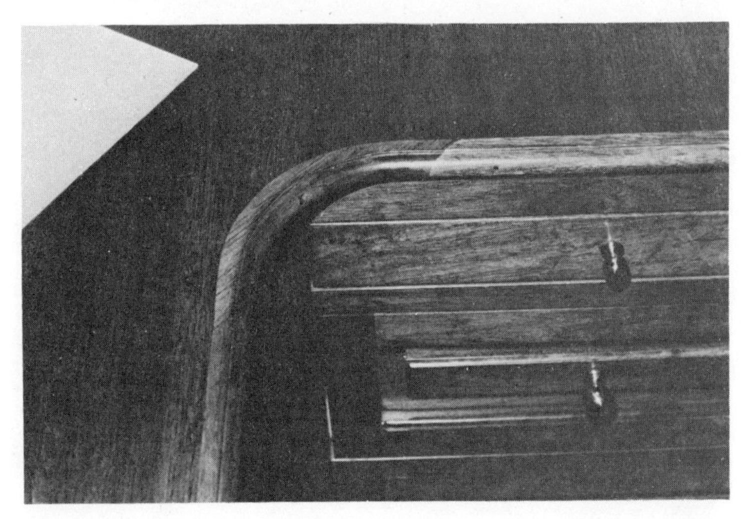

Fig. 14-27. Detail showing how fancy top edging was continued around the corner.

hole), the ends of the adjoining top panels should be splined together. All joints should be supported underneath on both sides by an end panel or some other structural member with direct bearing on the floor.

Fig. 14-28. The molding is continued across the top of the window wall, tying the cabinets on two sides of the room together. Library table between the two desks may be used from either side for additional planning room.

Fig. 14-29. Detail of smaller desk. Two drawers are letter/legal file drawers mounted on 100 lb. ball-bearing side slides. Notice the fancy heavy plinth molding.

TRIMMING TO THE CEILING

Cabinets that will completely fill the space from floor to ceiling cannot be made exactly to size and then moved into the room. They must be made smaller so that they can be physically brought into the room, swung or raised into place, and set to the wall. One of the most common mistakes made by beginners is to

Fig. 14-30. Skirt above knee hole conceals a pencil drawer.

Fig. 14-31. Detail of fancy top edging on the smaller desk.

Fig. 14-32. Detail showing how desk top joins television case next to it.

Fig. 14-33. Television cabinet. Television section and the shelf below it (for a VCR) extend through the wall into the garage. The rest of the case is 9½″ deep. VCR shelf is lighted.

Fig. 14-34. Television mounting in closed position.

not check the hypotenuse of the end panel (the distance from a bottom corner to a diagonally opposite top corner) to make sure that it is less than the height of the ceiling. If it is the same value or more the cabinet cannot be brought into the room and stood in place.

The slack required for jockeying the case into room position is generated by having a removable plinth or toe kick platform and by

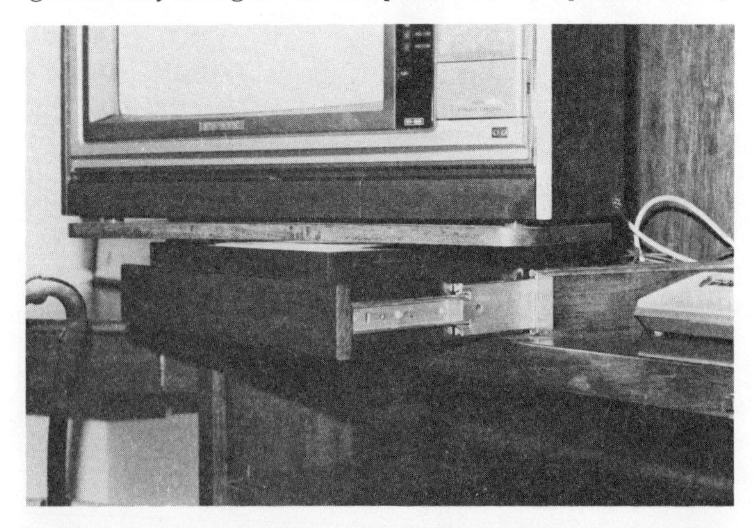

Fig. 14-35. Television mount pulled out and swiveled. Slides are 150 lb. rated. Television can be aimed directly at any point in the room, even along the same wall to which the cabinet is mounted.

Fig. 14-36. Television cabinet is connected by cornice molding over the fireplace to a bookcase in the opposite corner.

having at least an additional half inch gap at the top. The toe kick height should reduce the length of the end panel hypotenuse to less than the actual ceiling height, while the half inch at the top will allow for irregularities in the floor and ceiling (There are always some).

Fig. 14-37. Detail of fireplace mantle. The same molding was used for the mantle as for the plinth and cornice.

Fig. 14-38. Back around again to the cabinet containing the stereo receiver. The little boy is optional!

The top half inch gap is eventually trimmed with some sort of molding to complete the installation (Fig. 14-17). This molding should lap the top edge of the face frame by at least another half inch as previously described. This allows enough room to nail on the molding without splitting either it or the top of the face frame. If you have a half inch gap between the top of the cabinet and the ceiling, you will need a 1 inch wide scribe molding. If you plan to use the heavier cornice molding (see Chapter 13), then you will have to adjust the gap accordingly in your original design. For example, with a 3¼″ wide plinth and cornice molding, the top of the carcase set in position on top of the plinth assembly should be about 2¾″ from the ceiling.

The molding should be mounted with the point of the finish nails angled towards the floor. This uses gravity to help keep the molding tight on the cabinet. Lighter moldings can be tacked into place with spots of hot-melt glue.

SOME SPECIAL PROBLEMS WITH BUILT-INS

Since built-in stereo cabinets are mounted directly on the wall, there is no space behind the cabinet to exhaust the hot air from the ventilation currents. Under these circumstances, the ventilation exhaust is usually ducted up above the equipment section via a false back, with exhaust holes in the false back above a shelf placed over eye level. This method keeps the exhaust ports mostly hidden, but requires that the shelf directly under the exhaust holes not be completely filled up with books or other objects that might block the air flow.

Fig. 14-39. Office cabinet can conceal stereo unit in base section and reference books and catalogs in the upper section.

Fig. 14-40. Comprehensive bedroom built-ins offer book shelving, desk top surfaces, file drawers, and typewriter pullout. Stereo can be completely concealed in base sections with speakers set in opposite bookcases.

Intake ports are easier. Holes are usually bored in the bottom of the equipment compartment directly below the equipment. These are concealed by the low level of the equipment compartment bottom and the light baffle of the turntable compartment (if any).

Fig. 14-41. Detail of the file drawers shown in Fig. 14-40.

Wire accessibility is likewise more complicated in a built-in stereo system. You will have to rely more on fishing longer patch cords through the false backs. Since the equipment must be removed sooner or later to change a fuse or for other service, leave the patch cords long enough to completely remove the components from the equipment panel and set them down without having to unhook any lines. The turntable ground will probably have to be lengthened, and longer patch cords purchased.

Future equipment changes are handled just like in any other cabinet with an equipment panel (see Chapter 8). Make the bottom of the equipment compartment semi-permanently mounted. Use only cleats and screws so that it can swing down out of the way and allow the equipment panel to be removed and replaced.

Built-ins are dependent on the construction of the building they will be placed into. The special problems generated by room design and construction technique are nearly infinite.

Windows may not be placed for the convenience of the cabinetmaker (Fig. 14-18). A large bookcase may have a critical wall register behind it (Fig. 14-19). Shelves that have been left open with no case back (used for a less massive appearance) may need an occasional false back to conceal equipment wiring (Fig. 14-20).

The circumstances will vary for nearly every cabinet so much that generalized techniques or procedures would be useless.

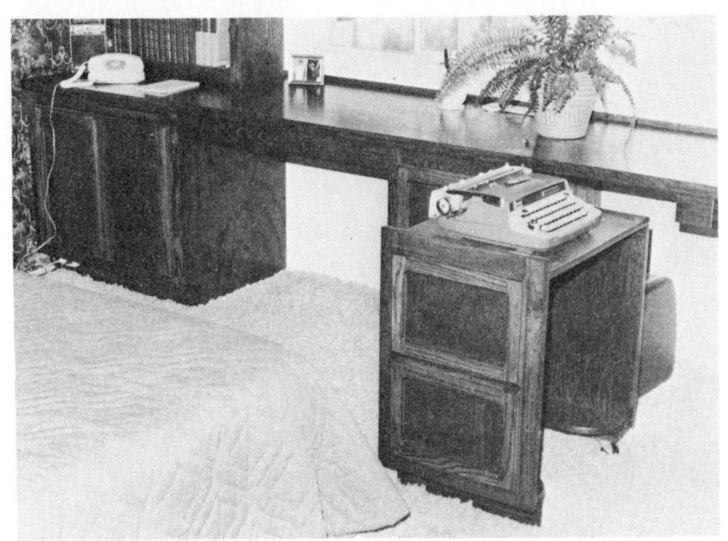

Fig. 14-42. The typewriter pullout as shown in Figs. 14-40 and 14-41.

Consider such problems as challenges to your skill and ingenuity. The secret is to work out any problem in great detail in the paper planning stages and to not wait until the cabinets are mostly built. Figures 14-21 through 14-42 show a variety of different designs to help stimulate your imagination.

Finishing

The term *finishing* is something of a misnomer. It implies that putting a protective chemical coating on a piece of furniture is the last thing you do. While a significant amount of this work *is* the last thing done (if you don't count final assembly of the subassemblies, mounting the decorative hardware, etc.), the bulk of the finishing work must be done as you proceed through each and every construction step.

For example, immediately after cutting out the face frame, door stiles, and rails is the very best time to sand the saw kerf marks from the edges of the pieces. This is nearly the very *first* thing done. How carefully you cut and fit your joints in the white will have a direct bearing on how much filling and sanding (and faking, if they are really bad) you will have to do before you apply the very first coat of stain. Even in the planning stages you must be thinking about how the very construction procedures will affect the eventual application of a stain or clear finish.

To completely build a stereo cabinet before starting the finish is expensive folly, and a major difference between the amateur and the professional. Forty or fifty percent of the labor time in a well made stereo cabinet will be sanding. The stain and the top coats do nothing more than amplify what is underneath them. Even if you simply paint the cabinet flat white or flat black, any mistakes or misfit joints will become more obvious. Therefore, the way to end up with the best looking cabinet is to spend the most time and care in smoothing and preparing the surface for that finish. If the basic surfaces are "perfection" in the white, then the finished cabinet will be "perfection amplified." This is the sum total of the "secret" of a superior finish.

BASIC PRECAUTIONS

Most of the problems that plague the amateur are completely avoidable. If you keep an eye on the way the grain runs in the wood while you are cutting out the parts, you can avoid garish graining combinations. These can cause optical illusions such as making a perfectly hung door appear to sag, or make the top rail of the face look like it is at a slant even though it is perfectly square.

Keep your work bench clean and smooth at all times. If you accidentally drip some glue on the surface wipe it up right away with a damp rag, and lightly sand the surface later. Even a nearly invisible bump of glue can leave a scratch across tender veneer. Likewise, keep your machine tables clean and polished. Put hand tools away immediately after using them and they can never fall onto a panel and puncture the veneer, or dent a freshly shaped top edging.

PROBLEMS WITH GLUE

Maintain a constant vigilance when working with glue. A small dab of glue anywhere on an exposed surface will absolutely refuse to take stain later. If you accidently get a spot of glue in the middle of a panel, wipe it up immediately if not sooner. After the wood surface has dried, lightly resand it with the finest finishing paper. If too much glue is used in a joint and bleeds out under clamping pressure, it is better to wait until the glue dries hard and scrape it off with a sharp knife.

Keeping a bowl of water with a rag in it for wiping any fresh glue spots is a good idea for any assembly procedure. But a strong word of caution: never keep the bowl of water (or cup of coffee, or anything liquid) on the same bench at which you are working. Knocking the liquid and spilling it all over everything is absolutely inevitable. If you place it on another table you won't spoil your work when you do knock it over.

All the above is written with the realization that no one will take it seriously until they have made all these stupid mistakes themselves. Since you will have to learn sometime (in company with those of us who have already learned the hard way), make your first stereo cabinet for a friend. By the time you have finished it you will take this advice seriously. You will then be ready to build a cabinet for yourself that you will be proud to keep forever. Good friends are hard to come by, but so are good stereo cabinets. Nothing worthwhile is without some cost.

REPAIRING MINOR SURFACE PROBLEMS

Besides introduced flaws, anything made out of a natural material will also have natural defects. All flaws and damage must be repaired as soon as they are noted before their location is forgotten.

Shallow dents in lumber or veneered surfaces can frequently be raised by steaming the dent with a wet cloth and a hot iron. Lay the cloth over the dent and set the iron on it for a few seconds. The steam should cause the wood to swell around and in the dent raising the grain. Let the area dry for a few hours before sanding the surface down flush.

Different species of wood react differently to steam and heat, so test this procedure on a scrap piece of wood before making a commitment on the actual cabinet. This also will give you a little practice with the technique. Be careful not to wet the veneers too much or they may blister or delaminate from the core.

Any other minor scratches, nail holes, and other blemishes should be filled with wood dough. Wood doughs or plastic wood come from many manufacturers with many different characteristics. Before filling the actual parts of the cabinet test the filler in a variety of scratches, holes and dents made in a scrap piece that matches the wood of the cabinet. Then finish sand and stain the scrap. This will give you a good idea of how the filler reacts to sanding and staining. You may have to try several different brands before finding one that is not very obvious.

Once you are content with the match of the filler, fill all nail holes, frame and case joints, and natural or introduced scratches that the parts may have. The filler should mound up slightly higher than the surface of the defect so that it can be sanded even with the surface when it dries.

SANDING CHRONOLOGY

Sanding is not something done at the end. The many different parts each have an ideal time for finish sanding that corresponds with the order of construction.

All lumber pieces should be belt sanded along the edges immediately after ripping and cross cutting to remove the saw kerf marks. Otherwise these marks will end up on the inside edge of a door or face frame. Inside corners are very hard to sand.

Lumber tops, implants, light baffles, etc., should be belt sanded to remove planer chatter marks on the faces immediately after they have received their final edge shaping, screw mounting holes, or other machining. Parts that will be difficult to finish later and will undergo no other machining should also be finish sanded at this stage.

Lumber frame and molding members should be completely machined and assembled before any belt or finish sanding (other than the previously noted kerf marks). Door frame faces and joints wait until after the door frame has been completely assembled and any shaping is done. Plinth and cornice molding wait until the frames are completely put together.

Internal faces of panel parts are finish sanded as soon as any dadoing or edging assembly is completed. It is far easier to sand these panels flat on a table than it is to try to sand inside corners later. External faces must wait until after the face frame is nailed on and the edges are flush-trimmed or belt sanded flush with the end panels. The front face of door or drawer panels is sanded before the panel is inserted into the door frame, but the back usually waits until the door assembly is complete.

End panels that will receive an end panel frame must be sanded before the frame is applied. The end panel frame inside edges must also be sanded before application, but the exposed face of the end panel frame is sanded flush with the edge of the face frame and finish sanded after it is mounted.

SANDPAPER

For removing saw kerf marks, planer chatter marks, or cutting lumber parts down flush with veneer surfaces, a belt sander with 60-X or 80-X is ideal. This coarse grit cuts very rapidly so care must be taken not to sand off any of the veneer. A belt sander will sand right through the 1/28" or thinner veneer in a matter of seconds.

The nicest finish sanding requires using a silicon carbide grit open coat sandpaper. Start with 100-C or 120-C to remove scratches left by the coarser belt sanding, and to cut the filler mounds down flush with the wood surface. After thorough sanding (again be careful not to sand through the veneers) with 120-C, use every step of sandpaper from 150-C to 220-A to remove the increasingly smaller scratches left by the previous paper. Diminishing returns for different wood species are reached at various times and must be determined by your own standards and experience. It is usually pointless to sand Oak or Pecan after about 180-C, but Walnut and Brazilian Rosewood demonstrate a noticeable improvement all the way to 400. A well sanded panel will show a good reflection of light and detail when the dust is blown or vacuumed off and the panel is sighted at an oblique angle.

Fig. 15-1. End panels are stained on the exposed interior surfaces before carcase assembly. This saves a lot of time since it eliminates the trouble of staining inside corners later. Care should be taken to not get stain in any glue area.

STAINING

The different subassemblies (carcase, plinth molding, cornice molding, doors, drawer fronts, adjustable shelves, semi-permanent shelves) are all stained and varnished or lacquered before the final assembly step. Any good quality store-bought stain may be used, but check out its effect on a piece of scrap before committing the cabinet.

The sanding dust must be completely cleaned off the parts with either a tack cloth, vacuum cleaner, or air nozzle before finishing begins. Carcase panels may be stained on any of the inside exposed surfaces before carcase assembly (Fig. 15-1). This saves the trouble of staining the inside corners later. Care must be taken not to stain any part that will be later glued since the glue will not adhere well to a finished surface. There is no need to stain any sections that will not be exposed in the finished cabinet (interior of the drawer sections, inside the equipment compartment, etc.).

Wipe or brush the stain the same direction as the grain runs (Fig. 15-2). The intensity of the color can be controlled a little by how long the stain is allowed to soak in. The excess is wiped off with a soft rag.

At this point any glue spots or scratches that might have been missed before should show up. These must be repaired and evenly restained before the parts are sealed. See Figs. 15-3 and 15-4.

SEALER AND TOP COATS

The type of sealer used will pretty much dictate the type of top coat put over it. The two must be compatible or else the finish will bubble up into the most awful mess. If any doubt exists test the combination on a piece of scrap.

In general, a lacquer base sealer must be used under a lacquer top coat, a varnish sealer under a varnish top coat, and polyurethane under a polyurethane top coat. The sealer is usually a much thinner version of the top coat, but sometimes has a completely different formulation. It is brushed or sprayed on relatively lightly and when dry, is polished with 0000 steel wool or 400 open grit sandpaper. The steel wool rubbing both roughens the surface slightly so that the top coat will adhere well and cuts down or smooths over miniscule bubbles and other surface irregularities.

The top coat is brushed or sprayed on more heavily than the sealer coat. It is better to plan to build up several light top coats than it is to try to put on a single heavy coat. The additional

Fig. 15-2. Stain is wiped or brushed on in the same direction as the grain runs. Otherwise lap marks may occur.

patience required will pay off handsomely in a much more even and deeper looking finish. All layers should be steel wooled or sanded lightly between coats. Be careful not to cut right through the finish on the corners or other sharp detail.

Fig. 15-3. Carcases and parts cleaned up, stained, and now ready to have the top coat put on.

Fig. 15-4. Sometimes a screw head will show. Flat black screws look more professional than ones with a zinc finish. If black screws are unavailable, paint your own with flat black paint. Only the heads need painting.

If a lacquer is used the waiting time between coats can be as little as an hour. However, lacquer dries so fast that it requires special spraying equipment to apply. Brushable lacquers like Deft are available. Polyurethanes need six to twelve hours between recoats, but absolutely must not go longer between coats than the manufacturer's recommendation. Polyurethane is funny stuff in that no chemical bonding will take place after eighteen to twenty-four hours. It won't even stick to itself. Varnishes should be allowed to dry thoroughly as long as two or three days between coats. In all cases follow the recommendation of the manufacturer. Each formulation has its own peculiarities.

OIL FINISHES

Walnut, Teak, and Brazilian Rosewood have a lot of inherent color and interest without staining the wood. These look great with a built up oil finish. Less dramatic and less expensive woods such as oak, maple, and pine can be finished this way as well, but generally look more interesting if stained first.

The easiest of all oil finishes is *Watco Danish Oil*. This is a special formulation that polymerizes within the surface of the wood itself, providing protection from within. After the parts have been completely sanded and the dust has been cleaned off, the surfaces are completely saturated with the oil to the point of it dripping off. Keep the surface wet with oil for fifteen or twenty minutes while the finish soaks into the wood. If spots dry out during this time, rewet them with more oil.

The excess is then wiped off with a lint-free rag and rechecked every ten minutes or so for about one hour for additional oil bleeding out of the corners and joints. After the finish has dried overnight the surfaces are thoroughly scrubbed with 0000 steel wool and another saturating coat of oil is applied. This coat is treated in the same manner as the first. After the second coat is dry the cabinet is waxed with a liquid carnauba satin wax applied with 0000 steel wool and buffed with soft rags.

Tung oil cut 50/50 with turpentine or mineral spirits can be applied pretty much the same way. Periodic re-application of the oil will maintain the cabinet in mint condition.

CABINET CARE

The cabinet should be placed in a room away from direct sunlight, but in a place where the controls of the stereo equipment are readily accessible. For best results the relative humidity of the house should be forty to sixty percent, and ideally fifty percent. The winter is the worst time for furniture. The heating system squeezes out almost every last bit of water vapor from the air. The wood in the cabinet will shrink slightly as it reaches a drier equilibrium with the room air. As it shrinks stress is put on the glue joints and the finish. After a few cycles of drying out, re-absorbing spring moisture, and drying out again, the glue lines begin to fail and the finish may start to check.

Applying a light coat of oil or wax to the cabinet will put yet another vapor barrier between the wood fibers and the room air and slow down internal moisture content changes. This layer of polish also serves to absorb physical abuse from outside. Something bumping a well polished cabinet is more likely to skid off to one side. An unpolished surface may permit a bump to dig in and dent or scratch the surface.

Use lemon oil for varnished or lacquered cabinets. This stuff is cheap and easy to apply, although it needs to be renewed every couple of weeks. Pay special attention to any horizontal surfaces, and any portion of the cabinet that may get touched frequently (doors, drawer fronts, equipment panels, etc.).

Oil finished surfaces can be waxed. Use a liquid carnauba satin wax and apply it with steel wool. Buff off with soft rags. This coating should be renewed a couple of times a year, more often in horizontal surfaces and handled areas.

Do not use a wax polish on any open grain woods (oak, pecan, ash, hickory, elm, etc.). The wax will build up in the pores and leave a series of white dots.

If you build your cabinet right and maintain it properly, you should be able to expect one to two hundred years of enjoyable service from it.

16

Commercially Available Stereo Furniture

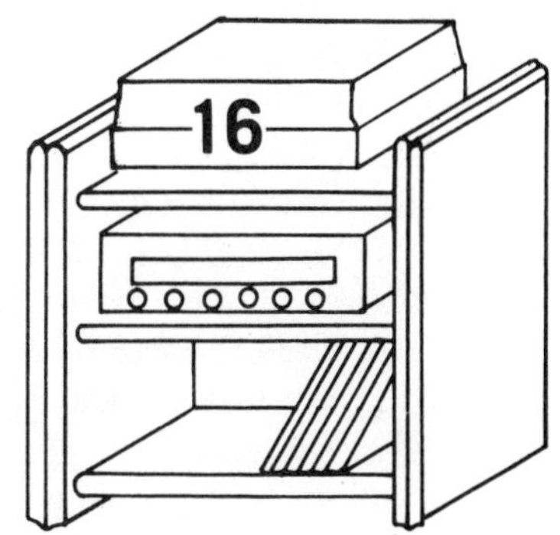

You may not wish to design and build your own stereo cabinet for any one of a wide variety of reasons. You may be a skilled woodworker, but simply lack the tools to do the job right. You may doubt your ability and be worried about buying several hundred dollars worth of fine quality lumber only to convert it into exotic fire wood. You may not even like woodworking (this is theory—people like that really don't exist). It is also possible that you may not be fussy about some of the finer points of stereo furniture such as concealed wiring and dust baffled ventilation.

In any case, there are many varieties of furniture on the market both designed specifically for stereo and adaptable to stereo. These break down into five general categories: built-ins by a local cabinet shop, general purpose casework adapted for stereo, shelving units and wall furniture, stereo "racks", and high-end stereo furniture.

No one type is categorically "better" than the others. Each type has a unique mix of appearance, function, life expectancy, and cost factors. The consumer will have to evaluate his own priorities with these in mind before coming up with a logical choice.

In general, all types of furniture tend to be a good investment if you buy the best and plan to keep it forever. "Buy the best and cry once," someone (probably wealthy) supposedly said. It is a fact that there is less and less good quality lumber and veneers available every year. With or without inflation lumber and furniture will become more expensive relative to the costs of other goods and services. This will occur for no other reason than the increase in population and demand coupled with a steady decrease in the supply and quality of materials.

Furniture tends to be a bad investment if you plan to replace it at a later date. Trying to unload your used furniture is exasperating. The used furniture market is nearly always hopelessly glutted, and liquidation prices are quite low. Specialty furniture is even worse in this regard than general-use furniture. Over the last twenty-five years countless excellent quality wood console television cases have been taken to the dump or converted into doll houses because of lack of standardization in the television business. Once the electronics gave out there was nothing else that would fit the case perfectly. This, by the way, is a good reason to make sure your stereo cabinet can be adapted for future equipment changes.

To buy the best is good general advice for nearly all durable goods, but if it were practical advice we would all drive Rolls-Royces. From a more pragmatic perspective, buy the very best you can afford, so that you maximize the value of your investment. Generally speaking, the more expensive the cabinet the longer it will last. Consider flexibility: the life of the case itself doesn't count for much if the cabinet will not adapt for future contingencies.

If you know you want a stereo cabinet but really are confused about what to buy, wait until you have looked all around before you finally make up your mind. Shop with your hands in your pockets and take no credit cards or cash with you. To the smooth, persuasive salesman you are just another moment of his day and a dollar in his pocket, but you could end up stuck with a cabinet you really don't like for a long, long time.

After you have seen everything that is available in your area buy exactly what you want. You may have to scrimp and save a little, but this sort of discipline is always good practice. Avoid getting talked into borrowing money to buy a stereo cabinet or financing it on a credit card. After all, in spite of the author's bias, a stereo cabinet is basically a luxury item you can live without for a while. Save your good credit for more necessary things. End of sermon.

LOCAL CABINET SHOPS

There probably is no company marketing custom-fit built-in stereo cabinets on a national basis, at least for a reasonable price. You will have to depend on a local cabinet shop. The quality of the product from these organizations can vary widely. You might luck out and get some of the best handcrafting done in the world by a modern day Chippendale or Hepplewhite, or you might get a piece of junk by a self-proclaimed cabinetmaker with two left thumbs who only recently read a book on cabinetmaking and bought a radial arm saw.

Discrimination is the key here. Insist on seeing several completed projects before even considering a commitment. Call the Better Business Bureau to see if any complaints have ever been filed against him. Ask around or check the local business license records to see how long he has been in business. Find out if he has ever done any stereo furniture before. Inspect his workmanship and the general appearance of his shop. Arrange to talk to owners of his work privately for a candid opinion.

It is a good idea to insist on a scaled isometric (perspective) drawing of the cabinet before signing a contract. Even an orthographic (three view) drawing is helpful, but it requires a little more imagination on your part to visualize the finished cabinet. This is important since most small cabinet shops in the country specialize in kitchen cabinets, and everything they make looks like a kitchen cabinet.

Don't settle for a scribbled sketch on a ripped out sheet of notebook paper. Although the cabinetmaker may personally be able to envision the finished product, it is part of his job to communicate this precisely to you. By the same token any contract should include a very thorough itemized and dimensioned specification sheet. The more precisely the cabinet is designed on paper the fewer nasty surprises you will encounter later. A word of caution when viewing cabinet drawings (or even photographs): the finished cabinet will always seem to look larger in real life than it appears on paper.

Once you have found a reputable and skilled cabinetmaker your options become nearly infinite. If you want something 103¼" wide, you can get the cabinet built 103¼" wide. If all the rest of your furniture is Padauk, you can have your stereo cabinet built of Padauk. At that point your main problem will be narrowing down your options into something that will reasonably go with the rest of your furniture and room layout and will not cost a small fortune.

Be prepared to wait a while before your cabinet is delivered. The best craftsmen are always swamped with business (beware of those who aren't), and your order may have to wait in line for weeks before even the first board is cut. Cabinetmaking is not like public construction projects. The product is a unique form of luxury, not a life or death situation. The cabinetmaker himself tends to be more of an artist with pride in his workmanship, but a very loose sense of time. He is less of a cold-blooded profit-monger, and is not likely to be precisely on time at the expense of very time-consuming quality considerations.

Most good craftsmen would rather you were mad at them for an extra week or two and then be happy with their creations for the rest of your life than to have you pleased with prompt delivery, only later to find evidence of hasty and careless workmanship. This is not to say wait forever. The best craftsmen usually have a little exhibitionism in their blood and don't mind you checking up on progress from time to time.

Check the design over to make sure the most important features of stereo furniture (dust proofing, wire concealment, panel mounting, etc.) are included. Definitely make sure adequate ventilation is provided for the amplifier. This has direct bearing on the life expectancy of your electronic equipment—and maybe your own. Heat deteriorated insulation within your unit can possibly short and cause a fire.

All panel mounting and other equipment compartment features should always be made changeable. Equipment wears out or becomes obsolete at a much faster rate than the cabinet.

ADAPTED GENERAL PURPOSE CASEWORK

You may have the good luck to bump into an antique or new armoire that seems just perfect for converting into a stereo cabinet. The main considerations are the inside width and depth of the case. Most general purpose casework is simply not deep enough to house a typical stereo system. When you measure be sure to include knobs, switches, patch cords, false backs, ventilation ducts, and wire runs in your figures.

Antiques raise an ethical question. Is it right to take a piece of furniture that has survived this long and arbitrarily desecrate it with irrepairable wire run holes? Are we not making a mockery of the original purpose of the work? A piece of furniture in its original form is a type of historical document that records levels of technology and skill, resources and function, and the wear and tear of day-to-day life over a period of time. To thoughtlessly take that record and exploit it for immediate circumstances would be like using the Declaration of Independence for wrapping fish, or the Dead Sea Scrolls for lining shelves. We would be confusing the record for future generations, and God knows they will need all the help they can get to interpret our present mess.

If you buy a brand new piece for the sole purpose of converting it into a stereo cabinet you should suffer no pangs of conscience. Indeed, by so doing you are helping to record the present circumstances for better or for worse. Again, make sure that any modifications allow for adequate ventilation and future equipment changes. This will be more difficult than with a cabinet especially constructed to accommodate stereo, but it is not impossible. Each cabinet will present its own problems.

Fig. 16-1. A complete organizing center for all home entertainment and storage needs. This fine piece is specifically sized for audio and video and constructed with solid Appalachian oak frames with exposed finger joint detailing. Shelves are scratch-resistant vinyl. Both sides are finished for use as a room divider (courtesy of Barzilay Company).

Fig. 16-2. Another shelf-type unit is also sized deep and wide enough for stereo. Only the highest grade of oak lumber and veneers have been selected for use in this cabinet, no particle board or plastic is used. Tempered smoked glass doors are featured (courtesy of Nomadic Furniture).

The main advantage of converting a general purpose cabinet into a stereo cabinet is one of style versus cost. The larger furniture manufacturers are better equipped to perform difficult operations such as ornate moldings and carvings, and can do these much more economically than a small shop. It is also possible to buy a piece that is part of a matched set with your present furniture.

WALL FURNITURE AND SHELVING UNITS

Various types of bookcases and general purpose shelving called "wall furniture" are currently on the market. See Figs. 16-1 through 16-3. These are usually made in modular sections so that different combinations of standard sized units will fit into most room layout situations.

There are several advantages to wall furniture. At present the units are quite easy to find in many furniture or wall furniture specialty showrooms. They are also frequently used as room dividers. Next, the modular sizes are designed to take advantage of normal raw material dimensions. This, coupled with mass-production makes most wall furniture a very good buy in quality per dollar.

The main disadvantages of wall furniture are style, depth, and exposed wiring. Although some very good traditional styles are available, most of the present units are lighter oak with more comtemporary styling. In many areas if you don't like this style you are out of luck. Sometimes wall furniture shelving is only 16″ to 16½″ deep. The logic behind this depth is that 16″ goes into a 48″ wide panel three times evenly. Any wider shelf would reduce the yield of the material by a third, and raise the price accordingly. Depending on your equipment, it is sometimes possible to work

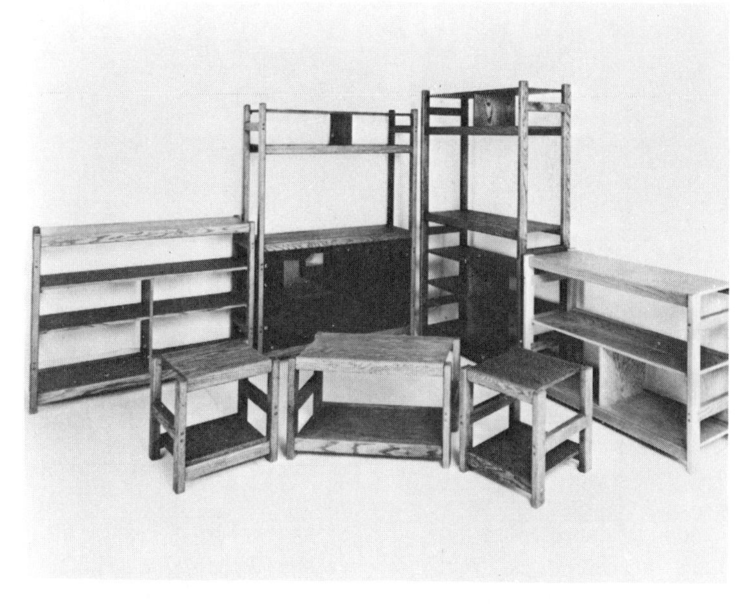

Fig. 16-3. Designed-for-stereo shelving units can come in all formats and sizes. These are crafted in solid 1½″ × 2½″ Appalachian oak. Metal to metal connectors are used to assure solid and accurate assembly (courtesy of Nomadic Furniture).

Fig. 16-4. One of the original stackable component "racks." This design along with slightly different variations are the mainstay of the stereo furniture commercially sold today. Finished with quality walnut vinyl, trimmed with matched edge moldings, this particular unit comes in two sizes that may be used independently or stacked (courtesy of Barzilay Company).

around this problem by letting part of the components hang a little over the shelf edges. Probably the worst aspect of open shelf units is that the wires and less attractive component features are fully exposed to view and dust. If the cabinets are at least 18″ or 20″ deep you can get around this problem by installing your own false backs where needed.

Fig. 16-5. An economical stereo component cabinet with built-in record racks. Quality walnut vinyl finish with black vinyl shelves. Doors are smoked tempered glass with full length bronze anodized hinges. Casters optionally available (courtesy of Barzilay Company).

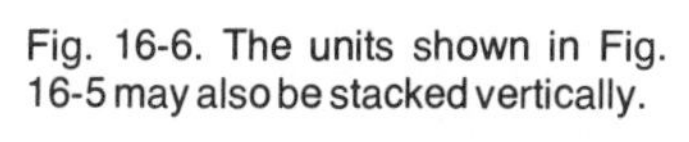

Fig. 16-6. The units shown in Fig. 16-5 may also be stacked vertically.

Fig. 16-7. Solid oak and oak veneer rack-type unit. Smoked tempered glass door latched with magnetic touch latches that open with a slight push (courtesy of Nomadic Furniture).

Fig. 16-8. Both vertical and horizontal formats are used in this line of audio and video cabinets. The cabinets come in both light and dark finishes which bring out the rich graining inherent to the solid oak and oak veneers. These cabinets are shipped knocked-down, but go together very quickly. The connections are located so that no fasteners or holes show from the outside (courtesy of Greenwood Forest Products, Inc.).

STEREO RACKS

Stereo racks are smaller shelving or special rack-type units designed specifically for stereo components. See Figs. 16-4 through 16-9. These were the first real industry-wide effort to improve upon the esthetics of the typical component stereo system. The first racks were metal or wooden cabinets open front and back with steel component mounting strips in the front for specially made high-end equipment. Later, the same cabinets were provided with shelves instead, to accommodate the more common popularly priced equipment. Nearly every rack-type design is a minor variation of the same theme: a minimal cabinet with various numbers of shelves to handle different sized systems.

Rack designs are gradually becoming more sophisticated. Some manufacturers offer wire concealment provisions, better equipment layouts, and furniture-quality materials and construction method. The trend is towards more of a furniture look away from the "sound studio" look.

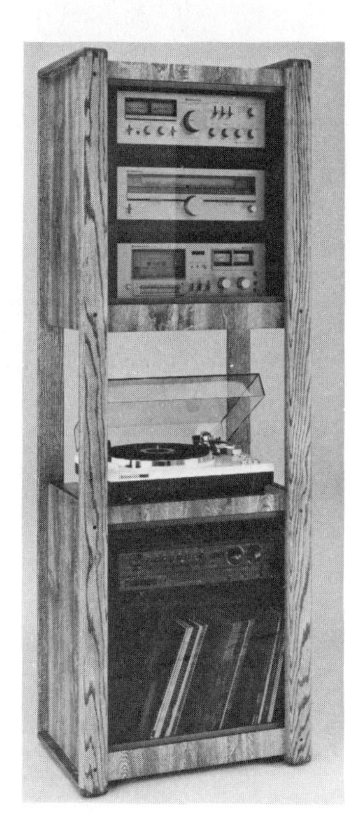

Fig. 16-9. Solid, oiled finish oak frames and vinyl panels are enclosed by tempered, bronzed glass doors with matching hinges and magnetic touch latches. This unit has indirect lighting over the turntable section and built-in record racks in the lower section. Upper section has two adjustable black vinyl shelves (courtesy of Barzilay Company).

Fig. 16-10. This is one of the finest quality stereo component enclosures made. The many outstanding features include gray tempered glass door, indirect light over turntable section, accessory drawer on self-closing ball-bearing slides, a touch latch door over both record storage and equipment compartments. Only premium quality woods are used. Available in either hand-oiled Applachian oak or American black walnut (courtesy of Brazilian Company).

The main advantage of racks is cost. A minimal cabinet designed and sized to house your stereo is mass-produced on a very large scale. It does the job well and costs less than almost any other method of equipment housing.

The main disadvantage is that appearance, convenience, and protective features have been compromised to hit the biggest possible market. You may have to put up with your receiver at knee height, or the view of the dangling patch cords, or the sound studio look in your French Provincial living room. Also consider theft. Some racks neatly package the stereo into one convenient unit thoughtfully provided with wheels. All a burglar has to do is to notice the expensive equipment through the showcase glass doors, roll it out to his car in one piece on the factory installed casters, and drive off (the perfect sixty-second crime). Larger units of course provide a little more security.

Another minor disadvantage is the common use of tempered glass doors, both clear and smoked. These do look very sharp, but the clearance required for glass door hinge mounting offers little barrier to dust infiltrating from the front. The glass requires more maintenance since it shows fingerprints a lot more than a semigloss finished wood door.

HIGH-END STEREO FURNITURE

Highly sophisticated stereo furniture (Figs. 16-10 through 16-15) is available commercially, but you have to look for it. For a partial list of companies see Appendix A. Most manufacturers of this type are small and regional. They rarely have the huge advertising budgets or marketing channels the rack manufacturers have. Prices will vary from one company to another depending on how complicated the product is, the quality of construction, and the wood species used.

Once you have found one of these companies, be prepared to buy a well designed piece of furniture with a full complement of features. Turntable lighting, dust proofing, forced air and convection ventilation systems, ergometrically placed equipment, panel mounting—the whole works are usually standard.

Fig. 16-11. Inside view of the fine enclosure shown in Fig. 16-10.

Fig. 16-12. This more traditionally styled cabinet is available in both six and seven foot tall versions. Both hold up to 150 albums in the easy gliding ball-bearing mounted record drawer, dozens of cassettes in the ball-bearing mounted accessory drawer, have indirect lighted turntable section, completely concealed wiring, dustproof ventilation system, and panel mounted equipment.

Likewise, the materials used are usually of better quality (and more expensive). You will find solid hardwood and hardwood veneers instead of less durable and more difficult to repair vinyl-clad particle board. You will find a variety of wood door styles, and mass that should discourage even a determined thief. For the most part you get what you pay for. Like the local cabinet shop be sure to inspect several completed models and check out the age and reputation of the company with the local Better Business Bureau.

Fig. 16-13. The shorter version holds two or three components besides the turntable, the taller one (Fig. 16-12) holds three to four (Figs. 16-12 and 16-13 courtesy of Presidential Industries).

If you cannot find a manufacturer in your area, some companies are willing to operate by correspondence and freight the

Fig. 16-14. This double-wide cabinet has all the features of the cabinet in Fig. 16-12, with the addition of nearly doubling record, tape, and accessory storage. Very few systems are too large for this unit. Available in the best oak, pecan, or walnut lumber and veneers (courtesy of Presidential Industries).

Fig. 16-15. Another armoire-type unit shows the versatility of style available only at high-end stereo furniture manufacturers. This unit is the same layout as Fig. 16-14. only in more of an Oriental motif. The proportions are nearly classical (1:1.618), and the bulk should be discouraging to several thieves (courtesy of Presidential Industries).

cabinet. But then, with this book in your hand, why not build your own?

A Small Full-Feature Stereo Unit

It is not expensive to build a stereo cabinet with a full set of features. The following plans show you how to build Presidential Industries' S-8 Minicabinet, complete with dust proofing, convection ventilation, and concealed wiring, without having to spend a fortune (see Fig. 17-1). The entire cabinet is constructed using little more than one full sheet of hardwood veneer plywood.

This unit is designed to accommodate a typical component stereo system: a receiver under fifty watts per channel output, a front loading cassette deck, and a turntable. The speakers will be located elsewhere in the room to maximize the stereophonic effect.

Panel mounting the equipment is the main missing feature. This was left off to simplify the design so that more people would be encouraged to attempt the project, and to maximize the cabinet flexibility for those who have very limited access to woodshop equipment. This design would be very easy to convert to panel mounting. Instead of one adjustable shelf, an equipment panel would be cut. The dadoed-in fixed shelf would be replaced by a shelf mounted with cleats and screws, and the false back would be behind only the record section. Actually, the false back could be deleted altogether, but by installing one you insure that you could conceal the wiring of auxiliary components you might get in the future.

S-8 MINI-CABINET SPECIFICATIONS

■ **Case dimensions**. 24″w, 37½″h, 20″d. Turntable mounts on top. Interior equipment section 22½″w, 18¼″h under top face rail, 19½″h behind front face rail, 17½″d. One adjustable shelf, ¾″ thick.

Fig. 17-1. Completed S-8 Minicabinet (courtesy of Presidential Industries).

Fig. 17-2. End panel dadoed, drilled for shelf clips, rabbeted for back, with solid lumber top edging splined into place.

■ **Equipment capacity**. Holds two average components, sometimes three short components with additional adjustable shelf.

■ **Storage capacity**. Holds about 150 single albums. Space could be converted to additional component space if desired.

Fig. 17-3. Actual cabinet back has one row of exhaust vent holes, two rows of intake holes and twelve inch square access port. False back has intake and exhaust holes for the amplifying unit on top shelf, with wire holes for the other components. ¾" by ¾" mounting cleats are glued to the back side of the false back.

■ **Features.** Concealed wire runs behind false back, simple dust-baffled ventilation can accept units with output of about 35 watts/channel with doors shut, unlimited if left open while playing. Doors have dust lips and are latched with magnetic touch latches.

Fig. 17-4. Parts completely cut out and labeled.

Fig. 17-5. Door frames assembled.

■ **Limitations.** Doors open to accept 19″ wide equipment set straight in, up to 20½″ wide equipment if adjustable shelf is dropped to top of the fixed shelf and equipment is angled when installed. Shelf would then be moved back up and set in place with shelf clips. Wider components would require making the cabinet wider. This can significantly increase material costs.

BILL OF MATERIALS

1 - 48 × 96 sheet ¾″ A-2 veneer core hardwood veneer plywood.

1 - 48 × 48 sheet ¼″ A-3 veneer core hardwood veneer plywood.

1 - 24 × 48 sheet ¼″ lauan plywood or hardboard.

About 10 board feet 5/4 FAS hardwood lumber planed to 1″.

About 1 board foot 4/4 FAS hardwood lumber planed to 25/32″.

2 - 1½″ × 32⅜″ brass plated piano hinges with screws.

2 - magnetic touch latches with screws and strike plates.

4 - plastic shelf clips with round ¼″ shank.

Glue, finish nails, sandpaper, stain, sealer, varnish, etc.

All above quantities include sufficient waste allowance.

Fig. 17-6. Face frame assembled (shown here upside-down).

Fig. 17-7. After the end panel top edging is splined together, the assembly is clamped for a couple hours while the glue dries. Notice that the clamps alternate sides to even out the clamping pressure.

Careful layout and material selection may allow you to get by with slightly less, particularly on the lumber.

CUT LIST

(All dimensions are in inches)

5/4 Lumber

■ 2 - 2 1/16 × 37 9/16 **Face stiles** S-8-1

Mortise the inside edges to receive rails. The top rail mortise should be 1 15/16 long, the bottom rail mortise should be 2 15/16

Fig. 17-8. Close-up of splined on end panel edge.

Fig. 17-9. Start of construction. It is a good idea to test fit all the joints before applying any glue.

long. All mortises should be 11/16 deep and matched with corresponding tenons.

■ 1 - 3 × 21¼ **Bottom face rail** S-8-2

Fig. 17-10. Shelves are set in bottom end panel dry, but top end panel is glued and nailed. Carcase is carefully held together while it is turned over, then the other end panel is glued and nailed.

Fig. 17-11. Partial carcase readied for face frame and top panel to be mounted. Work must be done rapidly while glue is still wet.

Tenon both ends and shoulder tenon on inside edge. Tenons should be cut ⅝ long and matched with corresponding mortises.

■ 4 - 2 × 32⅜ **Door stiles** S-8-3

Mortise the inside edges to receive door rails. Both top and bottom rail mortises should be 1 15/16 long.

■ 4 - 2 × 7 **Door rails** S-8-4

Tenon both ends of all rails. Shoulder all the inside edges to lap mortise end radius.

■ 2 - 2 1/16 × 21¼ **Top face rail** S-8-5

Tenon both ends and shoulder on inside edges.

4/4 Lumber

■ 2 - 2 1/16 × 19 **End panel top edging** S-8-6

Fig. 17-12. Face frame in place. The frame is nailed to the lower half first, then the top is inserted and the face is nailed to the top half of the case.

Fig. 17-13. After top and face have been assembled to carcase, top back cleating is glued and nailed into place.

Fig. 17-14. Glue and nail the bottom back cleat.

Groove edge to be joined with end panel for ¼″ spline. Groove may be 11/16 deep to use the same set-up as is used for mortising frame stiles.

■ 1 - ¼ × 23¼ **Fixed shelf edge** S-8-7

Glued and nailed to fixed shelf front edge (S-8-12).

■ 1 - ¼ × 22¼ **Adjustable shelf edge** S-8-8

Glued and nailed to adjustable shelf front edge (S-8-13).

■ 6 **Glue blocks** (cut from lumber scrap) SGB

¾ Hardwood Veneer Plywood

■ 2 - 19 × 35½ **End panel** S-8-9

Run a ¾ wide ⅜ deep dado for bottom panel 2¼/3 from bottom edge. Run another ¾ wide dado for fixed shelf 16½/17¼ from bottom edge. Run ¼ rabbet along back edge, blind at the very top (see Fig. 17-23). Drill series of ¼ diameter holes at 1 inch increments for adjustable shelf clips. Holes should only be ½″ deep (use drill press or stop block on hand drill so that holes don't go all the way through). Top edge is grooved to be splined to S-8-6. Use the same setup for grooving as for mortising frame stiles. See Fig. 17-2.

■ 1 - 19 × 22½ **Top panel** S-8-10

No machining required.

■ 1 - 18¾ × 23¼ **Bottom panel** S-8-11

No machining required.

■ 1 - 17¼ × 23¼ **Fixed shelf** S-8-12

Edge on front edge with S-8-7.

■ 1 - 17 × 22¼ **Adjustable shelf** S-8-13

Edge on front edge with S-8-8.

■ 1 - 2 × 22½ **Top back cleat** S-8-14

No machining necessary.

■ 1 - 2¼ × 22½ **Bottom back cleat** S-8-15

No machining necessary.

¼ Hardwood Veneer Plywood

■ 1 - 22½ × 33¾ **False back** S-8-16

Fig. 17-15. Entire carcase is thoroughly clamped so that all joints are prefectly tight with no gaps.

Fig. 17-16. After case has dried, glue blocks are glued into place.

Bore a row of one inch exhaust holes centered 2¾ from top edge, spaced two inches apart. Bore a row of intake holes centered 6¾ from top edge corresponding to the exhaust holes. Bore wire run holes centered 12¼ and 18⅝ from top edge. Cut two ¾ × 30¾ and one ¾ × 20 15/16 cleats from ¾ plywood scrap and glue and nail to the bottom edge and sides of the false back. The side cleats

Fig. 17-17. Top right hand corner of carcase after flush-trimming, rounding, and finish sanding. Mortise and tenon joint is shown.

are three inches shorter than the false back in order to clear the top back cleat when the cabinet is assembled. See Fig. 17-3.

■ 1 - 9⅜ × 32 7/16 **R. hand door panel** S-8-17

No machining required.

■ 1 - 10⅛ × 32 7/16 **L. hand door panel** S-8-18

No machining required.

¼ Lauan Plywood Or Hardboard

■ 1 - 23 3/16 × 11/16 **Cabinet back** S-8-19

Drill a row of one inch diameter exhaust holes centered 2⅜ from top edge. Make sure that these holes will be staggered with the false back exhaust holes when the case is completely assembled. This is a key part of the dust proofing system. Blind cut a 12″ square cutout 4¼ from the top edge of the back to 16¼. This cutout should be centered across the width. Drill two rows of one inch vent holes centered along lines 18 and 19½ from the top edge of the back panel. See Fig. 17-3.

■ 1 - 14 × 14 **Access panel** S-8-20

Drill four screw mounting holes, one in each corner spaced one-half inch from the edges.

CONSTRUCTION

1. Thoroughly read and understand all instructions before commencing to build the cabinet.

2. Cut out all parts to raw sizes (Fig. 17-4).

Fig. 17-18. Case and parts are stained and varnished before final assembly.

3. Mortise and tenon all frame members. Test fit the assemblies, then assemble the door frames (Fig. 17-5) and the face frame (Fig. 17-6).

4. With the ½″ radius bit, round over the front edges, both inside the door frame and out. Leaving the back edges square and sharp, completely finish sand the door frames.

5. Glue the door panels to the back of the door frames. The left hand panel should extend past the edge of the stile about 5/16″ on the side opposite the hinge. The right hand panel will come about ⅜″ short on the side opposite its hinge. The extension of the left hand panel will lap the recession of the right hand panel to form the

dust lip. This lip stops about 50% to 60% of the dust from sneaking in between the cracks around the doors. The door panel should slightly overlap the other edges of the frame. The joint will be filled and the excess sanded down, and the back of the door finish sanded.

6. Machine all panels as noted. Put the lumber edges on the shelves, and spline the top edgings on the end panels (Figs. 17-7 and 17-8).

Fig. 17-19. Piano hinges are usually installed with screws on opposite plates alternating to prevent head to head interference.

7. After the glue has cured, lay one end panel down on a padded table or the floor. Without glue, insert the fixed shelf and cabinet bottom panel (Fig. 17-9).

8. Put a bead of glue on the top edges of the upright shelves. Make sure that the lumber edges of the shelves are towards the front edge of the end panel. Then lay the second end panel on the

Fig. 17-20. The projecting heavy lumber of the door frames gives the cabinet a sculptured appearance.

assembly with the shelves in the appropriate gooves (Fig. 17-10). Check the alignment of the shelves with the end panel, and pin the end panel into place with two or three finishing nails per shelf.

9. Hold the case together with arm pressure and flip over. Remove the first (unglued) end panel, glue the shelf edges, and mount the end panel back in place with finishing nail pins the same way as the other side.

10. Flip the partial carcase over on its back (Fig. 17-11). Block up the cabinet bottom panel with a ¼″ shim, and the fixed shelf with a 1½″ shim. These blocks keep the shelf in place when nailing on the face frame.

Fig. 17-21. False back is nailed into place. Here we see why the mounting cleats are three inches shorter than the false back.

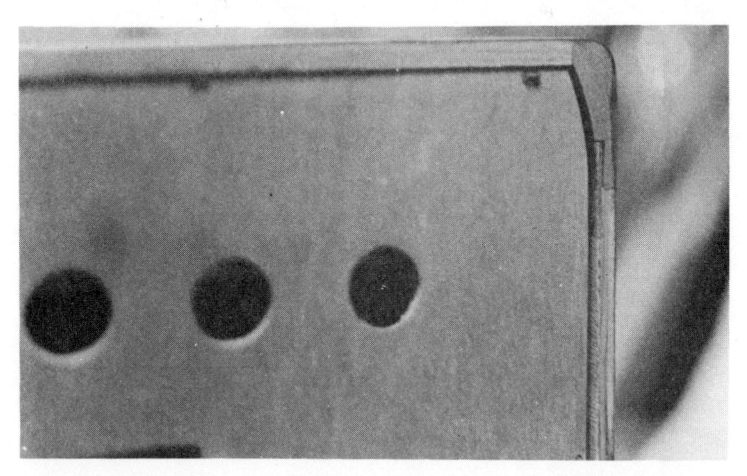

Fig. 17-22. Top right hand back corner shows blind end of rabbet and matching curve in cabinet back.

11. Apply a bead of glue to the panel front edges and lay the face frame in position on the partial carcase (Fig. 17-12). Align and nail the face frame to the lower half of the carcase. Glue the ends and front edge of the top panel and set in place by spreading the end panels slightly. The end panel top edging and the face frame should both extend beyond the top surface by about 1/16″. This will be trimmed off later. Nail the end panels to the top, then complete the nailing of the face frame for the top half of the cabinet.

12. Glue and nail the top back cleat into place (Fig. 17-13). Glue and nail the bottom back cleat into place (Fig. 17-14). Now quickly, while all the case glue joints are still wet, clamp all the joints until all gaps are pressed out (Fig. 17-15). Set the assembled carcase aside overnight for all the glue joints to set and cure.

13. Up-end the carcase on a pad and mount the glue blocks under the top and bottom panels (Fig. 17-16). Let it dry for several hours.

14. Belt sand or flush-trim all face frame to case joints flush. Round over the top and front outside edges with a ½″ radius rounding over bit (Fig. 17-17).

15. Mount the cleats to the false back. Set all nails in all assemblies, fill all joints with wood dough, and finish sand all parts.

16. Stain and varnish all parts before final assembly (Fig. 17-18).

17. Mount piano hinges on the doors and hang them (Figs. 17-19 and 17-20). The doors should project out in front of the face

frame about ½". This gives the cabinet a more sculptured look and reduces the boxy appearance.

18. Set the completed and finished false back assembly into place and nail to the cabinet sides through the mounting cleats (Fig. 17-21).

19. Hold the back in position and mark the radii at the blind end of the rabbet cut near the cabinet top (Fig. 17-22). Shape the top

Fig. 17-23. Cabinet back installed prior to mounting access panel. Wires are easily fished with access panel removed.

corners of the back accordingly with a sander or jig saw, and mount in place (Fig. 17-23).

20. If the back has been machined and lined up properly, the exhaust vent holes will be staggered between the back and the false back (Fig. 17-24).

21. Mount the magnetic touch latches. Mount the access panel with ½" #5 flat head screws.

22. Sign your work in pencil on the bottom (Fig. 17-25).

Fig. 17-24. Light through exhaust holes demonstrates how two sets of vents have been staggered to reduce dust infiltration.

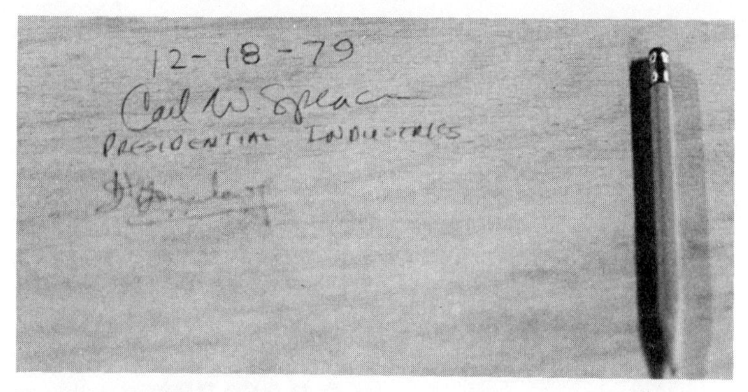

Fig. 17-25. Sign your completed cabinet on the bottom in pencil. It should be something to be proud of.

MOUNTING THE STEREO

1. Set turntable on top of the cabinet. Insert patch cords, power cord, and ground strap through the most convenient exhaust vent holes in both the back and the false back.

2. Set the cassette deck on the fixed shelf and run the wires out the closest wire run hole in the false back (the second wire run hole is in case you have three very short components to go inside. The hole is positioned for an additional adjustable shelf). Remove the access panel and pull the patch cords and power cord up between the false and real backs and out behind the receiver area through one of the intake holes in the false back.

3. Insert the speaker leads in through the intake holes of the real and false backs into the receiver area.

4. Insert the antenna lead in through the intake holes to the receiver area.

5. From the inside, fish the power plug of the receiver through the intake holes to the outside of the cabinet. Hook up all patch cords, speaker leads, and the antenna lead to the back of the receiver. Plug the cassette deck into the switched outlet and the turntable into the unswitched outlet.

6. Set the receiver on the adjustable shelf. Pull all extra wire lengths neatly out of the visible area within the cabinet and tuck down the sides between the false and real backs. Be careful not to obstruct the ventilation with a wire tangle or excess wires in the middle.

7. Plug the receiver plug into the wall, put on your favorite record or tape, and call all your friends over for a party.

A Vertical Format Cabinet

If you have confidence in your woodworking abilities, you may as well go all out and build a superior cabinet. This taller, vertical format cabinet (Fig. 18-1) can hold three to four components plus a turntable. It is ideal for people owning "separates" component systems, or a receiver, front loading cassette deck, and possibly an open reel deck.

The tall design takes full advantage of the chimney effect on convection ventilation flow, and has plenty of room in the cornice frame for dust-baffling the exhaust ports. The equipment is mounted at nearly ideal heights (see Fig. 18-2), and the records and tapes are conveniently located in pull-out drawers to eliminate having to get down on your belly on the floor just to pick a record.

The design takes up a minimal amount of floor space and blends in well with almost any style of furniture. The wood-species along with the door and drawer pulls chosen can make this cabinet look Country French, Italian, or Spanish without altering the plan one iota.

B-3 VERTICAL FORMAT CABINET SPECIFICATIONS

■ **Case dimensions.** 30″w, 84″h, 20½″d; Molding extremity dimensions: 32¾″w, 84″h, 21⅞″d; Turntable compartment dimensions: 28½″w, 18″h, 18″d; Equipment compartment (behind equipment panel): 28½″w, 29″h, 16½″d.

■ **Equipment capacity**. Holds three to four typical components, or two typical components plus a typical open reel deck.

■ **Storage capacity**. Holds about 150 single albums in bottom drawer. Top drawer can hold up to 90 cassettes, or is deep enough to store most headphones and record cleaning kits.

Fig. 18-1. Completed vertical format cabinet with doors shut. The design looks good with most furniture, yet doesn't attract a burglar's attention to the stereo as glass-doored cabinets will (courtesy of Presidential Industries).

Fig. 18-2. Vertical format allows near perfect equipment placement. The components are all at ideal heights while the record weight at the bottom helps to stabilize the tall cabinet (courtesy of Presidential Industries).

■ **Features.** Incandescent lighting over turntable (switched through receiver), false back with concealed wire runs behind turntable, dust-baffled convection cooling system (90 watts/channel convection, 300 watts/channel with 55 cfm forced air system), removable/replaceable equipment panel, magnetic touch latches for the doors, ideal equipment placement.

BILL OF MATERIALS

2 - 48 × 96 sheets ¾″ A-2 veneer core hardwood veneer plywood.

Fig. 18-3. End panels after all machining. It is easier to sand and stain interior panel surfaces before assembly.

Fig. 18-4. Fully machined lumber cabinet parts for several different cabinets. The four right hand stiles are for slightly different height cabinets as described in this chapter.

1 - 48 × 96 sheet ¼″ A-3 sound back veneer core hardwood veneer plywood (the back of the door panels will be exposed).

1 - 48 × 96 sheet ¼″ lauan or hardboard.

1 - 60 × 60 sheet ½″ Baltic birch plywood or equivalent.

Fig. 18-5. Door frames are filled and completely sanded before panels are glued to the back.

About 8 to 9 board feet 5/4 FAS hardwood lumber milled to 1″.

About 26 board feet 4/4 FAS hardwood lumber milled to 25/32″.

About 5 to 6 board feet ¾ FAS lumber planed to ½″.

3 - pair 2″ fancy butt hinges with screws (Amerock #2355 or equivalent).

Fig. 18-6. Starting carcase assembly of vertical cabinet.

Fig. 18-7. Glue bead along shelf end. Notice that the glue is spread slightly closer to the unfinished side of the panel. Any bleeding will occur where it won't be seen.

2 - magnetic touch latches with screws and strike plate.

1 - pair Grant 16″ 50 lb. #236 ball-bearing side slides or equivalent.

1 - pair Grant 16″ 100 lb. #329 ball-bearing side slides or equivalent.

2 - 6 or 7 watt night light bulbs.

2 - sockets for bulbs.

About 6 to 8 feet 18 ga. zip cord.

1 - male quick-connect plug end.

About three dozen 2½″ #10 wood screws, flat head.

About two dozen 1¼″ #10 wood or sheet metal screws, flat head.

Glue, finishing nails, sandpaper, stain, sealer, varnish, etc.

All above quantities include sufficient waste allowance. Careful layout and material selection may allow you to get by with slightly less, particularly on the lumber.

CUT LIST

(All dimensions are in inches)

5/4 Lumber

■ 2 - 2 1/16 × 77½ **Face stiles** B-3-1

Mortise the inside edges to receive the rails. The top rail mortise should be 1 15/16 long, as should the bottom rail mortise. The mortises for the rails above the two drawers should be 1⅜ long each. The mortise for the rail above the bottom drawer is 16 13/16 - 18 3/16 from the bottom end of the stile, the top drawer rail is 24 5/16 - 25 11/16 from the bottom end. All mortises should be 11/16 deep and matched with corresponding tenons.

■ 2 - 2 × 27¼ **Face rails** B-3-2

These are the top and bottom rails. Tenon both ends and shoulder tenon on the inside edge to lap the radius left in the mortise. Tenons should be cut ⅝ long and matched with corresponding mortises.

■ 2 - 1½ × 27¼ **Face rails** B-3-3

These are the rails above the two drawers. Tenon both ends and shoulder the tenon on both edges.

4/4 Lumber

■ 4 - 2 × 49⅝ **Door stiles** B-3-4

These should be cut out of the best looking and straightest stock. Mortise for top and bottom door rails to 1 15/16 from both ends. In addition, mortise for the two interior rails 15 15/16 - 17 13/16 from both ends.

Fig. 18-8. Carcase panel assembly complete.

Fig. 18-9. Blocking up carcase top and bottom panels to brace for face mounting.

■ 4 - 2 × 23 3/16 **Drawer rails** B-3-5

Tenon both ends and shoulder the tenon on the inside edge.

■ 2 - 2 × 5 15/16 **Drawer stiles** B-3-6

Mortise for top and bottom drawer rails to 1 15/16 from both ends.

■ 2 - 2 × 14⅝ **Drawer stiles** B-3-7

For bottom drawer. Mortise for top and bottom drawer rail to 1 15/16 from both ends.

■ 8 - 2 × 10⅛ **Door rails** B-3-8

Tenon both ends. On four, shoulder the tenons on one edge only. On four shoulder the tenons on both edges. The first four will be the bottom and top rails, the second four will be the interior rails.

■ 1 - 2 × 28½ **Light baffle** B-3-9

Bore two 3/16 holes, each 6″ from an end, from one edge to the other. Countersink for the two 2½″ #10 wood screws used to mount the baffle.

■ 2 - 3¼ × 96+ **Plinth cove molding stock** B-3-10

With a Sears or Rockwell molding head on the table saw, cut a ⅝″ radius cove along one edge of the good face. The cut should be ⅜″ deep and about 1″ wide. In hardwood, it is much safer and cleaner to make the cut in eight to ten passes.

■ 2 - 2 × 96+ **Plinth detail molding stock** B-3-11

Using a ½" radius rounding over bit on the router set with a ⅛" lip, shape along one edge of the good face.

■ 1 - ¼ × ¾ × 28½ **Equipment Shelf edge** B-3-12

■ 10 - **Glue blocks** (cut from lumber scrap). B-GB

¾ (½") Lumber

■ 6 - 6 15/16 × 11⅞ **Door panel implants** B-3-13

Using a ¼" radius rounding over bit on the router set with a 1/16" lip, shape along the four edges of the good face. Be sure not to use any cupped lumber.

■ 1 - 8 11/16 × 19 15/16 **Drawer panel implant** B-3-14

Using a ¼" radius rounding over bit on the router set with a 1/16" lip, shape along the four edges of the good face. Be especially sure not to use any cupped lumber.

¾ Hardwood Veneer Plywood

■ 2 - 19½ × 77½ **End panels** B-3-15

Run a ¾ wide ⅜ deep dado for the bottom and top panels 1¼ × 2 from both ends. Run another dado 25 - 25¾ from the bottom edge for the turntable shelf. Run a ¼ wide rabbet along the back edge. See Fig. 18-3.

Fig. 18-10. Blocking up turntable shelf.

Fig. 18-11. Carcase assembly complete. All joints not perfectly tight should be clamped to remove any gaps.

■ 1 - 19¼ × 29¼ **Top panel** B-3-16

Drill two rows of 1″ holes along lines centered 4 and 6 from the front edge spaced about 2″ apart. These are the convection exhaust holes. If you plan to use forced air ventilation, you will need to consult the instructions that came with your whisper fan and cut one large hole instead. Be sure that the circumference of the single hole is no closer than 4″ from the front edge of the shelf.

■ 1 - 18 × 29¼ **Turntable shelf** B-3-17

No machining required.

■ 1 - 28½ × 29 **Equipment panel** B-3-18

Cut out to fit your components, mount equipment supports made from scraps (see Chapter 8).

■ 1 - 17⅝ × 28½ **Equipment shelf** B-3-19

Edge with B-3-12. If you have upright sockets for your lighting, drill appropriately sized mounting holes 4″ from both ends of the shelf centered on a line 4″ from the front edge of the shelf with the lumber edging already applied. See Figs. 6-1 and 6-3. Cut two cleats 2 15/16 × 15¼ from ¾ plywood scrap stock and pre-drill the mounting holes (see Fig. 8-66) and glue and nail to the top side of the shelf flush with the back edge (Fig. 8-67).

The following part may be cut out of ¾″ particle board or other plywood scrap to conserve the expensive hardwood veneer plywood if desired.

■ 1 - 19¼ × 29¼ **Cabinet bottom panel** B-3-20

No machining required.

The following parts may be cut out of any species ¾″ plywood scrap to conserve the expensive hardwood veneer plywood if desired.

Fig. 18-12. Detail of carcase joint.

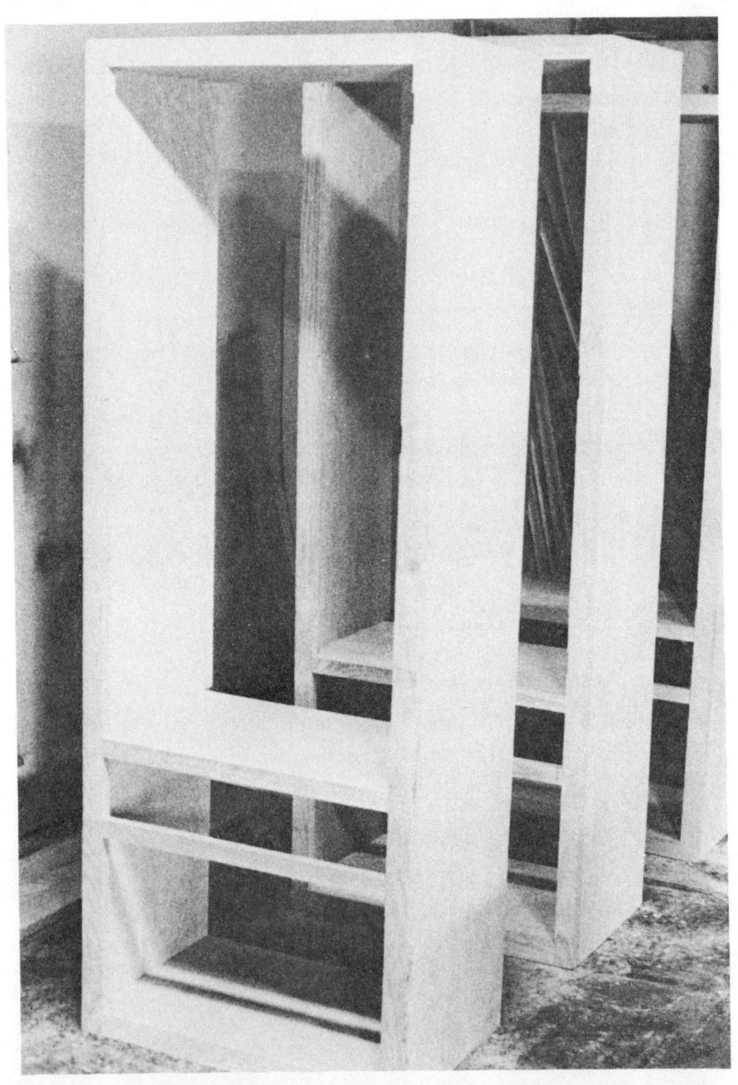

Fig. 18-13. Three vertical format carcases of slightly different height ready for finish sanding and staining.

■ 1 - 1¼ × 28½ **Bottom back cleat** B-3-21

No machining required.

■ 1 - 1¼ × 28½ **Dust baffle fence** B-3-22

Drill two 3/16″ holes 6″ from each end from one edge to the other. Counterbore with a ⅜ bit ½″ to allow the mounting with 1¼ #10 wood screws.

■ 4 - 2 15/16 × 29⅝ **Front & back molding cleats** B-3-23

■ 4 - 2 15/16 × 18¾ **Side molding cleats** B-3-24

■ 16 - **Corner blocks** (cut from 2 15/16 wide scrap stock) B-3-25

¼ Hardwood Veneer Plywood

■ 1 - 28½ × 21½ **False back** B-3-26

Cut two ¾ × 21½ cleats from ¾ plywood scrap, glue and nail to the side edges of the back face of the false back. Drill a 1″ hole for wires on a line 1½ from the bottom edge of the false back, and centered between the sides.

■ 1 - 6 × 26 **Top drawer panel** B-3-27

No machining required.

■ 1 - 14¾ × 26 **Bottom drawer panel** B-3-28

No machining required.

■ 1 - 12⅜ × 49¾ **R. hand door panel** B-3-29

No machining required. Make sure back is sound.

■ 1 - 13⅜ × 49¾ **L. hand door panel** B-3-30

No machining required. Make sure back is sound.

¼ Lauan or Hardboard

■ 1 - 29¼ × 76¼ **Cabinet back** B-3-31

Bore three rows of one inch intake ventilation holes on lines. 30, 32, and 34 from the bottom of the back panel (Fig. 18-19). Space the holes two inches apart. Cut a twelve inch square wire access port 11 to 23 ins. from the top of the back panel and centered across the width of the back.

Fig. 18-14. Completed plinth or cornice molding assembly.

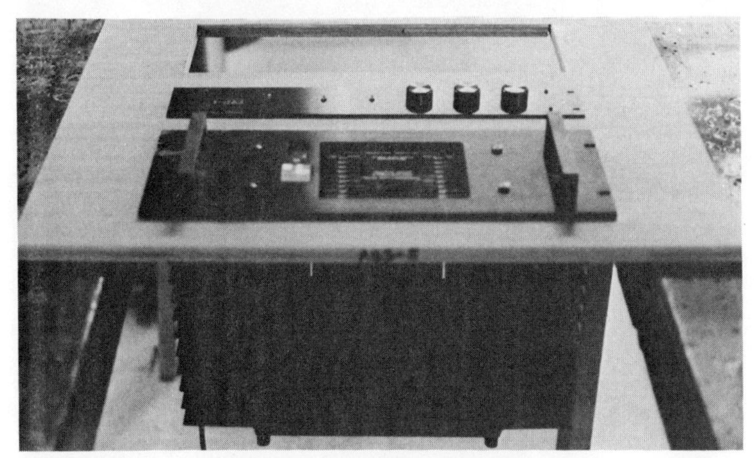

Fig. 18-15. Fitting the equipment in the equipment panel.

■ 1 - 14 × 14 **Access panel** B-3-32

Drill four screw mounting holes, one in each corner spaced one-half inch from the edges.

■ 1 - 14 × 24 7/16 **Bottom drawer bottom** B-3-33

No machining required.

■ 1 - 16 × 24 7/16 **Top drawer bottom** B-3-34

No machining required.

½ Baltic Birch

■ 2 - 13½ × 14 **Bottom drawer sides** B-3-35

Run dado ¼″ wide, ¼″ deep, ¼″ from bottom inside edge to receive drawer bottom.

■ 2 - 13 × 14 **Bottom drawer front-back & back** B-3-36

No machining required.

■ 2 - 5½ × 16 **Top drawer sides** B-3-37

Run dado ¼″ wide, ¼″ deep, ¼″ from bottom inside edge to receive drawer bottom.

■ 2 - 5 × 24 **Top drawer front-back & back** B-3-38

No machining required.

Miscellaneous Hardwood Lumber Scrap

■ 4 - 1¼ × 1¼ × 4 **Equipment panel mounting cleats** B-3-39

Bore screw mounting holes as shown in Fig. 8-71.

■ 4 - 1½ × 1¼ × 18 **Drawer mounting cleats** B-3-40

Bore three 3/16 mounting holes along the length of the 1½ wide face. Counterbore the front side with ⅜ diameter holes ½″ deep.

CONSTRUCTION

1. Thoroughly read and understand all instructions before commencing to build the cabinet. Doublecheck all dimensions to be sure you understand how the parts go together, and that your particular stereo will fit.

2. Cut out all parts to raw sizes.

3. Mortise and tenon all frame members (Fig. 18-4). Test fit all framed assemblies, then assemble with glue, clamps, and short brads or staples (see Chapter 11).

4. With a ¼ or 5/16 rounding over router bit set with a 1/16 lip, shape the inside front edges of the door frames. Leaving the back edges square and sharp, fill (Fig. 18-5) and completely finish sand the door frames.

5. Mark the edge of the door stiles for the hinge mortises. Set the table saw or router and mortise out a recess for the butt hinges. For most hinges, this will be ⅛″ deep, but check with your exact hinges.

6. After completely sanding the door and drawer panels, glue them to their respective frames. On the doors, the left hand panel should extend past the edge of the unhinged stile by about 7/16, while the right hand panel will come about ½ short along its unhinged stile. The extension of the left hand panel will lap the recession of the right hand panel to form the dust lip. The overlap

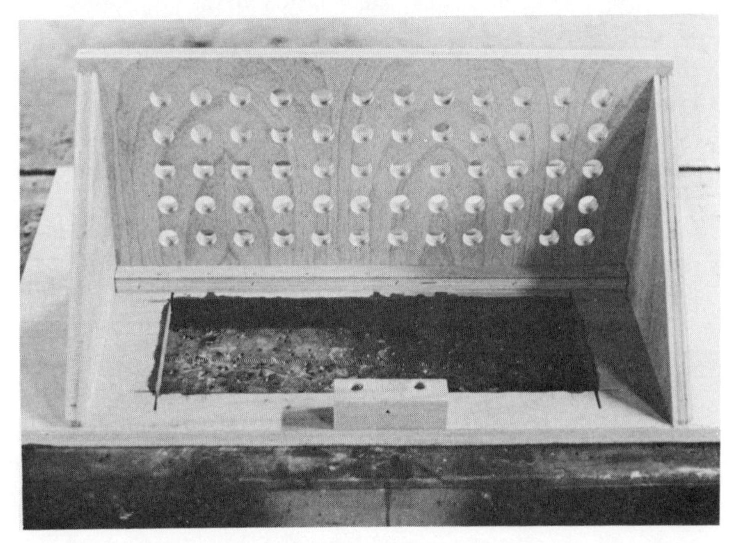

Fig. 18-16. Mounting the equipment supports to the back of the equipment panel.

Fig. 18-17. Completed vertical cabinet drawers awaiting mounting. The bottom drawer is sized to accept records while the top drawer is the full depth allowed by the drawer slides.

closes the center gap between the pair of doors and stops perhaps 50% to 60% of the dust that would otherwise enter from the front of the cabinet. The panels will also slightly overlap along the other three edges of the frame. This lip simplifies filling and sanding the joint, which is finally sanded down flush with the top, bottom, and hinged edges. Then the back of the door is filled and finish sanded.

7. Glue and nail door and drawer implants to the center of each framed panel section.

8. Machine all panels as noted. Check the fit of the dadoes and rabbets before attempting to glue the assemblies together. The fit should be snug, but not a force fit.

9. Lay one end panel down on a padded table or the floor. Without glue, insert the top and bottom panels, and the turntable shelf (Fig. 18-6).

10. Put a bead of glue on the top edges of the upright shelves (Fig. 18-7). Make sure that the top and bottom panels are positioned with their good sides to the carcase interior, and that the turntable shelf is flush with the front of the end panel. Then lay the second end panel on the partial assembly (Fig. 18-8), guiding the shelves into their appropriate grooves. This is easily done without smearing the glue if the shelf corners are first inserted into the front end of the dadoes while the end panel is held up like the half-opened cover of a book. The end panel is then lowered, and the

shelves will slip into place. Flush up all the front edges and pin with finishing nails.

11. The case is then held together with arm pressure and rolled over. The unglued end panel is removed, the shelf edges are glued, and the panel repositioned and pinned.

12. Flip the carcase panel assembly over on its back, and block up the gaps between the shelves and the floor. The top and bottom

Fig. 18-18. Installing equipment panel.

Fig. 18-19. Back of completed cabinet. Notice intake ventilation holes, screwed on access panel, and exhaust slot under cornice molding frame (courtesy of Presidential Industries).

panels will need ¼″ shims at both ends (Fig. 18-9), while the turntable shelf will require a 1½″ shim (Fig. 18-10). These bear the force of the hammer and nails and direct it to the floor rather than knocking the shelves out of position. The alternative is sloppy face frame-to-shelf joints and a lot of struggle.

13. Test fit the completed face frame assembly on the carcase. Apply a bead of glue all along the front carcase edges in such a fashion that any glue bleeding will occur on the least exposed side of the face frame-to-panel joint. Position and nail the face fame to the panel assembly. Now clamp all the joints that are even the slightest bit open (Figs. 18-11 and 18-12) and set the carcase aside to thoroughly dry (Fig. 18-13). A properly assembled case should have the face frame perfectly flush with all three horizontal panels, perfectly flush with the top and bottom of the end panels, but should hang over the front edges of the end panels by 1/16″ or so. This will be flush trimmed or sanded down after drying. It is much easier to make the solid lumber face frame slightly oversize and trim it down to the end panel than it is to try to fudge the construction and compress the end panels to fit a slightly undersize frame. Remember, your error in cutting parts to size doubles at each joint, and quadruples across the face frame. If you cut each part to a 1/32 tolerance, your face frame could net out as much as ⅛″ too wide or too narrow.

14. When the carcase glue has cured, mount the glue blocks above the top panel and below the bottom panel to double or triple the strength of the joints. Sand or flush trim the face frame overlap flush with the end panels. Set all the nail holes, fill all the holes and joints, and finish sand the entire case.

15. Fasten the plinth and cornice detail moldings to the plinth and cornice cove molding with glue and staples or screws. After drying, carefully cut out and miter 45°, the six main molding pieces: two fronts, two right hand pieces, and two left hand side pieces. The front pieces should measure 29¾ on their back side (the critical measurement), and approximately 32¾ at the widest extreme of the detail face. The side pieces should measure exactly 20⅜ on their back side, and about 21⅞ at the extreme on the molding detail side. Be sure to cut two "lefts" and two "rights" for the side pieces. Then construct as detailed in Chapter Thirteen. See Fig. 18-14.

16. Lay out, cut, and fit the equipment to be panel mounted in the equipment panel (Fig. 18-15) as detailed in Chapter 8. At this point, construct all necessary skids, shelving, and other bracing out of ¾ plywood scrap to support the equipment (Fig. 18-16). Fill and finish sand equipment panel.

17. Pre-sand and assemble the basic drawer box (see Chapter 10), and glue and fasten the drawer front in place (Fig. 18-17).

18. Touch-up sand any scuffs or minor blemishes that the case and parts may have accumulated during subassembly. Stain all exposed parts. Put on a coat or two of sealer compatible with your planned top coat, sanding lightly with 400 paper or 0000 steel wool between coats. Then apply two coats of varnish on top, sanding or steel wooling between coats. Be sure to follow application and curing procedures recommended by the manufacturer.

19. Screw the plinth and cornice molding to the carcase (see Chapter 13).

20. Mount equipment shelf and equipment panel in the carcase (Fig. 18-18) and see Chapter 8. The front of the finished edge of the equipment shelf should be ½″ from the back of the face frame to allow clearance for door swing when using touch latches. Screw light baffle to the underside of the equipment shelf, 1⅞″ from the front edge.

21. Screw the drawer mounting cleats in appropriate positions in the case to hold the case side of the drawer slides on a surface flush with the inside of the face frame. Mount the drawer slides on the cleats and the drawer sides, and install and adjust the drawers.

22. Gently tap the snug fitting false back into position behind the turntable section. Nail it to the turntable shelf and to the end panels through the ¾ × ¾ cleats. *Do not* nail it to the equipment shelf. This allows the shelf to pivot and allows you to change the equipment panel in the future.

23. Nail, but *Do not glue*, the cabinet back in position. Always install the false backs and cabinet backs before hanging the doors. The backs are the principle brace for the entire case. The cabinet can shift and rack the doors before the back is installed, but will not afterwards.

24. Screw the access panel over the wire access port with four ½″ #5 wood or sheet metal screws (Fig. 18-19).

25. Screw the butt hinges to the door mortises with the hinge knuckle on the front side of the door.

26. Lay the carcase on its back and hang the doors as described in Chapter 11.

27. Mount the magnetic door latches.

28. Mount light sockets, bulbs, and lighting power lead. See Chapter 5.

29. Screw dust baffle to top side of cabinet top, directly behind exhaust vent holes.

30. From scrap ¼ plywood, cut a panel the exact size to fit in the ¼″ recess of the cornice molding (see Figs. 3-5 and 13-39).

This may be simply dropped into place and held by gravity, or may be pinned with three or four ½″ #5 wood screws.

31. Sign your work in pencil on the bottom panel.

INSTALLING THE STEREO

1. Remove access panel. Drop a six or seven foot pull wire down between the backs, and poke one end through the wire run hole in the bottom of the false back. Be sure to hang onto the top end or hook it over the bottom edge of the access port.

2. Set turntable on turntable shelf and tape the wires to the pull wire.

3. Gently work the turntable wires up into the equipment compartment and untape from the pull wire.

4. Again, drop the pull wire down between the backs, only tuck the bottom end of the pull wire out the back of the cabinet through the intake vent holes.

5. Tape the leads from the speakers and the antenna to the pull wire and pull up gently between the backs into the equipment compartment. Untape the leads.

6. Slide the panel mounted components into position. Drop the receiver or amplifier power lead down between the backs and pull out through one of the intake vent holes.

7. Hook up the speaker lead to the anplifier, the antenna lead to the tuner, and the component patch cords to their appropriate receptacle. Hook up the component power leads to the outlets on the back of the receiver or preamplifier, the turntable plug going in an unswitched outlet, all others going into a switched outlet (see Chapter 5).

8. Plug the receiver or preamplifier power cord extending from the intake vent holes into the wall and check all your hookups before moving the cabinet against the wall.

9. Move the cabinet into position in the room making sure to leave it at least 1½″ to 2″ from the wall to insure proper ventilation.

10. Invite your friends over to share your happiness at creating a most worthwhile and functional piece of furniture.

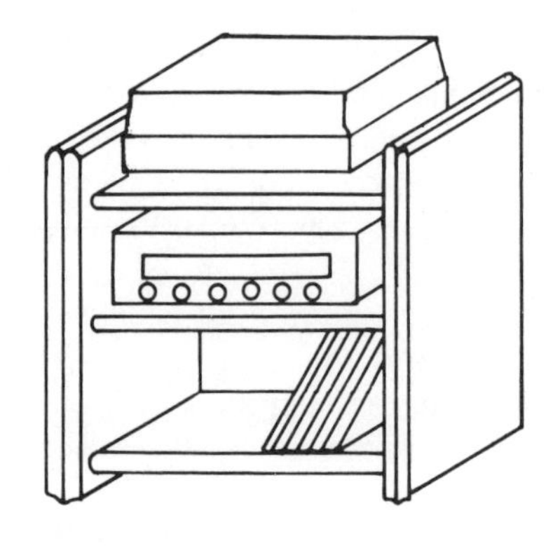

Glossary

access panel - A trap door over an access port on the actual cabinet back used to simplify wiring and servicing without having to remove the equipment from the case.

armoire - A large and ornate general purpose case, usually used for a wardrobe.

built-in - Casework permanently attached to a building, or cabinet that looks like permanent fixtures.

carcase - The basic structure of casework. Usually refers to the box or storage unit devoid of trim.

casework - Furniture designed for storage. Most casework is a variation of a simple wooden box, but may be very ornate (such as a china cabinet or buffet).

component - A piece of electronic equipment with one or more highly specialized functions that can be connected with other components with different functions to complete a sound system.

consoles - A manufacturers pre-packaged sound system complete with electronics and furniture. Most consoles are long, low sideboards with speakers at both ends, and turntable, tuner and amplifier inside.

convection - The circulation of air of nonuniform temperature because of the variation of density and the action of gravity. Hot air rises while denser cooler air falls.

cornice - The molding at the top of the upper carcase used to balance the appearance.

cove - To make in a hollow concave form.

cube tap - The real name for those funny little things that convert a single power receptacle into three or four power receptacles.

dado - Any groove cut across the grain of a board.

dust baffles - Obstructions placed in the path of ventilation currents to force them to negotiate turns. The heavier-than-air dust tends to settle out at each turn by force of gravity.

dust lip - The overhang of one of a pair of doors lapping a corresponding recess in the other to form a barrier against dust infiltration when closed.

entertainment center - A general term applied to shelving or casework used for a variety of electronic equipment, usually both audio and video.

equipment panel - Any panel into which stereo components are placed.

ergonomics - Efficiency of motion.

false back - A second back mounted a certain distance inside the actual cabinet back for the purpose of concealing wires and acting as one side of the ventilation duct. The false back is usually made from the same veneer as the case exterior and finished.

flush-trim - To trim off excess material by means of a router bit especially designed to follow the contour of the proper plane while removing any excess material. A flush-trimming bit is basically a straight cutting bit with a pilot bearing on the end.

freestanding - In this book the term usually refers to a cabinet that requires no additional support such as a wall. Any cabinet that is not a built-in is freestanding.

front-back - The front end of a basic drawer box that is fastened to the back of the finished decorative drawer front.

groove - Any channel or groove cut with the grain of a board.

hi-fi - Although the term "hi-fi" can be used to denote any high fidelity electronic sound equipment, it is typically used to describe older monaural set-ups.

hygroscopic - Induced by the absorption of moisture.

kerf - The slit made by a saw blade.
kerf marks - Marks and scratches left on the edge of a board by the act of sawing.

lift-lid - A hinged (and usually counter-balanced) cabinet top that is lifted to reveal the sound equipment under it.
light baffle - A board or other shade mounted directly in front of a light source to block the source from direct view. All light actually seen in then indirect.

modules - Cases built as standardized units to be used with other modules for flexibility and variety in room or equipment layout.
mortise - A slot cut to receive a tenon.

ogee - An S-shaped form.

panel mounting - Fitting sound equipment into a general front panel to seal out dust from the equipment compartment and to completely hide the wires.
parallel circuits - Circuits arranged so that all positive poles, electrodes, or terminals are joined to one conductor, and all the negative ones to another.
plinth - The enclosed base beneath the base carcase, traditionally a heavy molding.
pull wire - A temporary wire threaded through the concealed part of the cabinet during construction for the purpose of simplifying fishing the patch cords through later.

rabbet - A channel or recess cut out of the edge of a panel, usually to receive the edge of another panel.
racks - Racks were originally frames with metal mounting strips for ham radio or hi-fi equipment. Now the term is used to describe any low-priced vertical format simple shelving unit, usually mass-produced by a commercial manufacturer.
rack mounting - Method of mounting electronic equipment by bolting it on both sides of the face to two vertical mounting frames made either of wood or metal. Racks are generally prebored at standard increments. The chief difference between rack mounting and panel mounting is that panel mounting

surrounds the equipment face on four sides while racks are only on two. Rack mounting is not dust proof and may allow some wire visibility while panel mounting is dust proof and completely conceals the wires.

rail - Any horizontal member of a frame.

series circuit - The arrangement of parts or elements in a circuit so that the whole current passes through each element without branching.

shear bracing - A support that prevents two contiguous parts of a case from moving relative to each other in a direction parallel to their plane of contact. The case back acts as a shear brace in that it keeps the case a rectangle with 90° corners instead of allowing it to become a parallelogram.

shelf mounting - Merely setting the electronic equipment on a shelf.

stile - An vertical member of a frame.

tenon - A tongue of wood inserted into a mortise to make a joint.

tweeters - The speaker elements that reproduce only the higher acoustic frequencies.

waist molding - Molding around the cabinet part way up usually used to disguise the joint between the upper and base carcases, and to break up excessively long vertical lines.

zip cord - Lamp cord, usually 18 or 16 gauge. It gets its name because the two conductors may be pulled, "zipped," apart with ease.

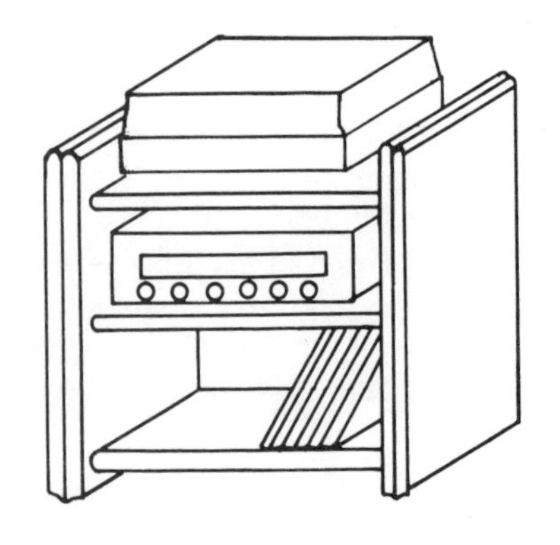

Appendix
Stereo Furniture Manufacturers

Apres Audio Ltd.
7 Revere Court
Suffern, NY 10901

rack-type framed in oak with oak veneers

Audio Works
840 Piner Road, Suite 14
Santa Rosa, CA 95401
(707)528-0422

Stack-rack equipment cabinets and speaker stands. Rack-type unit with unique "rack-mount" look, a variety of panel mounting.

Barzilay Company
18737 S. Reyes Ave.
Compton, CA 90221
(213)774-3321

the partiarch of stereo furniture, from simple-rack type units so well thought out furniture type. Excellent quality and construction.

Custom Electronics Display Cabinetry
1547 Birchwood
St. Germain, WI 54558

strictly custom units made to your specifications.

Custom Sound Corporation
8460 Marsh Rd.
Algonac, MI 48001

rack-type formica-clad units and record cabinets. Locking doors both front and back, ventilation allowance.

Danwood Design
21616 87th Ave. S.E.
Weedinville, WA 98072
(206)485-8524

Solid and veneer oak racks and shelving units.

Delphi Custom Stereo Cabinets
9826 Pice Blvd.
Los Angeles, CA 90035

high-end custom cabinets, both standard and adapted to order.

Ello
1034 Elm St.
Rockford, IL 61102

modular wall systems tv/hi-fi trolley, lowboys

Ethan Allen, Inc.
Ethan Allen Dr.
Danbury, CT. 06810

stereo cabinet bases and uppers for custom room plan wall furniture coordinates with the rest of their general furniture line.

Gusdorf Corporation
6900 Manchester
St. Louis, MO 63143

probably the largest manufacturer of stereo furniture, popularly priced KD (knock down) stereo racks and storage units. Rendura clad composite board with smoked glass doors.

Mariana by Greenwood Forest Products
8285 S.W. Nimbus, Suite 139,
Beaverton, OR 97005
(800) 547-1449

high-quality rack-type audio furniture made from solid oak and oak veneers. Although the cabinets come KD, no external screws are visible.

Nomadic Furniture
19505 Business Center Dr.
Northridge, CA 91324
(213) 885-5711

solid oak and oak veneer racks, wall units, and shelf systems designed strictly for stereo, Excellent workmanship.

O'Sullivan
19th and Gulf Streets
Lamar, MO 64759

popularly priced KD stereo racks and storage units.

Pennsylvania House
Lewisburg, PA
17837

freestanding oak stereo cabinet designed to coordinate with the rest of their general furniture line.

Presidential Industries
9710 Arlington Ave.
P.O. Box 2531
Riverside, CA 92516
(714) 354-2145

the author's company. High-end custom or production stereo furniture with full line of features (panel mounting, dust-baffled ventillation, concealed wiring, etc.) in oak, pecan, and walnut.

Sound Plus Wood
20 N. Federal
Boca Raton, FL 33432

high-end custom stereo furniture. Company specializes in a complete package, both equipment and cabinetry.

Timberline Products
3330 Brace Street
Burbank, CA 91504

real wood constructed rack-type furniture.

Index

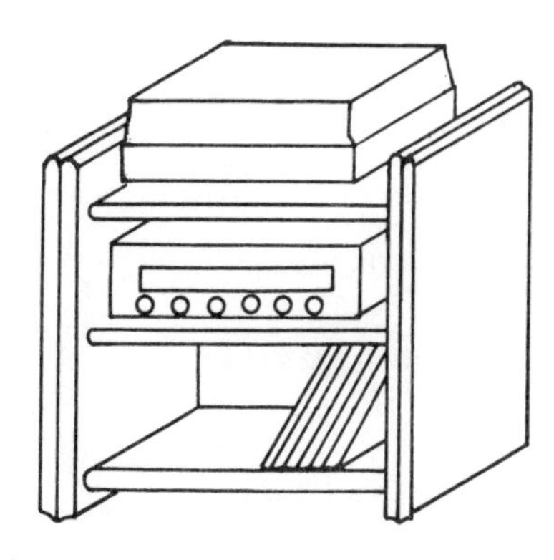

D

E

F

G

H